U0424800

北京理工大学“双一流”建设精品出版工程

SCI期刊论文阅读与写作（工科）

Reading and Writing SCI Journal Articles (Engineering)

闫鹏飞 ◎ 主编

北京理工大学出版社
BEIJING INSTITUTE OF TECHNOLOGY PRESS

图书在版编目(CIP)数据

SCI期刊论文阅读与写作:工科 = Reading and Writing SCI Journal Articles (Engineering) : 英文 / 闫鹏飞主编. -- 北京 : 北京理工大学出版社, 2024.3

ISBN 978-7-5763-3743-3

Ⅰ.①S… Ⅱ.①闫… Ⅲ.①科学技术-英语-论文-阅读教学-高等学校-教材②科学技术-英语-论文-写作-高等学校-教材 Ⅳ.①G301

中国国家版本馆CIP数据核字(2024)第065329号

责任编辑: 时京京　　**文案编辑**: 时京京
责任校对: 周瑞红　　**责任印制**: 李志强

出版发行 / 北京理工大学出版社有限责任公司
社　　址 / 北京市丰台区四合庄路6号
邮　　编 / 100070
电　　话 / (010) 68944439 (学术售后服务热线)
网　　址 / http: //www.bitpress.com.cn

版 印 次 / 2024年3月第1版第1次印刷
印　　刷 / 保定市中画美凯印刷有限公司
开　　本 / 787 mm×1092 mm　1/16
印　　张 / 23.25
字　　数 / 507千字
定　　价 / 68.00元

前　言

PREFACE

一、编写背景与教材定位

为主动应对新一轮科技革命与产业变革，教育部明确提出新工科建设，强调学科交叉融合，注重人才培养实效，突出创新创业实践，积极推进未来科技创新领军人才的前瞻性和战略性培养。拔尖创新人才培养就必须全面提升学生的国际视野和学术交际能力。显然，以语言基础和语言技能为主要培养目标的英语教学模式已不能适应新时期人才培养的定位与要求。与此同时，国内高校大幅削减英语课程的学时学分，英语教育教学不得不面对新形势、设计新思路、深耕新领域。在此背景下，学术英语充分体现学术性和工具性特点，又天然蕴含学术素养、学术思维和学术规范等人文内涵和思政要素，构成新工科背景下英语教育教学的重要方向。面向工科领域学生开展学术英语教学，充分利用有限的学时学分，高效培养其学术交际能力尤其是学术论文读写能力，可更好满足其汲取学科知识、开展学术交流以及发表科研成果的学业发展需求。

学术交际能力集中体现在学生能够从所在学科领域的国际期刊中汲取前沿信息，并将自身科研成果撰写成文发表。对理工科而言，使用英语开展学术交流并发表科研成果的国际化趋势凸显。其中，研究论文作为学术社团分享科研成果的主要形式，具有一定规约性的语篇结构、修辞策略、元话语功能和语言特征，体现学术话语知识构建、承前启后、问题导向、证据理据、读者交互等内涵属性，构成学术英语教学的重要切入点。当然，不同学科大类所秉持的本体论、认识论和方法论多有不同，研究对象和分析范式各有侧重，知识构建路径和认识发展规律也呈现不同倾向。具体到话语功能与形式，传统惯例、规约准则和方式策略以及更为具体的语篇结构、词汇语法和句法结构等也存有差异。除知识内涵和命题信息外，研究论文在文献引用、观点阐述、过程描述、结果解读、结论产出、劝说读者、规避责任等策略使用上也明显有别。因此，面向工科领域的拔尖创新人才培养，学术英语教学应聚焦这一领域权威国际期刊刊发的研究论文，积极引导学生观察语言使用、剖析文本特点并紧贴自身科研实践开展演练，以此提升语言输入与产出的系统性与针对性，一体化培养学生的语言技能、语用能力和语体意识，同时主动搭建语言知识和学科知识协同互促的有利平台。

鉴于上述思考，本教材定位为：融合前人学术话语研究尤其是语类分析、元话语研究和语体特征分析的重要成果，聚焦典型工科领域尤其是材料科学这一基础性、前沿性和交叉性学科，选取权威国际 SCI 期刊刊发的研究论文作为素材，引导学生观察剖析论文摘要、引言、方法、结果与讨论、结论等章节的交际意图、篇章结构、修辞策略以及显著多用的词汇语法、短语、句法、语义和语用特征，同时鼓励学生紧扣自身科研实践，将之逐步撰写成文并开展口头陈述，以此培养学生的语类意识、语体意识与语用能力尤其是学术论文读写能力，提升其学术技能、学术思维与学术素养。

二、教材特色

1. 立足学生学业发展需求，聚焦典型工科领域高水平 SCI 期刊论文

SCI 期刊论文作为学术知识构建和前沿信息传播的重要形式，早已成为拔尖创新人才学业发展过程中的必读必写文献类型。有效开展 SCI 期刊论文阅读与写作，可视为拔尖创新人才的必备技能和基本素养，也构成英语教育教学的重要抓手。本教材选取典型工科领域尤其是材料科学这一基础性、前沿性和交叉性学科的 SCI 期刊论文作为编写素材。具体而言，依据国务院学位委员会第六届学科评议组编写的《学位授予和人才培养一级学科简介》，对标这一领域所含材料物理与化学、材料学、材料加工工程、高分子材料与工程、资源循环科学与工程 5 大学科方向，同时考虑纳米材料科学这一热点方向；邀请专家学者推荐，确定多种权威国际 SCI 期刊，如 *Nature Materials*，*Advanced Materials*，*Chemistry of Materials*，*Biomaterials*，*Advanced Functional Materials*，*Journal of Alloys and Compounds*，*ACS Applied Materials & Interfaces*，*Acta Materialia*，*Journal of Materials Processing Technology*，*Composites Science and Technology*，*Polymer Testing*，*Polymer Composites*，*Energy & Environmental Science*，*Materials & Design*，*Nature Nanotechnology*，*ACS Nano*，*Nano energy*，从中采集较大批量的学术论文。

上述权威期刊刊发的学术论文，尤其是其中占比最高的研究论文，无论科研成果、学术规范以及文本写作，均代表这一领域的最高水平，可视为学生科研实践与成果发表的经典范本。本教材在文本选取、篇章分析、任务设计等各个环节，重点围绕研究论文这一典型语类展开。与此同时，鉴于近些年工科领域中国学者的论文发文数量、高被引论文数量、所获专利数量以及研究队伍规模均位于世界前列，本教材编写过程中选取了相当一部分中国学者撰写发表的高水平科研论文，以期激励引导青年学生学习科学家精神，树立远大理想、投身科研创新，抢占科技高地、助力科技强国。

2. 从宏观语篇结构到微观语言特征，基于语料库立体呈现典型工科领域 SCI 期刊论文特点

本教材旨在引导学生观察剖析典型工科领域 SCI 期刊论文的篇章结构、交际意图、修辞策略以及语言特征使用。科研人员撰写发表期刊论文目的在于，向业界同行介绍自身科研成果，以期推动学科知识发展并赢得相应学术声誉。相较于小说、新闻、信函等通用书面语篇以及专业教材、科普文章、专家访谈等特定学术语篇，SCI 期刊论文的正式程度、专业命题、读者群体、交互距离等情境要素清晰有别，决定了其篇章、修辞、措辞等层面的鲜明特性。但是，SCI 期刊论文的多维特征并非一目了然、静态呈现，须基于大量真实语料和文本分析技术进行深入探究。

本教材编写集中体现了这一理念，依托语言学理论和大量真实语料并借助语料库分析技术乃至自然语言处理技术，系统考察典型工科领域研究论文的文本特点，从而为教材编写提供了坚实保障。具体而言，首先，对照前文所述典型工科领域尤其是材料科学领域的多种权威国际期刊，从中采集近五年刊发的研究论文 3 100 余篇，建成库容 1 700 多万词

次的研究论文语料库；人工标注并批量提取研究论文的摘要、引言、方法、结果与讨论、结论等主要章节。其次，为精准描写研究论文各个章节的语篇组织，人工标注提取各个章节的共核语步结构，同时考察各自趋于使用的修辞手段以及所蕴含的认知图式。第三，借助语料库分析工具，自下而上挖掘研究论文各个章节乃至各个语步显著多用的词汇语法与句式特征如被动式、状语从句、定语从句。同时，鉴于名词短语构成学术语篇的重要特征，借助工业级自然语言处理工具包 spaCy 提取研究论文各个章节文本高频使用的名词短语，进而精准提取显著典型的动名搭配。此外，调用自然语言处理技术 word2vec 训练词向量模型，开展语义计算，纵深挖掘词汇与短语的语义（功能）相似项。基于上述研究论文的篇章结构标注和语言特征分析，本教材编写得以成功实现：精准提取语步结构颇具代表性的论文章节、论文章节所含主要语步的文本片段、传递相同交际意图的整句索引行、特定语步显著多用的词汇语法、短语、句法、语义和语用特征，以及特定章节乃至语步高频使用的动名搭配等。

在此基础上，本教材按照单元顺序，依次编写工科 SCI 研究论文的题目与摘要、引言、方法、结果与讨论、结论等主要章节的文本特点。以论文引言为例，首先基于语料库技术选取颇为典型的引言文本作为阅读材料，之后围绕引言的篇章结构、修辞策略和显著语言特征，并结合基于语料库提取出的文本示例，依次呈现教学内容、设计练习活动，引导学生观察归纳引言文本的多维特点。

综上所述，基于语言学理论和较大批量文本并借助语料库分析技术与自然语言处理技术，本教材深入考察了工科 SCI 研究论文从宏观语篇结构到微观语言特征不同维度的显著特性，有力支撑了教材编写理念，提供了丰富、典型的文本素材，确保了教学任务与练习活动设计的系统性、针对性与层次性，构建了高质量的语言输入，助力提升学生的学术交际能力。

3. 注重学术技能、学术思维和学术素养的一体化培养

本教材除多维度呈现典型工科领域 SCI 期刊论文的篇章结构和语言特征外，极为注重一体化培养学生的核心学术技能、学术思维和学术素养。SCI 期刊论文形式上包含摘要、引言、方法、结果与讨论、结论等章节，各个章节又进而呈现独特的篇章结构，但这些均建立于或蕴含着诸多共核的学术技能、学术思维和学术素养。学术技能诸如定义关键术语与概念、介绍公式及符号、描述图表及含义、作出综述或回顾、呈现系统与原理、梳理流程及步骤、强调特点与性能、描述实验设计等。学术思维涉及诸如问题导向、合乎规约、逻辑缜密、承前启后、证据理据、推理严谨、读者交互等。研究问题的提出、分析与解决无不体现主观与客观、整体与局部、普遍与特殊、抽象与具体、共性与个性、动态与静态、复杂与简约、事实与假设等多组关系的辩证统一。学术素养除叠加上述学术技能和学术思维外，还包括学术规范、学术诚信、创新意识、踏实作风、求真精神等重要元素。学术技能、学术思维和学术素养的提升不可能一蹴而就、立竿见影。本教材选取语言教学价值和思政教育价值均颇为典型的学术语篇作为切入点，引导学生体会、领会科学语言背后蕴含的核心技能、思维模式和基本素养，同时鼓励学生紧贴自身科研实践开展演练，如界定所在研究领域的关键术语、报道所读文献的研究成果并作出客观评判、描述实验设计并论述其科学性与可行性、结合图表数值解读其普遍意义，以此循序渐进地提升学生的学术交际能力、创造性思维水平和科学精神素养。

4. 突显语境化、任务型和项目制教学特点，紧贴学生专业学习和科研实践

本教材全方位凸显语境化、任务型和项目制教学特点，深度融合学生专业兴趣、学生自主学习与小组研讨活动，引导学生学中用、用中学，紧贴自身科研实践开展论文读写，最大程度地激发学生英语学习的内生动力。具体而言，本教材逐一剖析工科领域 SCI 期刊论文各个章节的篇章结构和语言特征之余，同步设置小组讨论、写作任务、口头展示、师生评价等环节，鼓励学生结合自身开展或参与的各类大学生创新项目、学科竞赛等，课外阅读自选或学业导师推荐的期刊论文。学生需应用课上所学学术技能，汲取归纳课外所读文献旨在传递的专业知识与前沿信息，并通过小组研讨、写作练习与课堂展示等，描述评价前人学术观点、研究目标、方法手段、研究结果等。与此同时，学生需描述自身所读文献的篇章结构与语言特征，讨论一致与差异、共性与个性，并在教师指导下探究成因。在此基础上，学生紧贴自身科研实践并依据教材单元进度，尝试界定核心概念、梳理前人文献、概括重要结果、明确尚存问题、提出研究目标、描述方法步骤等，逐一撰写论文题目、摘要、引言、方法、结果与讨论、结论等，开展小组陈述与互评，并对照拟投稿期刊要求进行修改完善，最终形成一篇结构完整、行文规范、表述精准的研究论文。简而言之，本教材极为注重教学材料、教学活动和产出导向的三位一体，凸显语境化、任务型和项目制特点，引导学生紧贴自身专业兴趣和科研实践读文献、写本子，同时充分发挥同伴互助作用，着力构建语言知识、交际能力和专业知识协同发展的有效平台，助力提升学术英语教育教学质量与效率。

三、教材使用

本教材适用于工科领域尤其是材料科学与工程、机械工程、环境科学与工程、兵器科学与技术、生物医学工程、冶金工程、力学学科以及相关数学、物理学、化学学科的本科生和研究生的学术英语课程，也可作为上述领域科研人员提升自身期刊论文撰写与发表能力的阅读材料。

本教材使用可设置为一学期 32、48 或 64 课时，面向本科生建议至少 48 课时。教学实践中，依据教材单元编排，依次讲授 SCI 期刊论文的题目与摘要、引言、方法、结果与讨论、结论等章节的文本特点。各个单元所需课时可平均配置，也可视实际情况灵活增减。除教材文本外，课程之初，引导学生尽快确定自身研究领域及具体话题，同时邀请学业导师为之推荐 SCI 期刊论文作为课外练兵场。依据教学进度，学生应用所学技能，剖析自身所在领域期刊论文的篇章结构与语言特征，并开展小组讨论。需要说明的是，学业导师推荐的学术论文本身就是学生专业学习的必读文献。因此，课外阅读不仅不会增加学生的学习负担，更有助于其学以致用、学用相长，进一步了解期刊论文写作特点，并充分汲取学科前沿信息。基于课内外文本阅读与分析，学生尝试将自身科研实践撰写成文，依据教学进度逐步撰写论文题目、摘要、引言、方法、结果与讨论以及结论等，并开展小组陈述与互评。

依据教材设计，课程考核方式建议采取形成性评价和终结性评价相结合的方式，各占课程总分的 50%。形成性评价体现在多种练习活动设计，如依据教材单元进度所开展的课内外文本阅读、小组讨论、写作练习以及口头陈述。终结性评价为学生提交一篇结构完整、行文规范、表述精准的学术论文，同时模拟学术会议进行 PPT 陈述并回答师生提问。

综上所述，本教材紧扣新工科建设对于拔尖创新人才培养提出的更高要求，紧贴学生学业发展需求和学术话语社团情境要素，聚焦典型工科领域 SCI 期刊论文文本特点，深度融合学术技能、学术思维和学术素养，引导学生围绕自身科研实践并通过语境化、任务型和项目制教学，以提升学生的学术交际能力尤其是学术论文读写能力，助力其汲取学科知识并传播自身成果，从而为增强国家科技实力和提升国家科技话语权做出贡献。

本教材入选北京理工大学“十四五”（2023 年）规划教材，同时得到第十一批“中国外语教育基金”项目“新工科背景下语类分析和学生科研实践相融合的学术英语教学模式构建研究”、北京理工大学校级教育教学改革项目“科技语篇分析与学生科研实践协同互促的学术英语教学模式优化研究”和“‘新四科’背景下面向学术英语教学的学术词汇及搭配列表研制”的大力支持。

本教材编写虽力求完美，但难免存有待商榷之处，敬请广大同仁批评指正。对于教材设计与使用若有疑问、对于课后习题答案如有需求，欢迎联系 readwritesci@126.com。

闫鹏飞

2023 年 10 月

Contents

Unit 1 Research Article Titles and Abstracts

Text Ⅰ Highly Stable Tungsten Disulfide Supported on a Self-Standing Nickel Phosphide Foam As a Hybrid Electrocatalyst for Efficient Electrolytic Hydrogen Evolution 8

Text Ⅱ Measuring the Competition Between Bimolecular Charge Recombination and Charge Transport in Organic Solar Cells Under Operating Conditions 35

Text Ⅲ Fabrication of Graphene Oxide and Hyperbranched Polyurethane Composite Via *in Situ* Polymerization with Improved Mechanical and Dielectric Properties 46

Unit 2 Introduction Sections

Text Ⅰ Architected Cellular Piezoelectric Metamaterials: Thermo-Electro-Mechanical Properties 61

Text Ⅱ Eddy Current Induced Dynamic Deformation Behaviors of Aluminum Alloy During EMF: Modeling and Quantitative Characterization 75

Text Ⅲ Full-field Measurement With Nanometric Accuracy of 3D Superficial Displacements by Digital Profile Correlation: A Powerful Tool for Mechanics of Materials 108

Unit 3 Methods Sections

Text Ⅰ Live Imaging of Label-Free Graphene Oxide Reveals Critical Factors Causing Oxidative-Stress-Mediated Cellular Responses 148

Text Ⅱ Lab Manual: Rockwell Hardness Test 170

Text Ⅲ Mixed Surface Chemistry: An Approach to Highly Luminescent Biocompatible Amphiphilic Silicon Nanocrystals 189

Unit 4 Results and Discussion Sections

Text Ⅰ Determination and Modeling of Mechanical Properties for Graphene Nanoplatelet/Epoxy Composites 208

Text Ⅱ A New Type of Vibration Isolator Based on Magnetorheological Elastomer 230

Text Ⅲ Improvement of S355G10+N Steel Weldability in Water Environment by Temper Bead Welding 267

Unit 5 Conclusion Sections

Text Ⅰ Formation, Photoluminescence and Ferromagnetic Characterization of Ce Doped AlN Hierarchical Nanostructures 290

Text Ⅱ High-Resolution Cryo-Electron Microscopy Structure of Block Copolymer Nanofibres with a Crystalline Core 315

Text Ⅲ A Mesocrystal-Like Morphology Formed by Classical Polymer-Mediated Crystal Growth 329

References 357

Unit 1
Research Article Titles and Abstracts

Objectives

- Identify typical content and various forms of research article titles in engineering disciplines;
- Acquaint yourself with generic structure of research article abstracts and their disciplinary variations;
- Distinguish between obligatory and optional structural elements of research article abstracts;
- Identify salient linguistic features characterizing each structural element of research article abstracts;
- Understand different rhetorical functions of common reporting verbs and the appropriate verb-noun collocations;
- Identify complex and compressed noun phrases conveying technical contents;
- Raise awareness of lexical and syntactic complexity in academic abstract writing.

Task 1: Read the research article titles below and discuss which of the following questions a typical research article title is expected to answer.

(1) Interplay of water and reactive elements in oxidation of alumina-forming alloys.

(2) Fabrication of graphene oxide and hyperbranched polyurethane composite via *in situ* polymerization with improved mechanical and dielectric properties.

(3) An autonomously electrically self-healing liquid metal-elastomer composite for robust soft-matter robotics and electronics.

(4) Electron delocalization and charge mobility as a function of reduction in a metal-organic framework.

(5) Small molecule acceptors with a nonfused architecture for high-performance organic photovoltaics.

(6) A Mg-doped high-nickel layered oxide cathode enabling safer, high-energy-density li-Ion batteries.
(7) Temperature-responsive magnetic nanoparticles for enabling affinity separation of extracellular vesicles.
(8) Metal-organic framework-derived Co_3O_4/Au heterostructure as a catalyst for efficient oxygen reduction.
(9) A robust process-structure model for predicting the joint interface structure in impact welding.
(10) Fabrication of porous NiAl intermetallic compounds with a hierarchical open-cell structure by combustion synthesis reaction and space holder method.
(11) A self-healing hydrogel as an injectable instructive carrier for cellular morphogenesis.
(12) Efficient magnetic enrichment of antigen-specific T cells by engineering particle properties.
(13) Flexible thermoelectric polymer composites based on a carbon nanotubes forest.
(14) Phototransistors with negative or ambipolar photoresponse based on as-grown heterostructures of single-walled carbon nanotube and MoS_2.
(15) A cobalt and nickel co-modified layered P2-$Na_{2/3}Mn_{1/2}Fe_{1/2}O_2$ with excellent cycle stability for high-energy density sodium-ion batteries.
(16) Three-dimensional printing hollow polymer template-mediated graphene lattices with tailorable architectures and multifunctional properties.
(17) Transparent and flexible piezoelectric sensor for detecting human movement with a boron nitride nanosheet (BNNS).
(18) Multilayered ferroelectric polymer films incorporating low-dielectric-constant components for concurrent enhancement of energy density and charge–discharge efficiency.
(19) Effects of state of charge on elastic properties of 3D structural battery composites.
(20) Concurrently improved dispersion and interfacial interaction in rubber/nanosilica composites via efficient hydrosilane functionalization.
(21) An encouraging recyclable synergistic hydrogen bond crosslinked high-performance polymer with visual detection of tensile strength.
(22) Influence of porosity and aspect ratio of nanoparticles on the interface modification of glass/epoxy composites.
(23) Structural design and mechanical response of gradient porous Ti-6Al-4V fabricated by electron beam additive manufacturing.
(24) Ultraviolet-light-driven charge carriers tunability mechanism in graphene.
(25) Validation of a deplasticizer-ball milling method for separating Cu and PVC from thin electric cables: a simulation and experimental approach.
(26) Oxygen vacancy creation energy in Mn-containing perovskites: an effective indicator for chemical looping with oxygen uncoupling.
(27) Effective and noneffective recombination center defects in Cu_2ZnSnS_4: significant difference in carrier capture cross sections.

(28) Finding the right bricks for molecular legos: a data mining approach to organic semiconductor design.
(29) Origins and control of optical absorption in a nondilute oxide solid solution: Sr(Ti,Fe)O_{3-x} perovskite case study.
(30) Molecular transmission: visible and rate-controllable photoreactivity and synergy of aggregation-induced emission and host-guest assembly.
(31) Three-dimensionally printed silk-sericin-based hydrogel scaffold: a promising visualized dressing material for real-time monitoring of wounds.
(32) Calcium phosphate nanoparticles as intrinsic inorganic antimicrobials: the antibacterial effect.
(33) Paving the way for K-Ion batteries: role of electrolyte reactivity through the example of Sb-based electrodes.
(34) Pt-embedded CuO_x-CeO_2 multicore-shell composites: interfacial redox reaction-directed synthesis and composition-dependent performance for CO oxidation.
(35) Smooth-shell metamaterials of cubic symmetry: anisotropic elasticity, yield strength and specific energy absorption.
(36) Grain size dependent physical properties in lead-free multifunctional piezoceramics: a case study of NBT-xST system.
(37) Tailored microstructures of gadolinium zirconate/YSZ multi-layered thermal barrier coatings produced by suspension plasma spray: Durability and erosion testing.
(38) Ultrasonic vibration-assisted metal forming: constitutive modelling of acoustoplasticity and applications.
(39) Internal decompression of the acutely contused spinal cord: differential effects of irrigation only versus biodegradable scaffold implantation.
(40) Hydrogels with an embossed surface: an all-in-one platform for mass production and culture of human adipose-derived stem cell spheroids.

Questions	Yes/No?
(1) Is theoretical background indicated?	
(2) Are test materials (or samples) stated?	
(3) Is research design (or method) mentioned?	
(4) Are experimental procedures reported?	
(5) Is experimental apparatus (or equipment) included?	
(6) Are research results highlighted?	
(7) Is the practical application value of the research indicated?	

Task 2: Read the above research article titles again and discuss which forms most titles prefer, either a single phrase (nominal or participial), a punctuation-partitioned multipart one (usually separated by a colon or a dash), or a full sentence (declarative or question). The following are examples with noun phrases consolidated.

(1) Interplay of water and reactive elements in oxidation of alumina-forming alloys.

(2) Fabrication of graphene oxide and hyperbranched polyurethane composite via *in situ* polymerization with improved mechanical and dielectric properties.

(3) An autonomously electrically self-healing liquid metal-elastomer composite for robust soft-matter robotics and electronics.

(4) Electron delocalization and charge mobility as a function of reduction in a metal-organic framework.

(5) Small molecule acceptors with a nonfused architecture for high-performance organic photovoltaics.

(6) A mg-doped high-nickel layered oxide cathode enabling safer, high-energy-density li-Ion batteries.

(7) Temperature-responsive magnetic nanoparticles for enabling affinity separation of extracellular vesicles.

(8) Metal-organic framework-derived Co_3O_4/Au heterostructure as a catalyst for efficient oxygen reduction.

(9) A robust process-structure model for predicting the joint interface structure in impact welding.

(10) Fabrication of porous NiAl intermetallic compounds with a hierarchical open-cell structure by combustion synthesis reaction and space holder method.

(11) A self-healing hydrogel as an injectable instructive carrier for cellular morphogenesis.

(12) Efficient magnetic enrichment of antigen-specific T cells by engineering particle properties

(13) Flexible thermoelectric polymer composites based on a carbon nanotubes forest.

(14) Phototransistors with negative or ambipolar photoresponse based on as-grown heterostructures of single-walled carbon nanotube and MoS_2.

(15) A cobalt and nickel co-modified layered P2-$Na_{2/3}Mn_{1/2}Fe_{1/2}O_2$ with excellent cycle stability for high-energy density sodium-ion batteries.

(16) Three-dimensional printing hollow polymer template-mediated graphene lattices with tailorable architectures and multifunctional properties.

(17) Transparent and flexible piezoelectric sensor for detecting human movement with a boron nitride nanosheet (BNNS).

(18) Multilayered ferroelectric polymer films incorporating low-dielectric-constant components for concurrent enhancement of energy density and charge-discharge efficiency.

(19) Effects of state of charge on elastic properties of 3D structural battery composites.

(20) Concurrently improved dispersion and interfacial interaction in rubber/nanosilica composites via efficient hydrosilane functionalization.

(21) An encouraging recyclable synergistic hydrogen bond crosslinked high-performance polymer with visual detection of tensile strength.

(22) Influence of porosity and aspect ratio of nanoparticles on the interface modification of glass/epoxy composites.

(23) Structural design and mechanical response of gradient porous Ti-6Al-4V fabricated by electron beam additive manufacturing.

(24) Ultraviolet-light-driven charge carriers tenability mechanism in graphene.

(25) Validation of a deplasticizer-ball milling method for separating Cu and PVC from thin electric cables: a simulation and experimental approach.

(26) Oxygen vacancy creation energy in Mn-containing perovskites: an effective indicator for chemical looping with oxygen uncoupling.

(27) Effective and noneffective recombination center defects in Cu_2ZnSnS_4: significant difference in carrier capture cross sections.

(28) Finding the right bricks for molecular legos: a data mining approach to organic semiconductor design.

(29) Origins and control of optical absorption in a nondilute oxide solid solution: $Sr(Ti, Fe)O_{3-x}$ perovskite case study.

(30) Molecular transmission: visible and rate-controllable photoreactivity and synergy of aggregation-induced emission and host-guest assembly.

(31) Three-dimensionally printed silk-sericin-based hydrogel scaffold: a promising visualized dressing material for real-time monitoring of wounds.

(32) Calcium phosphate nanoparticles as intrinsic inorganic antimicrobials: the antibacterial Effect.

(33) Paving the way for K-Ion Batteries: role of electrolyte reactivity through the example of Sb-Based Electrodes.

(34) Pt-Embedded CuO_x-CeO_2 Multicore-Shell composites: interfacial redox reaction-directed synthesis and composition-dependent performance for CO oxidation.

(35) Smooth-shell metamaterials of cubic symmetry: anisotropic elasticity, yield strength and specific energy absorption.

(36) Grain size dependent physical properties in lead-free multifunctional piezoceramics: a case study of NBT-xST system.

(37) Tailored microstructures of gadolinium zirconate/YSZ multi-layered thermal barrier coatings produced by suspension plasma spray: durability and erosion testing.

(38) Ultrasonic vibration-assisted metal forming: constitutive modelling of acoustoplasticity and applications.

(39) Internal decompression of the acutely contused spinal cord: differential effects of irrigation only versus biodegradable scaffold implantation.

(40) Hydrogels with an embossed surface: an all-in-one platform for mass production and culture of human adipose-derived stem cell spheroids.

Task 3: Analyze the noun phrase complexity in the research article abstract exemplified below by highlighting the head noun of each phrase and its possible pre- and post-modification. Examine the use of simple and complex noun phrases in Text I and distinguish technical jargon from general noun phrases.

Example[1]

Atomic layer deposition (ALD) enables the deposition of numerous materials in thin film form, yet there are no ALD processes for metal iodides. Herein, we demonstrate an ALD process for PbI_2, a metal iodide with a two-dimensional (2D) structure that has applications in areas such as photodetection and photovoltaics. This process uses lead silylamide $Pb(btsa)_2$ and SnI_4 as precursors and works at temperatures below 90 °C, on a variety of starting surfaces and substrates such as polymers, metals, metal sulfides, and oxides. The starting surface defines the crystalline texture and morphology of the PbI_2 films. Rough substrates yield porous PbI_2 films with randomly oriented 2D layers, whereas smooth substrates yield dense films with 2D layers parallel to the substrate surface. Exposure to light increases conductivity of the ALD PbI_2 films which enables their use in photodetectors. The films can be converted into a $CH_3NH_3PbI_3$ halide perovskite, an important solar cell absorber material. For various applications, ALD offers advantages such as ability to uniformly coat large areas and simple means to control film thickness. We anticipate that the chemistry exploited in the PbI_2 ALD process is also applicable for ALD of other metal halides.

(1) (Determiner) + (pre-modification) + **head noun**:

this **process** the **films**	determiner
various **applications** rough **substrates** smooth **substrates**	adjective
the starting **surface**	determiner, present participle
a $CH_3NH_3PbI_3$ halide **perovskite**	determiner, noun
an important solar cell absorber **material**	determiner, adjective, noun

(2) (Determiner) + **head noun** + post-modification:

the **deposition** of numerous materials in thin film form **temperatures** below 90 °C **exposure** to light **ALD** of other metal halides	prepositional phrase
conductivity of the ALD PbI_2 films which enables their use in photodetectors	prepositional phrase relative clause
advantages such as ability to uniformly coat large areas	prepositional phrase to-clause

(3) (Determiner) + pre-modification + **head noun** + post-modification:

Atomic layer **deposition** (ALD)	adjective, noun; noun phrase as appositives
no ALD **processes** for metal iodides an ALD **process** for PbI_2	determiner, noun; prepositional phrase
a metal **iodide** with a two-dimensional (2D) structure that has applications in areas such as photodetection and photovoltaics	determiner, noun; prepositional phrase, relative clause
a **variety** of starting surfaces and substrates such as polymers, metals, metal sulfides, and oxides	determiner; prepositional phrase
the crystalline **texture and morphology** of the PbI_2 films	determiner, adjective; prepositional phrase
porous PbI_2 **films** with randomly oriented 2D layers	adjective, noun; prepositional phrase
dense **films** with 2D layers parallel to the substrate surface	adjective; prepositional phrase
the **chemistry** exploited in the PbI_2 ALD process	determiner; past participial clause
simple **means** to control film thickness	adjective; to-clause

Text I

Highly Stable Tungsten Disulfide Supported on a Self-Standing Nickel Phosphide Foam as a Hybrid Electrocatalyst for Efficient Electrolytic Hydrogen Evolution[2]

The detrimental impacts on environment brought by the combustion of fossil fuel have urged scientists to search for various green substituents. In particular, molecular hydrogen is proposed as one of the competitive candidates among the wide spectrum of different energy sources due to the high gravimetric energy density, cleanliness and renewability. However, the batch production of H_2 via electrolytic route requires the cost-effective and high-performance electrocatalysts to minimize the energy loss during conversion. Tungsten disulfides (2H-WS_2) is an earth-abundant HER electrocatalyst which possesses high intrinsic catalytic activity and chemical stability. Nonetheless, the HER performance of WS_2 is limited by the sparse catalytic edge sites and the unsatisfactory on-plane electrical conductivity. On the other hand, the self-standing nickel phosphide (Ni_5P_4-Ni_2P) grown directly on the Ni foam is a compelling structure for water splitting due to its macroporosity and the metallic nature of 3D networks. Here, we report the hybridization of WS_2 and the 3D porous metallic Ni_5P_4-Ni_2P foam induced through a thermal process in a CVD furnace. The resultant hybridized WS_2/Ni_5P_4-Ni_2P electrode demonstrated intriguing synergistic effects that improve the overall electrode performances considerably. Experimentally, it was found that the surface of WS_2/Ni_5P_4-Ni_2P electrode contains a ternary $WS_{2(1-x)}P_{2x}$ constituent, bringing the advent of desired electronic perturbations. Notably, the 3D porous metallic WS_2/Ni_5P_4-Ni_2P electrode only requires 94 mV (vs. RHE) to drive a geometric current density of -10 mA cm^{-2} with a relatively small Tafel slope of 74 mV dec^{-1}. Such enhancements are correlated to the rational designs of WS_2/Ni_5P_4-Ni_2P electrode which (i) optimize the hydrogen adsorption Gibbs free energy ; (ii) increase the number of active sites of overall electrode; (iii) facilitate the electron transport from higher Fermi-level Ni_5P_4-Ni_2P foam to the lower Fermi-level WS_2; (iv) promote the access of electrolyte and the escape of hydrogen molecules. Additionally, such catalytic performances of the WS_2/Ni_5P_4-Ni_2P electrode can last for 22 h without significant degradations, clearly demonstrating the superior long-term operation stability in the acid environment (0.5 M H_2SO_4). This work provides a versatile platform for fabricating the 3D porous hybridized electrocatalysts for modulating the intrinsic catalytic activity of various transition metal dichalcogenides (TMDs) on nickel phosphide.

(1) (Determiner) + (pre-modification) + **head noun**:

____

____

(2) (Determiner) + **head noun** + post-modification:

____

____

____

(3) (Determiner) + pre-modification + **head noun** + post-modification:

____

____

____

Task 4: Analyze the structural patterns of the following research article abstracts sentence by sentence with reference to the examples below, and understand that research article abstracts can vary greatly in both structure and length.

Typical structural components of research article abstracts:

(1) Background (Setting the scene for the current study).

(2) Purpose (Stating the purpose of the study, research questions, and/or hypotheses).

(3) Methods (Describing test materials, equipment, theoretical framework, experimental procedures, etc.).

(4) Results (Reporting principle findings of the study).

(5) Discussion/Conclusion (Interpreting research findings, discussing implications/applications, and/or providing a summary).

Note: Not all the structural components will necessarily occur in a research article abstract. Each structural component could be realized by one main clause or subordinate clause, or one or more sentences. In addition, the embedding and cycling of the structural components may be observed, meaning that two or more structural components could be syntactically integrated into one sentence, functioning as the main clause or subordinate clause, while they may co-occur with each other in a particular sequence more than once.

Example (1)

Atomic Layer Deposition of PbI_2 Thin Films[1] 1)Atomic layer deposition (ALD) enables the deposition of numerous materials in thin film form, yet there are no ALD processes for metal iodides. 2)Herein, we demonstrate an ALD process for PbI_2, a metal iodide with a two-dimensional (2D) structure that has applications in areas such as photodetection and photovoltaics. 3)This process uses	 1): <u>Background</u> 2): <u>Purpose</u> 3): <u>Methods</u>

continued

lead silylamide Pb(btsa) and SnI_4 as precursors and works at temperatures below 90°C, on a variety of starting surfaces and substrates such as polymers, metals, metal sulfides, and oxides. 4)The starting surface defines the crystalline texture and morphology of the PbI_2 films. 5)Rough substrates yield porous PbI_2 films with randomly oriented 2D layers, whereas smooth substrates yield dense films with 2D layers parallel to the substrate surface. 6)Exposure to light increases conductivity of the ALD PbI_2 films which enables their use in photodetectors. 7)The films can be converted into a $CH_3NH_3PbI_3$ halide perovskite, an important solar cell absorber material. 8)For various applications, ALD offers advantages such as ability to uniformly coat large areas and simple means to control film thickness. 9)We anticipate that the chemistry exploited in the PbI_2 ALD process is also applicable for ALD of other metal halides.	4)-6): <u>Results</u> 7)-9): <u>Discussion/ Conclusion</u>

Example (2)

Bacterial resistance to silver nanoparticles and how to overcome it[3] 1)Silver nanoparticles have already been successfully applied in various biomedical and antimicrobial technologies and products used in everyday life. 2)Although bacterial resistance to antibiotics has been extensively discussed in the literature, the possible development of resistance to silver nanoparticles has not been fully explored. 3)We report that the Gram-negative bacteria *Escherichia coli* 013, *Pseudomonas aeruginosa* CCM 3955 and *E. coli* CCM 3954 can develop resistance to silver nanoparticles after repeated exposure. 4)The resistance stems from the production of the adhesive flagellum protein flagellin, which triggers the aggregation of the nanoparticles. 5)This resistance evolves without any genetic changes; only phenotypic change is needed to reduce the nanoparticles' colloidal stability and thus eliminate their antibacterial activity. 6)The resistance mechanism cannot be overcome by additional stabilization of silver nanoparticles using surfactants or polymers. 7)It is, however, strongly suppressed by inhibiting flagellin production with pomegranate rind extract.	1)-2): <u>Background</u> 3): <u>Purpose</u> 4)-5): <u>Results</u> 6)-7): <u>Discussion/ Conclusion</u>

Example (3)

Functionalized aramid nanofibers prepared by polymerization induced self-assembly for simultaneously reinforcing and toughening of epoxy and carbon fiber/epoxy multiscale composite[4]	
1)Aramid nanofibers (ANFs) are polymeric nanofibers that have been drawing tremendous attention as new nanoscale building blocks of advanced composites.	1): Background
2)Herein, we report a new strategy to rapidly synthesize functionalized ANFs (fANFs) through polymerization-induced self-assembly of poly(phenylene terephthalamide) (PPTA) using soluble sulfonated PPTA to tailor the size of fANTs, and the extraordinary reinforcing and toughening effect of fANFs on epoxy (EP) resin and multiscale carbon fiber (CF)/EP composites.	2): Purpose and methods
3)With a diameter of ~30 nm, and reactive sulfonic acid groups on the surfaces, the branched fANFs can be easily dispersed in EP and are reactive with the matrix, therefore, remarkably improve the interfacial interaction and mechanical properties (including both strength and toughness) of EP at low fANF contents. 4)The tensile strength, Young's modulus, toughness and elongation at break of fANFs/EP nanocomposites are increased by ~59%, ~19%, ~112% and ~36%, respectively. 5)The flexural strength, interlaminar shear strength, total energy dissipation and flexural strain at break of fANFs/CF/EP multi-scale composite are increased by ~57%, ~38%, ~65% and ~71%, respectively.	3)-5): Results
6)To the best of our knowledge, the enhancement in the strengths is comparative with those provided by graphene oxide, and the total fracture energy and flexural strain at break are more significantly increased by fANFs than almost all of the nanofillers ever reported.	6): Discussion/ Conclusion

(1)

An encouraging recyclable synergistic hydrogen bond crosslinked high-performance polymer with visual detection of tensile strength[5]	
1)We have successfully constructed a novel type of recyclable crosslinked protonated-polybenzotriazole ether sulfone ($PBTSH^+$), which possessed outstanding thermal stability and mechanical properties. 2)More importantly, the tensile strength of the resulting crosslinked films could be visually detected by taking advantage of the transparency decreasing effect of synergistic hydrogen bonds between protonated benzotriazole units.	1): _________ 2): _________

(2)

Serration dynamics in the presence of chemical heterogeneities for a Cu-Zr based bulk metallic glass[6]	
1)The serration flow behavior for $Cu_{47}Zr_{47}Al_6$ and $(Cu_{47}Zr_{47}Al_6)_{99}Sn_1$ (at. %) bulk metallic glasses (BMGs) was investigated through statistical analysis. 2)The statistical analysis reveals that the $(Cu_{47}Zr_{47}Al_6)_{99}Sn_1$ BMG has more stable self-organized critical (SOC) dynamics. 3)The chemical heterogeneities formation in the microstructure has a triple effect on the deformation dynamics, which leads to high ductility for the $(Cu_{47}Zr_{47}Al_6)_{99}Sn_1$ BMG.	1): _________ 2)-3): _________

(3)

P_4Se_3 as a new anode material for sodium-ion batteries[7]	
1)P_4Se_3 microparticles are successfully synthesized by simple heating of selenium and amorphous red phosphorus at low temperature. 2)These orthorhombic-phase microparticles exhibit a high reversible capacity of 654 mA h g^{-1} at 200 mA g^{-1} after 70 cycles and an excellent rate capability with a reversible capacity of 486 mA h g^{-1} at 3 A g^{-1} as an anode material in sodium-ion batteries. 3)This superior electrochemical performance is attributed to the formation of Na_2Se and Se during cycling with better conductivities ($\approx 10^{-6}$ and $\approx 10^{-5}$ S cm^{-1}) and the dominant position of capacitance behavior, which all accelerate kinetic behavior.	1): _________ 2): _________ 3): _________

(4)

Anode-Free Rechargeable Lithium Metal Batteries[8]	
1)Anode-free rechargeable lithium (Li) batteries (AFLBs) are phenomenal energy storage systems due to their significantly increased energy density and reduced cost relative to Li-ion batteries, as well as ease of assembly because of the absence of an active (reactive) anode material. 2)However, significant challenges, including Li dendrite growth and low cycling Coulombic efficiency (CE), have prevented their practical implementation. 3)Here, an anode-free rechargeable lithium battery based on a Cu||$LiFePO_4$ cell structure with an extremely high CE (>99.8%) is reported for the first time. 4)This results from the utilization of both an exceptionally stable electrolyte and optimized charge/discharge protocols, which minimize the corrosion of the in situly formed Li metal anode.	1)-2): ________ 3): ________ 4): ________

(5)

Local structural change and magnetic dilution effect in (Ca^{2+}, V^{5+}) co-substituted yttrium iron garnet prepared by sol-gel route[9]	
1)Co-substituted $Y_{3-2x}Ca_{2x}Fe_{5-x}V_xO_{12}$ samples ($x = 0$, 0.1, 0.2, 0.3, 0.4, 0.5, 0.6, 0.7, 0.8, 0.9, 0.95 and 1) were prepared by using citrate sol-gel method. 2)Synchrotron X-ray diffraction (SXRD) measurements combined with Rietveld refinement technique were used to investigate the crystallization, structural parameters and lattice distortion in the samples. 3)Magnetization and Curie temperature were studied by means of a vibrating sample magnetometer (VSM). 4)Molecular-field coefficients N_{aa}, N_{dd} and N_{ad} of the samples were determined by fitting the experimental thermomagnetization curves in the framework of two-sublattice ferrimagnetic model. 5)The information of local structure change was employed to analyze the influence of the substitution effects in the magnetic interactions.	1): ________ 2)-5): ________

(6)

Highly Efficient Electrocatalysts for Oxygen Reduction Reaction Based on 1D Ternary Doped Porous Carbons Derived from Carbon Nanotube Directed Conjugated Microporous Polymers[10]	
1)One-dimensional (1D) porous materials have shown great potential for gas storage and separation, sensing, energy storage, and conversion. 2)However, the controlled approach for preparation of 1D porous materials, especially porous organic materials, still remains a great challenge due to the poor dispersibility and solution processability of the porous materials.	1)-2): ________
3)Here, carbon nanotube (CNT) templated 1D conjugated microporous polymers (CMPs) are prepared using a layer-by-layer method.	3): ________
4)As-prepared CMPs possess high specific surface areas of up to 623 $m^2\ g^{-1}$ and exhibit strong electronic interactions between p-type CMPs and n-type CNTs.	4): ________
5)The CMPs are used as precursors to produce heteroatom-doped 1D porous carbons through direct pyrolysis.	5): ________
6)As-produced ternary heteroatom-doped (B/N/S) 1D porous carbons possess high specific surface areas of up to 750 $m^2\ g^{-1}$, hierarchical porous structures, and excellent electrochemical-catalytic performance for oxygen reduction reaction. 7)Both of the diffusion-limited current density (4.4 $mA\ cm^{-2}$) and electron transfer number ($n = 3.8$) for three-layered 1D porous carbons are superior to those for random 1D porous carbon.	6)-7): ________
8)These results demonstrate that layered and core–shell type 1D CMPs and related heteroatom-doped 1D porous carbons can be rationally designed and controlled prepared for high performance energy-related applications.	8): ________

(7)

Stability of La dopants in $NaTaO_3$ photocatalysts[11]	
1)Doping $NaTaO_3$ with La typically results in increased photocatalytic activity. 2)The photocatalytic activity of La-doped $NaTaO_3$ should therefore be sensitive to the photostability of the La species being doped. 3)Unfortunately, the photostability of the La species in $NaTaO_3$ is not fully known.	1)-3): ________
4)Herein, we perform an in-depth study on the photostability of the La species.	4): ________
5)Advanced characterization techniques, including X-ray photoelectron spectroscopy (XPS), X-ray absorption near edge structure (XANES), and extended X-ray absorption fine	5): ________

continued

structure (EXAFS), are employed to clarify the chemical state and local structure of the La species in $NaTaO_3$. 6)Our findings show that the examined La-doped $NaTaO_3$ can be effectively reused for six consecutive runs without a significant loss of its photocatalytic activity for the decomposition of recalcitrant organic compounds under UV light. 7)The reusability of La-doped $NaTaO_3$ is proposed to be due to the photostability of the La species. 8)The La species are photostable because their oxidation state is unaltered and their structure in the host $NaTaO_3$ is well preserved.	6): _________ 7)-8): _________

(8)

Sintering processes in direct ink write additive manufacturing: A mesoscopic modeling approach[12]	
1)Direct ink write (DIW) is an emerging additive manufacturing technique that allows for the fabrication of arbitrary complex geometries required in many technologies. 2)DIW of metallic or ceramic materials involves a sintering step, which greatly influences many of the microstructural features of the printed object. 3)Herein, we explore solid-state sintering in DIW through a mesoscopic modeling framework that is capable of capturing bulk and interface thermodynamics and accounting for various mass transport mechanisms. 4)Simulation results of idealized geometries identify regimes in materials parameter space, where densification rates are enhanced. 5)With the aid of several statistical and topological descriptors, the role of particle size distribution (PSD) on the microstructural evolution is explored and quantified. 6)More specifically, it is found that a bi-dispersed PSD enhances pore shrinkage kinetics. 7)However, bi-dispersity yields microstructures with pores that are highly eccentric, an effect that could be detrimental to the mechanical properties of the printed material. 8)On the whole, our modeling approach provides a capability to explore the phase space of DIW process parameters and determine ones that lead to optimal microstructures.	1)-2): _________ 3): _________ 4)-5): _________ 6): _________ 7): _________ 8): _________

(9)

High-Performance Nanowire Hydrogen Sensors by Exploiting the Synergistic Effect of Pd Nanoparticles and Metal-Organic Framework Membranes[13]	
1)Herein, we report the fabrication of hydrogen gas sensors with enhanced sensitivity and excellent selectivity. 2)The sensor device is based on the strategic combination of ZnO nanowires (NWs) decorated with palladium nanoparticles (Pd NPs) and a molecular sieve metal-organic framework (MOF) nanomembrane (ZIF-8). 3)The Pd NPs permit the sensors to reach maximal signal responses, whereas the ZIF-8 overcoat enables for an excellent selectivity.	1)-3) ________
4)Three steps were employed for the fabrication: (i) coating of a miniaturized sensor with vapor-grown ZnO NWs, (ii) decoration of these NWs with Pd NPs by atomic layer deposition, and (iii) partial solvothermal conversion of the tuned NWs surface to ZIF-8 nanomembrane.	4): ________
5)The microstructure and composition investigations of the ZIF-8/Pd/ZnO nanostructured materials confirmed the presence of both metallic Pd NPs and uniform ZIF-8 thin membrane layer. 6)The integration of these nanomaterials within a miniaturized sensor device enabled the assessment of their performance for H_2 detection at concentrations as low as 10 ppm in the presence of various gases such as C_6H_6, C_7H_8, C_2H_5OH, and CH_3COCH_3.	5)-6): ________
7)Remarkably high-response signals of 3.2, 4.7, and 6.7 (R_a/R_g) have been measured for H_2 detection at only 10, 30, and 50 ppm, whereas no noticeable response toward other tested gases was detected, thus confirming the excellent H_2 selectivity obtained with such a sensor design.	7): ________
8)The results obtained showed that the performance of gas sensors toward H_2 gas can be greatly increased by both the addition of Pd NPs and the use of ZIF-8 coating, acting as a molecular sieve membrane. 9)Furthermore, the presented strategy could be extended toward the sensing of other species by a judicious choice of both the metallic NPs and MOF materials with tuned properties for specific molecule detection, thus opening a new avenue for the preparation of highly selective sensing devices.	8)-9) ________

(10)

<table>
<tr><td>

Structural influence of proteins upon adsorption to MoS_2 nanomaterials: comparison of MoS_2 force field parameters[14]

1)Molybdenum disulfide (MoS_2) has recently emerged as a promising nanomaterial in a wide range of applications due to its unique and impressive properties. 2)For example, MoS_2 has gained attention in the biomedical field because of its ability to act as an antibacterial agent. 3)However, the potential influence of this exciting nanomaterial on biomolecules is yet to be extensively studied. 4)Molecular dynamics (MD) simulations are invaluable tools in the examination of protein interactions with nanomaterials such as MoS_2. 5)Previous protein MD studies have employed MoS_2 force field parameters which were developed to accurately model bulk crystal structures and thermal heat transport but may not necessarily be amendable to its properties at the interface with biomolecules. 6)By adopting a newly developed MoS_2 force field, which was designed to better capture its interaction with water and proteins, we have examined the changes in protein structures between the original and refitted MoS_2 force field parameters of three representative proteins, polyalanine (α-helix), YAP65 WW-domains (β-sheet), and HP35 (globular protein) when adsorbed onto MoS_2 nanomaterials. 7)We find that the original force field, with much larger van der Waals (vdW) contributions, resulted in more dramatic protein structural damage than the refitted parameters. 8)Importantly, some denaturation of the protein tertiary structure and the local secondary structure persisted with the refitted force field albeit overall less severe MoS_2 denaturation capacity was found. 9)This work suggests that the denaturation ability of MoS_2 to the protein structure is not as dire as previously reported and provides noteworthy findings on the dynamic interactions of proteins with this emergent material.

</td><td>

1)-5): ________

6): ________

7)-8): ________

9): ________

</td></tr>
</table>

(11)

<table>
<tr><td>

Electrochemical direct-writing machining of micro-channel array[15]

1)Bipolar plate is an important component in proton exchange membrane fuel cell. 2)The machining of micro-channel array is a key process for the fabrication of metallic bipolar plate. 3)This paper

</td><td>

1)-2): ________

3): ________

</td></tr>
</table>

continued

proposed a novel approach named electrochemical direct-writing machining for generating micro-channel array on thin metallic plate at one time, showing a high machining efficiency. 4)In this method, an insulated mask with a row of micro through-holes was integrated to a metallic nozzle, the electrolyte ejected from the nozzle reached to the workpiece through micro through-holes, and then the micro-channel array could be generated from dots to lines by controlling the movement of workpiece. 5)In the machining process, the non-processing zones were covered by the insulated mask, which avoided the stray corrosion. 6)By analyzing the results of simulation and experiment, it was found that due to the accumulation of electrolytic product in the ending point, the profile of this point was different from that of the starting point. 7)The parameters with the voltage of 20 V and pulse duty cycle of 20% were useful for preparing micro channel with good profile and low dimensional difference. 8)And compared with other pulse frequencies, the pulse frequency of 2 kHz could obtain a deeper micro channel. 9)With other parameters unaltered, the dimension of micro channel was decreased with the increasing moving speed of workpiece. 10)At last, with the optimized parameters (voltage = 20 V, pulse duty cycle = 20%, pulse frequency = 2 kHz, and moving speed = 20 $\mu m \cdot s^{-1}$), ten micro channels with the length of 60 mm were generated at one time on a metallic plate with the thickness of 0.5 mm. 11)The width was 302 ± 3.53 μm (Mean ± SD) and the depth was 95.9 ± 1.34 μm, showing a satisfied dimensional consistency.	4)-5): ________ 6): ________ 7): ________ 8): ________ 9): ________ 10): ________ 11): ________

Task 5: Determine what linguistic features (words, expressions, and sentence structures) are commonly used in the structural component "Background" of an abstract to state primary research interest, highlight practical needs, and/or point out the research gap in an academic field.

- Linguistic features <u>stating primary research interest</u>:

____

____

____

- Linguistic features <u>highlighting practical needs</u>:

____

____

____

- Linguistic features pointing out the research gap:

____

____

____

(1) These molecularly thin layers are expected to present unique properties with respect to the bulk counterpart, due to increased lattice deformations caused by interface strain.

(2) Reduced-dimensional metal halide perovskites (RDPs) have attracted significant attention in recent years due to their promising light harvesting and emissive properties.

(3) The excellent photocatalytic properties of titanium oxide (TiO_2) under ultraviolet light have long motivated the search for doping strategies capable of extending its photoactivity to the visible part of the spectrum.

(4) A large class of metal dodecaborides (MB12) is currently raising great expectations as multifunctional materials, but their refined structures are not fully resolved, which severely limits the understanding of structure-property relationships.

(5) The ability to modulate reaction rates like controlling speeds by the transmission of vehicles is highly desirable and challenging.

(6) Trigonal-planar units with high physicochemical stability and large polarizability anisotropy are one kind of promising fundamental building block (FBB) for constructing novel nonlinear optical (NLO) materials. Though great achievements have been made in the ultraviolet/deep ultraviolet (UV/DUV) region with trigonal-planar units, little attention has been paid to them in the infrared region owing to the lack of enough representatives.

(7) Atomic layer deposition (ALD) enables the deposition of numerous materials in thin film form, yet there are no ALD processes for metal iodides.

(8) In the global transition to a sustainable low-carbon economy, CO_2 capture and storage technology plays a key role in reducing emissions. Metal-organic frameworks (MOFs) are crystalline materials with ultrahigh porosity, tunable pore size, and rich functionalities, holding the promise for CO_2 capture. However, the intrinsic fragility and depressed processability of MOF crystals make the fabrication of the flexible MOF nanofibrous membrane (NFM) rather challenging.

(9) Direct ink write (DIW) is an emerging additive manufacturing technique that allows for the fabrication of arbitrary complex geometries required in many technologies.

(10) Bipolar plate is an important component in proton exchange membrane fuel cell. The machining of micro-channel array is a key process for the fabrication of metallic bipolar plate.

(11) Current alloy development efforts in High Entropy Alloys call for a better understanding of solution hardening in high-concentration chemically-complex alloys.

(12) Wearable sensing technologies have received considerable interests due to the promising use

for real-time monitoring of health conditions. The sensing part is typically made into a thin film that guarantees high flexibility with different sensing materials as functional units at different locations. However, a thin-film sensor easily breaks during use because it cannot adapt to the soft or irregular body surfaces, and, moreover, it is not breathable or comfortable for the wearable application.

(13) Colloidal quantum dots (QDs) have attracted scientific interest for infrared (IR) optoelectronic devices due to their bandgap tunability and the ease of fabrication on arbitrary substrates.

(14) Elemental phosphorus, one of the most promising high-capacity anode materials for sodium-ion batteries (SIB), suffers from the low practical capacity and severe cycling degradation, due to its low conductivity and significant volume change (>300%) upon electrochemical sodiation/desodiation.

(15) Doping $NaTaO_3$ with La typically results in increased photocatalytic activity. The photocatalytic activity of La-doped $NaTaO_3$ should therefore be sensitive to the photostability of the La species being doped. Unfortunately, the photostability of the La species in $NaTaO_3$ is not fully known.

(16) The electrolysis of water to produce clean, non-byproduct hydrogen energy is considered to be the most effective method. Therefore, it is of far-reaching significance that hydrogen evolution electrocatalysts with lower development costs and good performance.

(17) Silver nanoparticles have already been successfully applied in various biomedical and antimicrobial technologies and products used in everyday life. Although bacterial resistance to antibiotics has been extensively discussed in the literature, the possible development of resistance to silver nanoparticles has not been fully explored.

(18) The engineering of cooling mechanisms is a bottleneck in nanoelectronics. Thermal exchanges in diffusive graphene are mostly driven by defect-assisted acoustic phonon scattering, but the case of high-mobility graphene on hexagonal boron nitride (hBN) is radically different, with a prominent contribution of remote phonons from the substrate.

(19) In organic electronics the functionalization of dielectric substrates with self-assembled monolayers is regarded as an effective surface modification strategy that may significantly improve the resulting device performance. However, this technique is not suitable for polymer substrates typically used in flexible electronics.

(20) Biosurfaces with geometry-gradient structures or special wettabilities demonstrate intriguing performance in manipulating the behaviors of versatile fluids.

(21) Measuring molecular binding to membrane proteins is critical for understanding cellular functions, validating biomarkers, and screening drugs. Despite the importance, developing such a capability has been a difficult challenge, especially for small-molecule binding to membrane proteins in their native cellular environment.

(22) Controllable phase transitions are highly desirable for exploring exotic physics and

fabricating devices.

(23) As future generations of wearable electronics are expected to be directly worn, fiber-based electronics are expected to become increasingly more important in the coming years, as they can be weaved into textiles to provide higher comfort, durability, and integrated multi-functionalities.

(24) Wearable displays are considered as a bilateral communication tool in the hyperconnected era. Although several electronic clothing display was demonstrated, high power consumption issue still remained.

(25) Generally, triboelectric generators (TENGs) demonstrate a considerably lower output current than output voltage; this has largely limited their performance enhancement. Thus, enormous research efforts have been made to address this problem.

(26) The use of all "green" composite materials is a great environmental option for several applications such as construction or automotive.

(27) The properties of ethylenepropylenediene monomer (EPDM) insulations are often inadequate for solid rocket motor (SRM) applications. These materials exhibit relatively high erosion rates during the operation of SRM unless they are reinforced with suitable fiber fillers.

(28) Organic/inorganic hybrid particles with various morphologies play a critical role in the development of advanced materials.

(29) The rational design of new high performance materials for organic photovoltaic (OPV) applications is largely inhibited by a lack of design rules for materials that have slow bimolecular charge recombination. Due to the complex device physics present in OPVs, rigorous and reliable measurement techniques for charge transport and charge recombination are needed to construct improved physical models that can guide materials development and discovery.

(30) Potassium-ion batteries (KIBs) are an attractive energy storage system due to their advantages of low cost and abundant potassium resources. However, their rapid development is greatly impeded by their low capacity and poor stability, due in particular to their much larger volumetric expansion compared to transition metal lithium-ion batteries.

(31) Urban haze is a multifaceted threat. Foremost a major health hazard, it also affects the passage of light through the lower atmosphere.

(32) Hybrid composites are made by incorporating two or more different types of fillers in a single matrix that is highly tailorable. Carbon fiber (CF) reinforced composites have been well developed for certain industries such as aerospace and sporting goods. However, the high cost of carbon fiber, as well as lack of cost effective processing technologies for mass production, prevents its penetration to many different markets. Wood fiber (WF), an environmentally sustainable bio-fiber, has been used widely in making wood/plastic composites (WPCs) for building products and automotive applications, due to its low cost and lightweight. Nevertheless, WPCs have very limited structural applications where strong

mechanical properties are required. Incorporating CF and WF into a polymer matrix to make hybrid composites through injection molding, would be a path to expanded applications for both.

(33) Tuning the charge carrier concentrations of graphene is a fundamental feature to obtain highly efficient electronic and optoelectronic devices.

(34) Processing of end-of-life products has become essential in the rare earth elements (REEs) recovery field because the demand for these metals has increased over the last years due to their intensive use in advanced technologies. Fluorescent lamp wastes are considered one of the most interesting end-of-life products for investigation due to their high REEs content, mainly yttrium and europium.

Task 6: Extract the commonly used sentence patterns in a research article abstract to explicitly report research purposes, with the aid of the underlined words and phrases in the following sentences.

- (Here/Herein/In this work/In this paper/In this article,) *we* + reporting verb in the present tense + noun phrase.

  e.g. Herein, we report the fabrication ...

  ____

  ____

  ____

- (Here/Herein/In this work/In this paper/In this article,) *we* + reporting verb in the present tense + that complement clause.

  e.g. We report that ...

  ____

  ____

  ____

- (Here/Herein,) we + reporting verb in the present perfect tense + noun phrase.

  e.g. Here, we have synthesized perovskites ...

  ____

  ____

  ____

- (Here/Herein,) noun phrase + reporting verb in the past/present tense and passive voice.

  e.g. ... three frameworks were constructed...

____
____
____

- (In this study/paper/article/work/essay,) noun phrase + reporting verb in the past/present perfect tense and passive voice.

  e.g. The effects … have been systematically investigated…

  ____
  ____
  ____

- This paper/study + reporting verb in the present/past tense + noun phrase.

  e.g. This paper presents a … model…

  ____
  ____
  ____

(1) <u>Here we report that</u> layered intermediate complexes formed with the solvent provide a scaffold that facilitates the nucleation and growth of RDPs during annealing.

(2) <u>Here, we have synthesized</u> centimetre-sized, pure-phase single-crystal RPP <u>perovskites</u> $(CH_3(CH_2)_3NH_3)_2(CH_3NH_3)_{n-1}Pb_nI_{3n+1}$ (n = 1–4) from which single quantum well layers have been exfoliated.

(3) <u>Here we elucidate</u> the formation mechanism and structural <u>features</u> of black TiO_2 using first-principles-validated reactive force field molecular dynamics simulations of anatase TiO_2 surfaces and nanoparticles at high temperature and under high hydrogen pressures.

(4) <u>Here we report</u> <u>an</u> ultrafast water transport <u>process</u> in the surface of a Sarracenia trichome, whose transport velocity is about three orders of magnitude faster than those measured in cactus spine and spider silk.

(5) <u>Here, we report that</u> the tetragonal tI26 structure is the thermodynamic ground state of ScB_{12}, <u>and we predict</u> the tetragonal YB_{12}, ZrB_{12}, and HfB_{12} to be metastable, whereas the cubic cF52 structure is the high-temperature phase of ScB_{12} and represents the thermodynamic ground state of YB_{12}, ZrB_{12}, and HfB_{12}.

(6) With the utilization of a "bifunctional ligand-directed strategy", three isostructural indium-organic <u>frameworks</u> based on dual secondary building units (SBUs) <u>were successfully constructed</u> with targeted structures.

(7) <u>Here, we show that</u> lithium migrates along the solid/liquid interface without leaving the particle, whereby charge carriers do not cross the double layer.

(8) Herein, we demonstrate an ALD process for PbI_2, a metal iodide with a two-dimensional (2D) structure that has applications in areas such as photodetection and photovoltaics.

(9) Herein, we report the fabrication of hydrogen gas sensors with enhanced sensitivity and excellent selectivity.

(10) Here, we present a new proof-of-concept multilayered LSPR sensor design that incorporates both a sensing layer and an encapsulated reference layer within the same region.

(11) The effects of He particles coming from the solar wind impinging on a gold thin film have been systematically investigated, considering absorbed doses compatible with the duration of the European Space Agency Solar Orbiter mission.

(12) Herein, we report a template-free approach to polydopamine nanocapsule formation in the presence of resveratrol, a naturally occurring anti-inflammatory and antioxidant compound found in red wine and grapes.

(13) Herein, we demonstrate an effective strategy for the versatile preparation of self-supported and flexible HKUST-1 NFM with ultrahigh HKUST-1 loading (up to 82 wt %) and stable and uniform HKUST-1 growth through the combination of electrospinning, multistep seeded growth, and activation process.

(14) In this work, [$AgSe_3$] and [$HgSe_3$] are rationally proposed as NLO-active FBBs.

(15) In this study, AgCl-KCl lamellar eutectic was solidified within lithographically fabricated pillar templates and the results were compared with phase-field simulations.

(16) Here we propose a general scheme for assessing the overall solute-dislocation interaction, independent of concentration, stress, and temperature.

(17) The orientations, locations, and sizes of approximately 2,500 grains in a Ni polycrystal were measured at six points in time during an interrupted annealing experiment by synchrotron x-ray based, near field high-energy diffraction microscopy.

(18) This paper presents a new physics-based constitutive model to accurately describe the deformation behavior of metals during ultrasonic vibration assisted (UVA) forming.

(19) A defect-free welded joint of 100 mm-thick SUS 304 steel plates is produced by ultra-narrow gap laser welding with filler wire in the laser conduction mode in 42 layers.

(20) This paper proposed a novel approach named electrochemical direct-writing machining for generating micro-channel array on thin metallic plate at one time, showing a high machining efficiency.

(21) We develop an efficient chemo/photothermal therapy system and propose an alternating photothermal strategy.

(22) As a proof-of-concept, the reactive oxygen species (ROS)-activatable thioketal (TK) bond was explored as the linkage between doxorubicin (DOX) and polyphosphoester (PPE-*TK*-DOX).

(23) A near-infrared (NIR) light-triggerable thermo-sensitive hydrogel-based drug reservoir that can realize on-demand antibiotics release and hyperthermia-assisted bacterial inactivation

was prepared to combat bacterial infection and promote wound healing.

(24) Herein, we engineered for the first time a chylomicron-pretended mesoporous silica nanocarrier that utilizes the digestion, re-esterification, and lymphatic transport process of dietary triglyceride to promote lymphatic transport of oral drugs.

(25) Herein, a new and general strategy of making electrochemical fabric from sensing fiber units is reported.

(26) Here, the use of laser-activated nanosealants (LANS) with gold nanorods (GNRs) embedded in silk fibroin polypeptide matrices is demonstrated.

(27) Here, an all-solid-state electrochemical transistor made with Li ion–based solid dielectric and 2D α-phase molybdenum oxide (α-MoO_3) nanosheets as the channel is demonstrated.

(28) Herein, we developed a smart nano delivery system STD-NM, showing tumor microenvironment responsive targeting, efficient drug delivery and precise evaluation of therapeutic effect in vivo.

(29) In this work, short-wave IR photodetectors based on lead sulfide (PbS) QDs with high detectivity and low dark current is demonstrated.

(30) Co-substituted $Y_{3-2x}Ca_{2x}Fe_{5-x}V_xO_{12}$ samples ($x = 0$, 0.1, 0.2, 0.3, 0.4, 0.5, 0.6, 0.7, 0.8, 0.9, 0.95 and 1) were prepared by using citrate sol-gel method.

(31) Here, we report a ball-milled phosphorus/reduced graphene oxide (rGO) nanocomposite for high-performance anode in SIBs.

(32) P_4Se_3 microparticles are successfully synthesized by simple heating of selenium and amorphous red phosphorus at low temperature.

(33) Herein, we perform an in-depth study on the photostability of the La species.

(34) In this essay, the different effects of Electrocatalytic Hydrogen Evolution of $FeNi_xS_2$-RGO Electrocatalysts are studied with different nickel-doped content added.

(35) Here, we present a tip-enhanced nano-optical approach to induce, switch and programmably modulate the XD emission at room temperature.

(36) Here, we demonstrate that topological defect-driven magnetic writing—a scanning probe technique—provides access to all of the possible microstates in artificial spin ices and related arrays of nanomagnets.

(37) Here, we report organic modifiers based on a paraffinic tripodal triptycene, which self-assembles into a completely oriented two-dimensional hexagonal triptycene array and one-dimensional layer stacking structure on polymer surfaces.

(38) Herein a bimetal oxide dual-composite strategy based on two-dimensional (2D)-mosaic three-dimensional (3D)-gradient design is proposed.

(39) By mimicking natural species, that is, the cactus spine with a shape-gradient morphology and the Picher plant with a lubricated inner surface, we have successfully prepared an asymmetric slippery surface by following the processes of CO_2-laser cutting, superhydrophobic modification, and the fluorinert infusion.

(40) Here we show that the binding of both large and small molecules to membrane proteins can be quantified on single cells by trapping single cells with a microfluidic device and detecting binding-induced cellular membrane deformation on the nanometer scale with label-free optical imaging.

(41) We report a direct observation of a controllable semiconductor-metal phase transition in bilayer tungsten diselenide (WSe_2) with potassium (K) surface functionalization.

(42) Herein, we demonstrate an intrinsically stretchable multi-functional hollow fiber capable of harvesting mechanical energy and detecting strain.

(43) Here, we present wireless powered wearable μLEDs (WμLEDs) with excellent stability.

(44) This paper demonstrated a novel energy utilization that thermoelectric generation (TEG) device can achieve self-powering by radiative cooling (RC) continuously.

(45) In this work, we present a simple method to enhance the triboelectric output current by burying an indium zinc oxide (IZO) layer under the triboelectric polymer friction layer.

(46) In this paper, bio-inspired triboelectric nanogenerator (BITENG) has been designed to mimic the motion of kelp, which gently sways along with wave and harness the energy in this process.

(47) This work aims to quantify the effect of a diamond-like carbon coating (DLC) treatment of aramid fibres and to reveal the conversion of a fibre-level performance leap on the macroscale mechanical behaviour.

(48) In the current investigation, BMI self-healing (SH) resin based on Diels-Alder (DA) reaction mechanism (identification code: BMI pp) was integrated into high performance aerospace carbon fiber reinforced plastics (CFRPs).

(49) This study proves the ability of polyvinyl butyral (PVB) nanofiber to improve the mode I and mode II interlaminar fracture toughness of out of autoclave phenolic-based composites.

(50) This paper provides results from a comprehensive experimental characterization on two gelatins extracted from golden grey mullet skin (GGSGs) with (5 U/g of skin) and without acid protease from the same specie.

(51) The pyroelectric properties of polyvinylidene fluoride/trifluoroethylene doped with graphene oxide were studied.

(52) This paper investigated the injection molding of CF-WF/polypropylene hybrid composites and their mechanical properties.

(53) In this article, we report the findings of comprehensive mechanical and ablative studies of EPDM insulations blended with polyacrylonitrile (PAN) fiber, pre-oxidized polyacrylonitrile fiber and carbon fiber.

(54) The microscopic mechanism of potassium permanganate (KMnO4) modification and the properties of ramie fiber/polypropylene (RF/PP) composites were investigated.

(55) In this work, the synthesis of raspberry-like hybrid poly(styrene-co-2-vinyl pyrrolidone)/nanosilica via dispersion polymerization have been reported, 2-vinylpyridine involved added

to improve the interaction of polymer and nanoparticles.

(56) In this study we investigate two polyoxometalates (POMs), $[SiW_{12}O_{40}]^{4-}$ and $[PV_{14}O_{42}]^{9-}$, as nano-sized electron shuttles.

(57) Here, we develop a new technique called impedance-photocurrent device analysis (IPDA) to quantitatively characterize the competition between charge extraction and charge recombination under steady state operational conditions.

(58) Herein, we report a new etching route to fabricate a new class of zero-strain potassium fluoromanganate hollow nanocubes (KMnF-NCs) for boosting the performance of KIBs in terms of capacity, rate capability and cycling stability.

(59) In this paper, we present a study addressing the impact of haze on the performance of photovoltaic installations in cities.

(60) In this study, novel Pb-Bi-X alloys, as improvement LBE, have been designed for a broad temperature range.

Task 7: Underline the passive verbs and their subject nouns (e.g. a system was developed; a model is optimized) in the following three sets of methods description of research article abstracts. Discuss the reasons for the dominant use of the past tense in Set (1), the present tense in Set (2), and different tenses in Set (3).

Set (1)

(1) The volume changes were compared to the geometric characteristics of the grains in both the original microstructures and microstructures in which adjacent twin related domains were merged.

(2) Dislocation dynamics and acoustic energy transformation mechanisms in materials under ultrasonic vibration were considered in the modelling. A user defined subroutine was also developed for modelling the UVA forming processes using the finite element method. To assess the proposed model, upsetting forming, press forming, and incremental forming were simulated separately, and the predicted results were compared with their corresponding experiments.

(3) In this method, an insulated mask with a row of micro through-holes was integrated to a metallic nozzle, the electrolyte ejected from the nozzle reached to the workpiece through micro through-holes, and then the micro-channel array could be generated from dots to lines by controlling the movement of workpiece. In the machining process, the non-processing zones were covered by the insulated mask, which avoided the stray corrosion.

(4) Synchrotron X-ray diffraction (SXRD) measurements combined with Rietveld refinement technique were used to investigate the crystallization, structural parameters and lattice distortion in the samples. Magnetization and Curie temperature were studied by means of a vibrating sample magnetometer (VSM). Molecular-field coefficients N_{aa}, N_{dd} and N_{ad} of the

samples were determined by fitting the experimental thermomagnetization curves in the framework of two-sublattice ferrimagnetic model. The information of local structure change was employed to analyze the influence of the substitution effects in the magnetic interactions.

(5) An ultrathin TEG with a multilayer thermal emitter was fabricated to convert the heat from the environment into electricity directly by using radiative cooling. The TEG device was consisted of more than 46,000 P-N modules in series, and each two TE modules were connected by air bridge.

(6) The raw industrial humins were cross-linked with and without catalyst. The effect of the process conditions on the final properties of the composites was studied. Good tensile properties were measured by tensile test and Dynamic Mechanical Analysis.

(7) CNF were first prepared through enzymatic and mechanical treatment of bleached hardwood kraft pulp. The bio-nanocomposites- were then fabricated through ring opening polymerization (ROP) of L-lactide, in the presence of various amounts of fibrils. Molecular weight, thermal properties, surface morphology, mechanical and wettability properties of the PLA-CNF nanocomposites were evaluated.

(8) A custom-designed wireless power supply system operated a 30 × 30 WμLED array on a fabric. The WμLED stability was intensively investigated under bending, stretching, 85 °C/85% relative humidity and artificial sunlight conditions.

(9) For investigation fracture behavior of glass-phenolic composites, a mode-I and mode II fracture test were carried out standard UD laminated and nanoweb interleaved specimens. PVB nanoweb with 25 μm thickness was placed in the laminate mid-planes which was subsequently subjected to a double cantilever beam and end-notched flexure tests.

(10) Four $CaCO_3$ added LLDPE polymer systems were compounded in a 24 mm diameter, co-rotating, closely intermeshed twin-screw extruder. The compounds differed from each other by the weight percent of $CaCO_3$ added. A laser beam system installed at the die exit of a capillary rheometer was used to determine the diameter of the swollen extrudate.

(11) The pyroelectric current measurements were performed for polarized and non-polarized samples with different amount graphene oxide: 5%, 10%, 15%, 20%, 25%.

(12) CIGS thin films with various compositions (defined by $x = Ga/(In + Ga)$) were prepared *via* a three-stage method employing a vacuum evaporation system.

Set (2)

(13) The laser beam concurrent heating of the groove side walls and bottom ensures the adequate side wall fusion, with the angle distortion belowed 1°, and the average fusion ratio of the whole weld of about 11%. The microstructure is composed of cellular grains adjacent to the fusion line and fine equiaxed grains.

(14) Combinations of different elastomers and carbon nanotube electrodes are investigated and optimized to meet performance characteristics appropriate to tactile display applications,

namely operation up to 200 Hz with a combination of a 1 N blocked force and free displacement of 1 mm, all within a volume of less than 1 cm^3. Lives in excess of 50 000 cycles have been obtained.

(15) Advanced characterization techniques, including X-ray photoelectron spectroscopy (XPS), X-ray absorption near edge structure (XANES), and extended X-ray absorption fine structure (EXAFS), are employed to clarify the chemical state and local structure of the La species in $NaTaO_3$.

(16) For energy harvesting, we have utilized a stretchable ferroelectric layer composed of P(VDF-TrFE) in a matrix of elastomer, sandwiched between stretchable electrodes composed of multi-walled carbon nanotubes and PEDOT:PSS.

(17) The DLC-based coating is applied directly to the reinforcement and laminates are infused with an epoxy matrix. After characterisation of the coated surfaces, the performance of the composite is analysed via interlaminar shear testing, fatigue testing and damage tolerance testing, microbond tests, and 3D finite element simulation using a cohesive zone model of the interface.

(18) The measurements are performed on actual lab scale solar cells, have mild equipment requirements, and can be integrated into normal device fabrication and testing workflows. We perform IPDA tests on a broad set of devices with varying polymer:fullerene blend chemistry and processing conditions.

Set (3)

(19) The sensor device is based on the strategic combination of ZnO nanowires (NWs) decorated with palladium nanoparticles (Pd NPs) and a molecular sieve metal-organic framework (MOF) nanomembrane (ZIF-8). The Pd NPs permit the sensors to reach maximal signal responses, whereas the ZIF-8 overcoat enables for an excellent selectivity. Three steps were employed for the fabrication: (i) coating of a miniaturized sensor with vapor-grown ZnO NWs, (ii) decoration of these NWs with Pd NPs by atomic layer deposition, and (iii) partial solvothermal conversion of the tuned NWs surface to ZIF-8 nanomembrane.

(20) The resulting gelatins were structurally characterized by optical microscopy, differential scanning calorimetry (DSC), as well as by X-ray diffraction (XRD) and nuclearmagnetic resonance (NMR) spectroscopies. According to optical micrographs, DSC analysis, and XRD spectra, GGSGs are semi-crystalline polymers.

Task 8: Classify the underlined words, expressions, and sentence structures utilized to report significant results in research article abstracts.

- Sentence structures: (e.g. it is found that …; the results demonstrate that …)

____

- Finite verbs (expressing causation, occurrence, existence, etc.): (e.g. enable the fabrication; exhibit a distribution; changes are proved)

- Past/present participle as nominal pre-modifier: (e.g. the optimized composite)

- Comparative, superlative, or positive evaluative adjective (e.g. softer, highest, excellent)

- Degree/attitudinal adverb (e.g. significantly; effectively delivered)

(1) RDPs exhibit a thickness distribution (with sizes that extend above n = 5) determined largely by the stoichiometric proportion between the intercalating cation and solvent complexes.

(2) By changing the polarizing substituents in situ with different halogens (Cl^-, Br^-, and I^-), three obtained isostructural MOFs show different channel characteristics, such as alkalinity of the polarizing substituents, acidity of the polarized open indium sites, extended channel sizes, and increased pore volumes (from -I to -Cl).

(3) The restriction of intramolecular motions of AIEgen activates the fluorescence of the transmission with the rate of $\sqrt{2} > \sqrt{1} > \sqrt{3} > \sqrt{4} \approx 0$ and enhances the fluorescent contrast before and after photoirradiation.

(4) Rough substrates yield porous PbI_2 films with randomly oriented 2D layers, whereas smooth substrates yield dense films with 2D layers parallel to the substrate surface. Exposure to light increases conductivity of the ALD PbI_2 films which enables their use in photodetectors.

(5) Experiments demonstrate that both $Ag_6HgSiSe_6$ and $Ag_6HgGeSe_6$ show strong second harmonic generation (SHG) responses, valuable phase-matchable features, and congruent-melting thermal behaviors.

(6) We demonstrate the suitability of this sensor for sucrose concentration measurements and for the detection of biotin-avidin interactions, while also showing that the sensor can self-correct for drift.

(7) Structural and morphological changes have been proved to be dependent not only on the dose but also on the irradiation flux.

(8) The loading rate of MOF is the highest level among the reported analogues. Significantly, the HKUST-1 NFM exhibits a prominent CO_2 adsorption capacity of 3.9 mmol g^{-1}, good CO_2/N_2 selectivity, and remarkable recyclability.

(9) Self-consistency of the method is shown by computing various strengthening parameters near critical percolation stress states at various system sizes, concentrations, and temperatures.

(10) More specifically, it is found that a bi-dispersed PSD enhances pore shrinkage kinetics. However, bi-dispersity yields microstructures with pores that are highly eccentric, an effect that could be detrimental to the mechanical properties of the printed material.

(11) It was found that the application of the ultrasonic vibration can significantly decrease the material flow stress, making the material softer in forming. A large vibration amplitude results in a large reduction in flow stress. Relevant experimental measurements showed that the model has captured the material behaviour and the major mechanics in UVA forming.

(12) By analyzing the results of simulation and experiment, it was found that due to the accumulation of electrolytic product in the ending point, the profile of this point was different from that of the starting point.

(13) Experimental results validate that the tube can be formed into the desired shape by cross-section compression under a certain internal pressure supporting.

(14) The cured RSWB joints achieved approximately 20% and 80% improvement in the peak load and energy absorption, respectively, in coach peel tests.

(15) Results show that the photothermal depth by ILAA NHCs is 2.1-fold than other common photothermal agents (PTAs), and the irradiated region exhibits a lower surface temperature.

(16) As a result, the chylomicron-pretended nanocarrier afforded 10.6-fold higher oral bioavailability compared with free LNV and effectively delivered LNV to gut-associated lymphoid tissues, where HIV persists and actively evolves.

(17) Based on this operating mechanism, the essential functionalities of synapses, such as short- and long-term synaptic plasticity and bidirectional near-linear analog weight update are demonstrated. Simulations using the handwritten digit data sets demonstrate high recognition accuracy (94.1%) of the synaptic transistor arrays.

(18) It is proved that the charge carrier diffusion length in the QD layer is negligible such that only photogenerated charges in the space charge region can be collected.

(19) The optimized P/rGO1000 composite, containing rGO reduced by hydrothermal reaction and heat-treatment, achieves high specific capacity (2032.4 mAh g^{-1} at 100 mA g^{-1}), excellent

rate capability (1306.6 mAh g^{-1} at 1C rate), and stable cyclability (98.7% capacity retention after 300 cycles at 780 mA g^{-1}).

(20) Our findings show that the examined La-doped $NaTaO_3$ can be effectively reused for six consecutive runs without a significant loss of its photocatalytic activity for the decomposition of recalcitrant organic compounds under UV light.

(21) It was observed that the boundary effects from the template drive changes in the orientation of lamellae within the template.

(22) The results demonstrate that the Ni element incorporated with a molar fraction of 10% ($FeNi_{0.10}S_2$-RGO) has a lower Tafel slope (71.96 mV dec^{-1}) and good electrochemical kinetics.

(23) We show that the ZKT triggers a new cooling pathway due to the emission of hyperbolic phonon polaritons in hBN by out-of-equilibrium electron-hole pairs beyond the super-Planckian regime.

(24) The triptycene films significantly improve the crystallinity of an organic semiconductor and the overall performance of organic thin-film transistors, therefore enabling the fabrication of high-performance organic complementary circuits on polymer substrates with high oscillation speeds and low operation voltage.

(25) As a result, a high reversible capacity (1,010 mA h g^{-1}) and areal capacity (1.48 mA h cm^{-2}) are attained, while ultrastable cyclability is obtained during high-rate and long-term cycles, rending great potential of our 2D-mosaic 3D-gradient design together with facile synthesis.

(26) We have demonstrated voltage and current generation under stretching and normal pressure, with output voltage and current as high as 1.2V and 10 nA, respectively.

(27) Maximum temperature drop of 4K was achieved. The output voltage of the TEG-RC reached up to 0.5 mV, and the TEG-RC exhibited a continuous average 0.18 mV output for 24 h.

(28) An output voltage and current density of ~ 140 V and ~ 180 μA/cm^2 were obtained, which are 4-fold and 9-fold higher, respectively, than a TENG without an IZO layer.

(29) The energy harvesting has been proven as stable and efficient, which could work under a vibration frequency as low as to 1 Hz.

(30) According to quasi static experimental results, it was shown that the incorporation of the BMI pp SHA did not deteriorate the in-plane mechanical properties of the entire composite.

(31) DSC analysis demonstrated the effect of CNF on crystallization and crystalline morphology of PLA.

(32) It was found that PVB nanofiber thanks to its excellent adhesion property with phenolic resin and unique features of nanoscales, has a tunable effect on improving mode I and mode II interlayer fracture toughness.

(33) During the course of this investigation, PAN fiber-filled EPDM insulations showed much superior mechanical and ablative properties than other fiber-filled EPDM insulations.

(34) Good consolidation qualities were obtained with configurations (ii) and (iii).

(35) The results show that the coating treatment improves the fatigue life and the S-N curve slope for the laminates, while the residual strength after impact damage and environmental conditioning (water immersion at 60 °C) remains high.

(36) Results from the IPDA technique exhibit significantly improved reliability and self-consistency compared to the open-circuit voltage decay technique (OCVD).

(37) After applying the compressive impact loads on the vertical and horizontal samples, it was found that the dynamic loading behavior of the two samples was almost similar despite the crucial differences in the initial microstructures.

(38) The key role of the surfactants in the washing was evidenced because the ORR values with the surfactants were significantly higher than the value with the control with no surfactant solution.

(39) Our results showed that the leakage rate over a long time period (50-1,000years) is 10 times higher than that of either a short (0-10years) or medium (10-50years) period.

(40) Charge carrier modulation of graphene by UV irradiation under gas environment show no degradation effect in mobility of device.

Task 9: Choose the right verbs to fill in the blanks in the following "discussion/conclusion" sentences from research article abstracts, Use the same verb more than once and change the verb forms when necessary. Then discuss whether each statement is hedged or emphasized.

[A] provide	[B] indicate	[C] suggest	[D] speculate
[E] imply	[F] open	[G] apply	[H] reveal
[I] render	[J] offer	[K] represent	[L] anticipate
[M] point	[N] would	[O] could	[P] can

(1) Our results ________ that, thanks to their dynamic structure, atomically thin perovskites enable an additional degree of control for the bandgap engineering of these materials.

(2) The results ________ a means to control the distribution, composition and orientation of RDPs via the selection of the intercalating cation, the solvent and the deposition technique.

(3) Besides confirming that the hydrogenated amorphous shell has a key role in the photoactivity of black TiO_2, our results ________ insight into the properties of the disordered surface layers that are observed on regular anatase nanocrystals under photocatalytic water-splitting conditions.

(4) We ________ that the superior catalytic efficiencies of the three MOF catalysts could be ascribed to the synergistic effect of open indium sites as Lewis acid with different halide ions as weak base sites.

(5) We ________ that the chemistry exploited in the PbI_2 ALD process is also applicable for ALD of other metal halides.

(6) The presented strategy ________ be extended toward the sensing of other species by a judicious choice of both the metallic NPs and MOF materials with tuned properties for specific molecule detection, thus ________ a new avenue for the preparation of highly selective sensing devices.

(7) Thus, the low-cost and scalable production pathway is able to convert MOF particles into self-supported and flexible NFMs, and thereby, they are better ________ to the efficient postcombustion CO_2 capture.

(8) Different forming processes, due to their difference in tool-workpiece contact condition, ________ have different acoustic energy and stress transformation efficiency.

(9) The microstructure morphology difference in the weld joint ________ the uneven distribution of microhardness which values in the equiaxed grain zone significantly exceed those the cellular zone.

(10) These results ________ an insight into the application of 2D oxides for large-scale, energy-efficient neuromorphic computing networks.

(11) Considering the facile synthesis of ball milling process, our approach ________ be highly promising for high-capacity anode in practical SIB.

(12) Our approach provides a facile way to harness excitonic properties in low-dimensional semiconductors, ________ new strategies for quantum optoelectronics.

(13) These features ________ our multi-functional fiber highly suitable for wearable electronic applications in the near future.

(14) Such a bio-inspiration ________ a new strategy for harvesting wave-energy with promise for new energy sources based on TENGs.

(15) The performed experiments ________ the basis to future applications of these blends as membranes and confirm their performance in industrial field.

(16) The microstructural analysis of the deformed samples ________ entangled networks of dislocations.

(17) The work ________ to the need for further research into the compatibility with structural materials at high temperatures.

(18) By combining an appropriate hydrometallurgical process with the present supergravity separation and concentration of precious metals, this clean and efficient process ________ a new pathway to recycle valuable metals and prevent environmental pollution by PCBs.

Task 10: Read Text Ⅱ and identify the similarities and differences in structural components between the abstract and conclusion sections.

Text Ⅱ

Measuring the Competition Between Bimolecular Charge Recombination and Charge Transport in Organic Solar Cells Under Operating Conditions[16]

Abstract: The rational design of new high-performance materials for organic photovoltaic (OPV) applications is largely inhibited by a lack of design rules for materials that have slow bimolecular charge recombination. Due to the complex device physics present in OPVs, rigorous and reliable measurement techniques for charge transport and charge recombination are needed to construct improved physical models that can guide materials development and discovery.	Background
Here, we develop a new technique called impedance-photocurrent device analysis (IPDA) to quantitatively characterize the competition between charge extraction and charge recombination under steady-state operational conditions.	Purpose
The measurements are performed on actual lab-scale solar cells, have mild equipment requirements, and can be integrated into normal device fabrication and testing workflows. We perform IPDA tests on a broad set of devices with varying polymer:fullerene blend chemistry and processing conditions.	Methods
Results from the IPDA technique exhibit significantly improved reliability and self-consistency compared to the open-circuit voltage decay technique (OCVD).	Results
IPDA measurements also reveal a significant negative electric field dependence of the bimolecular recombination coefficient in high fill factor devices, a finding which is inaccessible to most other common techniques and indicates that many of these techniques may overestimate the value that is most relevant for describing device performance. Future work utilizing IPDA to build structure-property relationships for bimolecular recombination will lead to enhanced design rules for creating efficient OPVs that are suitable for commercialization.	Discussion/Conclusion

...

IV. Conclusions

1	In order to increase the accuracy and reliability of charge transport and recombination measurements for OPVs, we have developed the impedance-photocurrent device analysis (IPDA) technique for measuring real devices under standard, steady-state operating conditions, and we apply the technique to a wide range of polymer:fullerene blends prepared with various fabrication conditions. As a comparison point, we also apply a more common transient photovoltage technique called open-circuit voltage decay (OCVD) to the same set of devices. We find that IPDA gives a much more reliable quantification of the recombination kinetics than the OCVD method, and that altogether, IPDA characterization of both charge transport and recombination correlates very well with the device fill factor when using the Bartesaghi model.[5] The IPDA measurement results compiled here on many different devices with a variety of blend chemistries, processing conditions, and active layer thicknesses represent one of the most self-consistent and comprehensive charge transport and recombination datasets available for OPVs. Our results also provide further supporting evidence that common transient photovoltage based techniques do not give reliable measurements of the bimolecular recombination kinetics in thin film solar cells, as shown recently by Kiermasch et al.[22] Furthermore, we find that very often, the bimolecular recombination coefficient is significantly less at maximum power operating conditions than at open-circuit, even after correcting for the charge carrier density dependence. As a result, even charge extraction based techniques that measure recombination at open-circuit may often overestimate the recombination coefficient. Further analysis reveals that most of the optimized devices tested here exhibit a recombination coefficient with a significant negative electric field dependence, which is not probed by most of the common recombination characterization techniques.	Purpose Method① Method② Result① Discussion① Result② Discussion②
2	Overall, we find IPDA to be a powerful technique that will allow researchers to characterize the primary device properties that ultimately determine the fill factor in most high performance	Concluding

continued

	OPVs. The reliability and reduced measurement uncertainty afforded by the IPDA technique could allow researchers to identify and more rigorously study materials that can maintain a high fill factor with a thick active layer. Recently, an increasing number of fullerene and non-fullerene blends have been able to achieve a high fill factor in thick films,[72,89,94–99] so there is great opportunity for elucidating the physical mechanisms that drive this behavior. While we have only investigated polymer:fullerene blends in this study, there should be no reason that the technique cannot also be applied to the wide variety of small molecule:fullerene blends, non-fullerene acceptor blends, polymer–polymer blends, co-evaporated BHJ blends, and even ternary blends that all continue to be developed. As long as the blends have a BHJ architecture and meet the assumptions laid out in this study, IPDA should be equally applicable. Future work utilizing IPDA to build structure-property relationships for bimolecular recombination could lead to enhanced design rules for creating efficient OPVs that are suitable for commercialization.	remarks Significance of research outcomes ① Potential applications①
3	IPDA could also play a role in increasing the rate of OPV materials discovery and optimization. In this vein, most work has been focused on using *ab initio* electronic structure methods to screen for appropriate chemical structures, largely based on their optical bandgap.[100–102] While this can be a useful method to narrow the search, *in silico* predictions of the fill factor are extremely difficult due to the highly complex series of factors that determine its value, and such methods cannot currently capture processing and morphology effects. Recent work has highlighted the importance of the thermodynamic parameter χ on the fill factor and suggested higher throughput miscibility experiments could be used to accelerate materials discovery,[90] but knowing χ alone is not enough to accurately predict the fill factor nor the fabrication conditions required to reach a particular level of performance. Instead of replacing device experiments completely, reliable higher throughput experimental characterization techniques that can generate information-rich databases could enhance recent efforts to create predictive data-driven models using machine learning.[103,104]	Significance of research outcomes ②

continued

IPDA measurements are performed on actual lab-scale solar cells, have mild equipment requirements, can be integrated into existing device fabrication and testing workflows, and yield a range of detailed metrics, thereby greatly increasing the scientific value of each experimental device fabricated. A starting example of such an information-rich dataset containing detailed starting materials properties, film fabrication details, and IPDA device characterization results is provided in the ESI. Combining this approach with various materials screening efforts could be a promising strategy for significantly reducing trial-and-error approaches to OPV materials development.	Potential applications②

Task 11: Fill in each blank below with the appropriate form of the verb provided. Each of the following sentences presents at least two of the structural components (namely background, purpose, methods, results, and discussion/conclusion) of a research article abstract with embedded present/past participial adverbial clause(s). Then group together the participial adverbial clauses describing each structural component.

(1) The magnetic properties of nanocomposites were studied ________ (use) the vibration sample magnetometer.

(2) Furthermore, a wireless system that can detect touch, temperature and pressure is successfully demonstrated ________ (use) a nanomesh with excellent mechanical durability.

(3) Their material and biological properties were characterized ________ (use) multiple techniques.

(4) The governing differential equation of motion of the free vibrated sandwich panel is obtained ________ (use) the classical Hamilton's principle and transformed to the set of algebraic form with the help of suitable finite element steps.

(5) Damage evolution was monitored ________ (use) thermal imaging, while post-failure modes were evaluated using scanning electron microscopy (SEM).

(6) Three kinds of grinding wheels, including octahedron structure wheel, honeycomb structure wheel and solid structure wheel, are designed and fabricated ________ (use) this novel method.

(7) The present approach is validated ________ (take) gold as bench mark material as results for 1D and 2D cases are already reported in literature.

(8) The waste paper was treated ________ (use) three different techniques, namely pulping, flotation and washing, after which it was subjected to an ultrafine grinding process to produce CNFs.

(9) The output is measured by scanning tunneling spectroscopy ________ (follow) the shift in energy of the electronic tunneling resonances at the end of the short branch of the molecule.

(10) Excellent rotational commensurability, van der Waals gap at the interface and moiré pattern are observed ________ (indicate) good registry between the $ZrTe_2$ epilayer and the substrate through weak van der Waals forces.

(11) This processing resulted in exfoliated chiral 2D MoS_2 nanosheets ________ (show) strong circular dichroism signals, which were far past the onset of the original chiral ligand signals.

(12) The latter indicates massless Dirac Fermions which are maintained down to the 2D limit ________ (suggest) that single-layer $ZrTe_2$ could be considered as the electronic analogue of graphene.

(13) Endoscopic treatment of Barrett's oesophagus often leads to further damage of healthy tissue ________ (cause) fibrotic tissue formation termed as strictures.

(14) Finally, a simple mechanical model was used to estimate the sub-nanometer thickness of the interfacial shell ________ (apply) the stress on the crystalline core.

(15) Heat aging caused properties to deteriorate due to chain scission reactions that eliminated network chains ________ (create) a more heterogeneous system.

(16) High Schottky barriers, on the other side, distort device characteristics ________ (make) extraction impossible.

(17) To address these issues, we designed Janus-structured gold-mesoporous silica nanoparticles ________ (use) a modified sol-gel method.

(18) Also, tan max of the DR phase decreases with increasing carbon black concentration ________ (indicate) mainly localization of filler in rubber phase.

(19) The amorphization and diffusion of ionized oxygen in nanorods is controlled by optimizing the voltage window, ________ (result) in the great increase of capacity retention from 26% to 80%.

(20) We propose that the formation of these fast ion transport "highways" improves accessibility to interior sites, ________ (lead) to significantly improved overall rate performance in the amorphous films.

(21) ________ (benefit) from the close cooperation of these two types of cores, the micromotors were imparted with a strong propulsion and prominent recyclability for the delivery of both microscale and macroscale objects.

(22) ________ (use) this database, we develop machine learning models capable of predicting shape persistence with an accuracy of up to 93%, ________ (reduce) the time taken to predict this property to milliseconds, and ________ (remove) the need for specialist software.

(23) Furthermore, we find that for a specific shell to diameter ratio, corresponding to the transition between core and shell, the stress concentration in the nanoparticles is apparently hindered, ________ (lead) to a delayed plastic deformation.

(24) Moreover, the conductive sponges displayed absorption-dominant mechanism to alleviate secondary radiation, ________ (make) them promising candidates as high EMI shielding materials.

(25) No appreciable cracks were observed in the four fusion zones after the surfacing process, ________ (suggest) that good fusion had occurred between the H13 steel substrate, sublayer and wear layer.

(26) Generally, magnetostriction elastomer were based on magnetic field-induced strains of elastomer matrix, ________ (result) from the rotation of magnetic particle in elastomer matrix.

(27) Neutron Bragg edge imaging is a novel technique, ________ (allow) for two-dimensional mapping of the Bragg edge broadening parameter, indicative of bulk plastic deformation.

(28) The unique structures impart intimate structural interconnectivities, highly opened freeway for ionic diffusion, large accessible surface area, as well as high structural stability, ________ (open) up a wide horizon for electrochemical applications, for example, high-energy, long-life lithium-ion batteries and lithium–sulfur batteries highlighted in this work.

(29) ________ (use) these fingerprints, layers composed of structurally related proteins with differing geometries can be discriminated.

(30) Pre-strained material emits light on straining perpendicularly, but not parallel to the original tensile direction, ________ (demonstrate) that covalent bond scission is highly anisotropic.

(31) Overall, the developed hierarchical structured and programmed vehicles load, protect, transport and release drugs locally to inflamed sites of intestine, ________ (contribute) to superior therapeutic outcomes.

(32) ________ (start) with a full description of the concept, the equations describing the control of the optical elements are deduced ________ (use) Jones calculus formalism.

(33) ________ (operate) in single-electrode mode at 2.5 Hz, the WP-TENG with an area of $6 \times 3\ cm^2$ produces an open-circuit voltage of 180 V, short-circuit current of 22.6 μA, and average power density of $4.06\ mW\ m^{-2}$.

(34) ________ (attribute) to the well-dispersed state and layered structure, incorporated OH-BN presented a barrier function to suppress the delivery of thermal degradation products of PVA matrix, thus ________ (enhance) thermal stability and fire safety.

(35) In addition, the PCE is maintained at 70% or more of its original value even after 1000 bending cycles at a bending radius of 8 mm, ________ (promise) for practical applications of the FPSCs as next-generation semi-transparent solar energy power sources.

(36) ________ (take) this into account, the critical stress for growth of twins in pure magnesium is found to be 7 MPa which is consistent with previously published measurements on macroscopic single crystals.

(37) ________ (exploit) those advantages, the HES-based spheroids were used for 3D bioprinting, and the spheroids within the 3D-printed construct showed improved retention

and VEGF secretion compared to the same 3D structure containing single cell suspension.

(38) ________ (be) easily reused, ammonium chloride is found to be efficient and posing minor environmental impacts during the overall process.

(39) Moreover, the graphene-based framework includes only a very small amount of edge sites, thereby ________ (achieve) much higher stability against oxidation than conventional porous carbons such as carbon blacks and activated carbons.

(40) All topological patterns of the optimized PhCs are reported and have regular and smooth features, ________ (mean) they can be readily fabricated.

(41) Strong field enhancement is exhibited by the modes of this extended structure, which is able to excite a wealth of high-order surface plasmons, ________ (enable) deeply subwavelength focusing of incident THz radiation.

(42) For the first time, nanoparticles serve the dual role of an emulsion stabilizer and a pore template for the shell, directly ________ (utilize) in situ generated CO_2 bubbles as template for the core.

(43) Magneto-transport measurements show dramatic improvements in performance, including a record-high Hall mobility reaching 34,000 $cm^2\ V^{-1}\ s^{-1}$ for six-layer MoS_2 at low temperature, ________ (confirm) that low-temperature performance in previous studies was limited by extrinsic interfacial impurities rather than bulk defects in the MoS_2.

(44) The future success of semiconductor technology relies on the continuing reduction of the feature size, ________ (allow) more components per chip and higher speed.

(45) ________ (base) on the in situ TEM observation and calculation, particle size and distribution affects the sintering process.

(46) The method is based on consecutive processes involving radical polymerization and hydrolytic polycondensation, ________ (follow) by ultralow-cost, highly scalable, ambient-pressure drying directly from alcohol as a drying medium without any modification or additional solvent exchange.

(47) ________ (compare) with the traditional methods, laser sintering technology has the advantages of shorter sintering time and retaining the surface microstructure which is more conducive to the adhesion of the catalyst.

(48) ________ (compare) to thermal reduction, chemical reduction contributes to maintaining the flexibility and strong intensity of G-Fs.

(49) ________ (take) together, the light-activatable hydrogel-based platform allows us to release antibiotics more precisely, eliminate bacteria more effectively, and inhibit bacteria-induced infections more persistently, which will advance the development of novel antibacterial agents and strategies.

(50) ________ (inspire) by natural mosaic dominance phenomena, $Zn_{1-x}Co_xO/ZnCo_2O_4$ 2D-mosaic-hybrid mesoporous ultrathin nanosheets serve as building blocks to assemble into a 3D Zn–Co hierarchical framework.

(51) The changes in their surface color and mechanical properties were tested, ________ (accompany) by characterizations using UV-Vis, SEM, and ATRFTIR.

(52) This study displays the first time-resolved 3D imaging of decompression failure in high-pressure gas exposed polymers, ________ (obtain) from in-situ X-ray computed tomography.

(53) The simultaneous introduction of Co and Ni effectively improves the cycle stability of the electrode, ________ (indicate) by the increase of the capacity retention rate from 51.5% for MF to 87.4% for MFCN over 100 discharge-charge cycles at the same current density of 130 mA g^{-1}.

(54) Direct experimental evidence of these states has been limited so far to their local thermodynamic and magnetic properties, ________ (determine) by the competing effects of edge topology and electron-electron interaction.

(55) ________ (apply) to uniaxial-stress EMF process, the present model exhibits preferable prediction accuracy by comparing the predicted results of analytical calculation with experimental ones.

(56) Additionally, a dense Fe shell forms on the exterior surface of most samples, ________ (cause) by lamellae contacting and sintering during oxidation, ________ (follow) by formation of an impermeable Fe layer during reduction.

(57) ________ (summarize) over all measurements, the refractive index of PMMA increases by 0.002,2/1,000 bar with the pressure in the injection molding process during the solidification of the material.

(58) ________ (grow) on a conductive substrate, the nanowire arrays can be operated in a well-defined electrochemical working point with high sensitivity and stability.

(59) ________ (assist) with thermal thrombolysis, the present formulated system shows a high efficiency, on-demand drug release, and thus a safer protocol for thrombolytic therapy, which fits the developing trends of precision medicine.

(60) Proofs of concept have been reported with semiconductor quantum dots, yet ________ (limit) by inefficient atom-photon interfaces and dephasing.

(61) Subsequent studies confirmed a considerable protective role of PDPIA in a model of severe hepatitis induced by concanavalin A, ________ (evidence) by reduced hepatocellular injury and evaded immune response.

(62) Moreover, the thermal conductivity of Al_2O_3/graphene/PU composite reaches to 0.502 W m^{-1} K^{-1}, ________ (enhance) by 141% compared with pure matrix, ________ (show) very high enhancement efficiency, which is attributed to the multilevel aligned structure and multi-contact conductive pathway inside the composite.

(63) This simple strategy does not involve any solvothermal and hydrothermal processes, ________ (pave) a new avenue toward the design of robust non-noble electrocatalysts for hydrogen production, ________ (aim) at commercial water electrolysis.

(64) This paper reports the fabrication of an electrochemical supercapacitor (ES) with high gravimetric and areal capacitances, ________ (achieve) at a high mass ratio of active material to current collector.

Task 12: Arrange the sentences in each set logically and coherently into a research article abstract.

I

(1) Despite recent successful efforts to imbue synthetic materials with some of these remarkable functionalities, many emerging properties of complex adaptive systems found in biology remain unexplored in engineered living materials.

(2) Hydrogels loaded with the fungus Ganoderma lucidum are three-dimensionally printed into lattice architectures to enable mycelial growth in a balanced exploration and exploitation pattern that simultaneously promotes colonization of the gel and bridging of air gaps.

(3) Biological living materials, such as animal bones and plant stems, are able to self-heal, regenerate, adapt and make decisions under environmental pressures.

(4) To illustrate the potential of such mycelium-based living complex materials, we three-dimensionally print a robotic skin that is mechanically robust, self-cleaning and able to autonomously regenerate after damage.

(5) Here, we describe a three-dimensional printing approach that harnesses the emerging properties of fungal mycelia to create living complex materials that self-repair, regenerate and adapt to the environment while fulfilling an engineering function.

Sequence of sentences: ______________________________

II

(1) A wheel was considered as a final target application of the composite material.

(2) We showed the possibility of applying ML particles embedded composites to real engineering products.

(3) As an ML component, ZnS:Cu was employed in the composite.

(4) The morphological, rheological, thermal, and mechanical properties of the composites were analyzed experimentally.

(5) The aim of this study is to demonstrate the fabrication process of mechanoluminescent (ML) composites.

(6) The manufacturing process and resulting mechanical behavior of the wheel were modeled numerically.

(7) ML powders embedded thermoplastic polyurethane (TPU) composites were prepared using injection molding.

(8) As the particle content and mechanical stimulus imposed on the specimen increased, the induced ML was enhanced significantly.

Sequence of sentences: ______________________

Ⅲ

(1) The reversible deformation of the flexible MOF-NS and the vertical interlamellar pathways were captured with electron microscopy.
(2) MOF-NS and polydimethylsiloxane synergistically contribute to the separation performance.
(3) The controlled growth followed by a surface-coating method effectively produced flexible and defect-free superhydrophobic MOF-NS membranes.
(4) We report a strategy to create highly flexible metal-organic framework nanosheet (MOF-NS) membranes with a faveolate structure on polymer substrates for alcohol-water separation.
(5) Molecular simulations confirmed the structure and revealed transport mechanism. The ultrafast transport channels in MOF-NS exhibited an ultrahigh flux and a separation factor of 8.9 in the pervaporation of 5 weight % ethanol-water at 40°C, which can be used for biofuel recovery.
(6) High-performance pervaporation membranes have potential in industrial separation applications, but overcoming the permeability-selectivity trade-off is a challenge.

Sequence of sentences: ______________________

Ⅳ

(1) This work highlights the importance of material quality in achieving the long-term operational stability of perovskite optoelectronic devices.
(2) In this work, we introduce a high-temperature dimethyl-sulfoxide-free processing method that utilizes dimethylammonium chloride as an additive to control the perovskite intermediate precursor phases.
(3) A population of encapsulated devices showed improved operational stability, with a median T80 lifetime (the time over which the device power conversion efficiency decreases to 80% of its initial value) for the steady-state power conversion efficiency of 1,190 hours, and a champion device showed a T80 of 1,410 hours, under simulated sunlight at 65 ℃ in air, under open-circuit conditions.
(4) Achieving the long-term stability of perovskite solar cells is arguably the most important challenge required to enable widespread commercialization.
(5) By controlling the crystallization sequence, we tune the grain size, texturing, orientation (corner-up versus face-up) and crystallinity of the formamidinium (FA)/caesium $(FA)_yCs_{1-y}Pb(I_xBr_{1-x})_3$ perovskite system.
(6) Understanding the perovskite crystallization process and its direct impact on device stability

is critical to achieving this goal.

(7) The commonly employed dimethyl-formamide/dimethyl-sulfoxide solvent preparation method results in a poor crystal quality and microstructure of the polycrystalline perovskite films.

Sequence of sentences: ______________________________

V

(1) Ex-situ measurements of SEI resistivity at microscopic scales are also lacking.

(2) The results show relatively uniform lateral resistivity distribution of the SEIs but steep decreases in resistivity in the vertical direction.

(3) Silicon is a promising candidate for the lithium ion battery (LIB) anode because of the order-of-magnitude improvement in capacity over current state-of-the-art graphite anodes.

(4) In addition to resistance mapping, this method also provides an alternative technique for locating buried interfaces, where mechanical or electronic properties differ sufficiently between layers.

(5) We report on a nanometer-resolution three-dimensional technique that enables ex-situ mapping of electronic resistivity of SEI formed on a model single-crystalline wafer Si anode.

(6) Resistivity vs. depth profiles are highly dependent on cycling conditions, but they generally show a resistivity decrease from the most superficial levels of SEI and a thickness increase with continued cycling prior to SEI stabilization.

(7) In systems featuring both C and Si anodes, electronic resistivity of the solid-electrolyte interphase (SEI) layer is a critical factor for preventing continuous electrolyte-decomposition reactions at the electrode/electrolyte interface.

(8) The most prominent resistivity increase was observed on SEI formed in Gen2 electrolyte (EC:EMC [3:7 by wt.] + 1.2 M $LiPF_6$) with 10 wt% fluoroethylene carbonate additive; this result may partially explain the significant improvements of sustained electrochemical cycling and coulombic efficiency observed with this electrolyte additive.

(9) Further validation of this method was obtained by resistance mapping of a reference sample with a designed α-Si:H layer stack of different doping concentrations.

(10) However, the in-situ measurement of SEI electronic resistance has been complicated by ion transport and electronic contributions from other parts of the battery circuit.

(11) Our approach provides a novel and unparalleled three-dimensional approach in characterizing electronic resistivity, which contributes significantly to understanding SEI formation and the intrinsic properties critical to battery performance.

(12) Our novel experimental approach uses scanning spreading-resistance microscopy resistance imaging and mechanical depth profiling.

Sequence of sentences: ______________________________

Task 13: Summarize the main points of each paragraph in Text Ⅲ and identify the paragraph(s) providing the basis for each structural component in the abstract.

Text Ⅲ

Fabrication of Graphene Oxide and Hyperbranched Polyurethane Composite Via *in Situ* Polymerization with Improved Mechanical and Dielectric Properties[17]

Herein, we report a simple and one-pot route for the synthesis of graphene oxide (GO)/hyperbranched polyurethane (HBPU) composite by *in situ* polymerization technique.	Purpose
The polyurethane (PU) chains formed covalent linkages with the exposed hydroxyl and carboxylic acid groups on the GO. The composite was characterized by optical microscopy, scanning electron microscopy (SEM), transmission electron microscopy (TEM), nuclear magnetic resonance (NMR) spectroscopy, and infrared (IR) spectroscopy.	Methods
The characterization results revealed that the GO sheets were well dispersed within the HBPU matrix, resulting in very good enhancement in tensile strength and Young's modulus of the composites.	Result①& discussion
Differential scanning calorimetry (DSC) suggests that the GO has a nucleating effect for the HBPU.	Result②& discussion
The composites also show a mild increase in stability towards thermal degradation, and significantly, higher dielectric constant compared with neat HBPU at low frequencies.	Result③& discussion

INTRODUCTION

1 Graphene, which is a single layer of graphite comprising sp^2-hybridised carbon atoms in a hexagonal lattice[1], has attracted considerable interest owing to its unique electronic properties[2, 3], high thermal conductivity[4], exceptional Young's modulus and mechanical strength[5], together with good thermal stability. Therefore, graphene is regarded as promising candidate as fillers in high-performance polymer composites[6-10]. A major problem associated with graphene is the tendency of the graphitic layers to aggregate and poor compatibility with

solvents and polymer matrices. This leads to difficulties in processing and creates impediments in the way of successfully harnessing the superlative properties of graphene. This problem of graphene is generally overcome to a great degree by its derivative, graphene oxide (GO), which has oxygen-containing functional groups along the edges, as well as distributed over, the graphitic basal plane[11]. Consequently, GO has significantly better solubility in common solvents including water, lesser tendency to aggregate, and is relatively easier to process. GO can also be synthesized in gram quantities from graphite by potentially scalable liquid-phase reactions[12, 13].

2 Hyperbranched polymers (HPs) are highly branched polymers prepared from multi-functional monomers. In addition to their easy synthesis[14, 15], their three-dimensional dendritic structure confers them with many advantageous properties over traditional linear polymers such as their non-aggregation, high solubility[16], and low viscosity. Considerable research attention has been devoted to hyperbranched polymers in recent years owing to their interesting properties, which makes them potential candidates for applications ranging from coatings[17, 18], non-linear optical materials[19], light-emitting diodes[14, 20] and sensors[14, 21] to catalysis[22]. Recently, Mahapatra et al.[21] reported the synthesis of hyperbranched polyurethane (HBPU)-GO via click coupling of azide-functionalized GO and alkyne-terminated polyurethane (PU). Although click chemistry has the advantages of high selectivity and yields, and good improvement in mechanical properties was observed, their procedure had the drawback of involving multiple reaction steps. Here, we report an extremely simple, one-pot method for the *in situ* synthesis of an HBPU-GO composite. The isocyanate component of the PU forms covalent bonds with the hydroxyl and carboxylic groups in the GO to form carbamate and amide functional groups respectively[23, 24]. The covalent linkage between the filler and matrix results in good dispersion and subsequent load transfer. The composites were characterized by nuclear magnetic resonance (NMR) spectroscopy, infrared (IR) spectroscopy and thermogravimetric analysis (TGA), and the morphology and distribution of GO in the polymer matrix was studied by scanning electron microscopy (SEM) and transmission electron microscopy (TEM). The composites show good improvement of mechanical properties with increasing content of GO. The effect of incorporation of GO in the HBPU was also investigated by differential scanning calorimetry (DSC). The composites also have higher dielectric constant at low frequencies as compared with pure HBPU.

EXPERIMENT

Materials

3 Graphite and 4,4′-methylenebis(phenylisocyanate) (MDI) were purchased from Sigma Aldrich and used without further purification. Polyol (polytetramethylene glycol, PTMG, MW = 1,800 g mol^{-1}) was purchased from Korea PTG, Ltd. and 1,4-butanediol (BD) was

obtained from Junsei Chemical, Japan. *N,N'*-dimethylformamide (DMF) was used after distillation by the conventional technique and stored with molecular sieves.

Preparation of GO/HBPU Composites by In Situ Polymerization

4 The graphene oxide was synthesized according to the literature procedure[25]. Hyperbranched polyurethane was synthesized by a two-step procedure. Firstly, in a 500 mL four-neck cylindrical vessel equipped with a mechanical stirrer and nitrogen inlet, 6 mmol (10.8 g) of polyol(tetramethylglycol) (MW = 1,800 g/mol) was dissolved in 30 mL of dry DMF. Thereafter 16 mmol (4 g) of 4,4′-methylenebis(phenylisocyanate) in 20 mL of DMF solution was slowly injected into the vessel at room temperature. The reaction temperature was slowly increased to 75°C and the reaction was allowed to proceed for 4 h. In the second step, after the completion of prepolymer synthesis, the system was cooled to 0°C and 10 mmol (1.49 g) of triethanol amine 0.1 g of dibutyltindilaurate and graphene oxide (GO) solution with different concentration (0.01 to 2) in 20 mL of DMF were added to it. The temperature was then increased slowly to 60°C, and the reaction was evaluated by monitoring the disappearance of the absorption peak due to the –NCO group in the IR region of 2,250-2,270 cm^{-1}. After completion of the reaction (3 h), the final viscous product was dried in a hot air oven at 50°C for 48 h to obtain the polymer films. ^{1}H NMR (600 MHz, DMSO-d_6, d): 9.47 (s, 2H, NH), 7.34 (d, 4H, ArH), 7.05 (d, 4H, Ar H), 3.77 (s, 2H, CH2–Ar), PTMG residue 4.3 (d, 2H, N–CH), 4.12, 4.01, 3.32, 3.34 (d, 2H, –CH–O), 1.48, 1.2 (d, 2H, –CH– CH).

Measurements

5 ^{1}H-NMR spectra of HBPU and HBPU/GO were recorded in a Bruker 600 MHz NMR spectrometer using tetramethylsilane (TMS) as the internal standard and DMSO-D_6 as a solvent. Fourier transform infrared (FT-IR) spectra for GO, HBPU and HBPU/GO were recorded on a Jasco FT-IR 300E device. The dispersion of GO in the composite was measured using a field emission scanning electron microscope (FE-SEM, S-4300SE, Hitachi). The surface morphologies of HBPU/GO composite was observed by transmission electron microscopy (TEM, JEM 2100F, JEOL). The mechanical properties of the nanocomposites were measured at an elongation rate of 10 mm min^{-1} at room temperature using a tensile testing machine (Instron 4468). Differential scanning calorimetry (DSC) measurements were carried out using a TA instrument 2010 (Du Pont) thermal analyzer in a temperature range of −50 to 250°C at a heating rate of 10°C min^{-1} in nitrogen flow. Thermogravimetric analysis was carried out using a TA Q50 thermal analyzer with a nitrogen flow rate of 30 mL min^{-1} and a heating rate of 10°C min^{-1}.

RESULTS AND DISCUSSION

6 The IR spectra of the GO, HBPU and HBPU-GO composite are shown in Figure 1. The carbonyl peak of GO at around 1,730 cm^{-1} appears to be lost in the slightly lower energy carbamate carbonyl stretching peaks (~1,725 cm^{-1}), however this change is not very obvious. In addition, the presence of other characteristic peaks, such as the amide carbonyl stretch

(~1,645 cm^{-1}) and C-N stretch (~1,540 cm^{-1}) in the composite, indicate the coupling of GO with the HBPU[26]. The ^{1}H-NMR data also support the successful synthesis of HBPU-GO. The secondary –NH protons appeared at 9.47 and aromatic peak of MDI segment appeared at 7.34-7.04. The aliphatic $-CH_2$ protons appeared at 5.1–1.2 ppm due to different chemical environment. The protons for the aliphatic PTMG residue appeared at 4.3–1.2 ppm. Comparing with pure HBPU, the spectra of HBPU-g-GO are well agreement with hyperbranched polyurethane structure[26].

7 Figure 2a shows the optical micrograph of the GO/HBPU-4 composite. The surface is corrugated, as a result of uniformly dispersed GO in the polymer matrix. The SEM image of the fracture surface of the GO/HBPU-4 composite (Figure 2b) is very rough, indicative of the good interfacial interaction between the GO and the polymer, which may be attributed to covalent bond formation between the GO and polymer chains. Also, the GO sheets appear to be well dispersed in the matrix. TEM images of the fracture surface for this composite (Fig 2c and d) show wavy and crinkled GO sheets, suggesting that they retain their dispersion and do not re-aggregate in the polymer. Our observations are in line with previous reports for GO/polymer composites[27-29]. Thus, good enhancement in mechanical properties can be expected in the GO/HBPU composites.

8 The tensile strength of the HBPU-GO composites increased to 27.4 MPa for 4 wt% GO, compared with 18 MPa for neat HBPU, thus an enhancement of 52% was observed (Figure 3 and Table 1). The Young's modulus of the 4 wt% GO-HBPU composite was 20 MPa, which represents a four-fold increase over that of neat polymer. The strain at break decreased from 1,610% for neat polymer to 1,180% for the 4 wt% GO composite, hence a reduction of 27%. The excellent improvement in mechanical properties may be attributed to the very good dispersion of the GO in the polymer matrix, as well as covalent bond formation between the hydroxyl and carboxylic acid groups on GO, and the isocyanate[23], resulting in effective load transfer[25].

9 DSC shows that the crystallization temperature (T_c) of the GO-HBPU composites increases slightly, while the melting temperature (T_m) is unchanged (Figure 4). This has been observed previously for graphene-polyurethane composites[9, 30], and is probably owing to nucleation by GO. Indeed, the broadening of the curve at crystallization supports the hypothesis that the GO provides additional nucleation sites. The increasing trend in T_c is indicative of affinity between GO and the HBPU matrix.

10 The incorporation of GO in the HBPU achieves a slight improvement in the thermal stability of the composites, as has been reported earlier[31-33]. The thermal degradation of neat HBPU and the GO/HBPU composites are shown in Figure 5. The two-step decomposition corresponds to the soft and hard segments of the PU. The degradation temperature at about 70% residual weight increases slightly from 380°C in the case of neat HBPU to 395°C for GO/HBPU-4. This enhancement in thermal stability has been attributed to the barrier effect

of the GO platelets, which prevents oxygen from diffusing into the polymer, thereby slowing down its decomposition[32].

11 The incorporation of graphene, GO as well as their related materials, carbon nanotubes have been previously reported to increase the dielectric constant of polymers[34, 35]. Here too, we observe an increase in the dielectric constant of the GO-HBPU composites with increasing GO content (Figure 5). The dielectric constant of the composites is much higher at lower frequencies than at higher frequencies, similar to our findings for functionalized carbon nanotubes (CNT)-polymer composites[35], as well as for graphene- and GO-polymer composites reported by other groups[36]. This observation has been attributed to the setting up of large number of polymer-filler interfaces, which block the charge carriers – as substantiated by SEM and TEM (Figure 2) and the predominance of interfacial polarization at these interfaces at low frequencies[37]. The blocking of charge carriers at the GO/HBPU interfaces is further enhanced by the electrophilic oxygen-containing functional groups on GO[35, 37].

SUMMARY

12 We have described a simple, one-pot method for the synthesis of graphene oxide-hyperbranched polyurethane. The graphene oxide undergoes covalent bonding with the isocyanate component of the polyurethane, and is well-dispersed in the polymer matrix. As a result, excellent load transfer is achieved, and the composites show very good enhancement in mechanical properties. The composites also show high dielectric permittivity compared with neat HBPU at low frequencies, owing to the blocking of charge carriers at the GO/HBPU interfaces, enhanced by electronegative oxygen-containing functional groups on the GO. The incorporation of GO also improves the thermal stability of the resulting composites owing to the barrier effect of the GO sheets. GO/HBPU composites with high mechanical and dielectric properties can find application in electronic devices, artificial muscle materials, and actuators.

(Figures and tables omitted)

References

…

Main points of each paragraph	Abstract
Para. 1: ____________________	Purpose
Para. 2: ____________________	
Para. 3: ____________________	Methods

continued

Main points of each paragraph	Abstract
Para. 4: ______________________________	
Para. 5: ______________________________	Result①& discussion
Para. 6: ______________________________	
Para. 7: ______________________________	
Para. 8: ______________________________	Result②& discussion
Para. 9: ______________________________	
Para. 10: ______________________________	
Para. 11: ______________________________	Result③& discussion
Para. 12: ______________________________	

Task 14: Transform the following high-frequency verb-noun collocations used in research article abstracts into passive voice. Determine typical verb-noun collocations clearly describing background, purpose, methods, results, and discussion/conclusion in an abstract.

No.	(Phrasal) Verbs	Noun collocates	Examples
(1)	achieve	density efficiency	achieve a high energy density achieve significantly higher encapsulation efficiencies
(2)	address	challenge issue	address the notorious challenges address the efficacy and safety issues
(3)	affect	property	affect the physical and mechanical properties
(4)	attract	attention interest	attract numerous attention attract considerable theoretical interest
(5)	boost	performance	boost the photovoltaic performance
(6)	conduct	study	conduct fundamental nanoscience studies
(7)	control	property	control the material properties

continued

No.	(Phrasal) Verbs	Noun collocates	Examples
(8)	deliver	capacity	deliver an ultrahigh capacity
(9)	demonstrate	application approach feasibility performance potential strategy	demonstrate their potential applications demonstrate structural interface engineering approach demonstrate the feasibility demonstrate excellent antistatic performance demonstrate the great potential demonstrate a synchronous reduction strategy
(10)	develop	approach method model device material strategy	develop instrumental imaging approaches develop a software-based drift correction method develop a general multiscale model develop high-efficiency photovoltaic devices develop novel electrode materials develop a simple, rapid, and robust strategy
(11)	discuss	mechanism	discuss the photoelectric mechanism
(12)	draw	attention	draw tremendous attention
(13)	enhance	activity conductivity efficiency performance property stability	enhance the water splitting activity enhance their electrical conductivity enhance charge separation efficiency enhance electrochemical energy storage performances enhance the electrical transport properties enhance the operational stability
(14)	establish	model	establish an appropriate thermal cycle processing model
(15)	evaluate	effect	evaluate their inhibition effect
(16)	exhibit	activity behavior capability capacity conductivity effect efficiency performance property resistance response stability strength structure	exhibit much higher photocatalytic activity exhibit stress - stiffening behavior exhibit excellent Li+ storage capability exhibit a reversible capacity exhibit extremely low diffusive thermal conductivity exhibit nonlinear magnetoelectric effects exhibit a high luminous efficiency exhibit excellent rate performance exhibit robust mechanical properties exhibit an excellent sheet resistance exhibit fast and stable responses exhibit excellent long-term stability exhibit excellent tensile strength exhibit 3D nanoscale pore structure

continued

No.	(Phrasal) Verbs	Noun collocates	Examples
(17)	find	application	find promising applications
(18)	form	structure	form an obvious core-shell structure
(19)	have	advantage application capacity conductivity effect impact implication influence performance potential property stability structure	have the structure advantage have potentially widespread applications have large loading capacity have an anisotropic thermal conductivity have a synergistically positive effect have long-lasting and fundamental impacts have practical implications have a decisive influence have stronger fatigue performance have a great environmental improvement potential have unique physical and chemical properties have a long-term cycling stability have a simply disordered atomic structure
(20)	highlight	importance	highlight the paramount importance
(21)	hold	potential promise	hold great potential hold a great promise
(22)	improve	conductivity efficiency performance property stability strength	improve electrode conductivity improve the computational efficiency improve the overall electrode performances improve the mechanical properties improve the mechanical stability improve the interfacial shear strength
(23)	increase	content density temperature	increase the acrylonitrile content increase the density increase phase transition temperatures
(24)	investigate	mechanism effect influence property	investigate tensile fracture mechanisms investigate the individual and associated effects investigate the influence investigate the crystal and magnetic properties
(25)	limit	application	limit the commercial application
(26)	offer	opportunity	offer new opportunities
(27)	open	avenue door	open a cost-effective avenue open a new door

continued

No.	(Phrasal) Verbs	Noun collocates	Examples
(28)	optimize	performance	optimize the multifunctional performance
(29)	overcome	limitation	overcome intrinsic efficiency limitations
(30)	pave	way	pave a smart way
(31)	perform	analysis test	perform a comprehensive statistical analysis perform calibrated responsivity tests
(32)	play	role	play a pivotal role
(33)	possess	property	possess seemingly incompatible properties
(34)	prepare	composite	prepare graphene reinforced metal matrix composites
(35)	present	approach method strategy	present an experimental single-molecule approach present a simple and efficient method present a novel composite strategy
(36)	promote	formation	promote rapid formation
(37)	propose	mechanism strategy	propose a reaction mechanism propose an alternating photothermal strategy
(38)	provide	approach avenue information insight method opportunity platform route site strategy understanding way	provide a conceptually novel approach provide a new venue provide key biological information provide fresh and much needed insights provide a simple and intuitive method provide an additional opportunity provide a promising structural platform provide a new route provide sufficient accessible reaction sites provide a sustainable and scalable strategy provide an insightful understanding provide an effective way
(38)	remain	challenge	remain a great challenge
(40)	report	approach method strategy synthesis	report a template-free approach report a general post-annealing method report a tailored microwave-aided synthetic strategy report a new large-scale colloidal synthesis

continued

No.	(Phrasal) Verbs	Noun collocates	Examples
(41)	show	activity behavior capacity conductivity decrease effect efficiency improvement increase performance potential property resistance stability trend	show remarkable high OER catalytic activity show strong rectifying behavior show much higher reversible capacity show a highest thermal conductivity show a resistivity decrease show a significant radio-sensitizing effect show a high efficiency show a significant improvement show a linear increase show unprecedented electrochemical performance show realistic potential show inferior photocatalytic properties show higher erosion resistance show enhanced operational stability show similar qualitative trends
(42)	solve	problem	solve the quadratic optimization problems
(43)	study	property effect	study both the electrical and photovoltaic properties study the electrophysiological effects
(44)	tune	property	tune the hydrogel network properties
(45)	understand	mechanism	understand the underlying molecular mechanism
(46)	use	method model analysis approach combination diffraction film microscopy process simulation spectroscopy system technique	use a one-pot template-free method use a developed numerical model use cohesive element-based finite element analysis use a multimodal imaging approach use carefully selected combinations use wide-angle-ray diffraction use noncompact films use high-speed atomic force microscopy use both vacuum and solution processes use molecular dynamics simulations use electron energy loss spectroscopy use a single-stage system use liquid solution-based techniques

Verb-noun collocations describing **background**:

____

____

Verb-noun collocations describing **purpose**:

____

____

Verb-noun collocations describing **methods**:

____

____

Verb-noun collocations describing **results**:

____

____

Verb-noun collocations describing **discussion/conclusion:**

____

____

Task 15: Discuss in groups what appropriate verbs could be used to fill in each blank in the following research article abstracts. Change the verb forms if necessary.

I

Microscopic Segregation Dominated Nano-Interlayer Boosts 4.5 V Cyclability and Rate Performance for Sulfide-Based All-Solid-State Lithium Batteries[18]

To implement the growing requirement for higher energy density all-solid-state lithium batteries (ASSLBs), further increasing the working voltage of $LiCoO_2$ (LCO) is a key to (1) __________ the bottleneck. However, $LiCoO_2$ severe structural degradation and side reactions at the cathode interface (2) __________ the development of high-voltage sulfide-based ASSLBs (⩾4.5 V). Herein, a nano-metric $Li_{1.175}Nb_{0.645}Ti_{0.4}O_3$ (LNTO) coated LCO cathode where microscopic Ti and Nb segregation at the interface during cycling potentially stabilizes the cathode lattice, and minimizes side reactions, simultaneously, is (3) __________. Advanced transmission electron microscopy (4) __________ that the stable spinel phase minimizes the micro stress at the cathode interface, avoids structure fragmentation, and hence significantly (5) __________ the long-term cyclic stability of LNTO@LCO @ 4.5 V. Moreover, the differential phase contrast scanning transmission electron microscopy (DPC-STEM) visualizes the nano-interlayer LNTO to boost Li^+ migration at the cathode interface. Electrochemical impedance spectroscopy (EIS) (6) __________ that sulfide-based cells with the LNTO nano-layer effectively (7) __________ the interfacial resistance to 140 Ω compared to $LiNbO_3$ (235 Ω) over 100 cycles. Therefore, 4.5 V sulfide-based ASSLBs (8) __________ gratifying long-cycle stability (0.5 C for 1,000 cycles, 88.6%), better specific capacity, and rate performance (179.8 mAh g^{-1} at 0.1 C, 97 mAh g^{-1} at 2 C).

II

Structural evolution of titanium dioxide during reduction in high-pressure hydrogen[19]

The excellent photocatalytic properties of titanium oxide (TiO_2) under ultraviolet light have long (1) __________ the search for doping strategies capable of extending its photoactivity to the visible part of the spectrum. One approach is high-pressure and high-temperature hydrogenation, which (2) __________ in reduced "black TiO_2" nanoparticles with a crystalline core and a disordered shell that absorbs visible light. Here we (3) __________ the formation mechanism and structural features of black TiO_2 (4) __________ first-principles-validated reactive force field molecular dynamics simulations of anatase TiO_2 surfaces and nanoparticles at high temperature and under high hydrogen pressures. Simulations (5) __________ that surface oxygen vacancies created upon reaction of H_2 with surface oxygen atoms diffuse towards the bulk material but encounter a high barrier for subsurface migration on {001} facets of the nanoparticles, which initiates surface disordering. Besides (6) __________ that the hydrogenated amorphous shell has a key role in the photoactivity of black TiO_2, our results (7) __________ insight into the properties of the disordered surface layers that are (8) __________ on regular anatase nanocrystals under photocatalytic water-splitting conditions.

III

3D Graphene Films Enable Simultaneously High Sensitivity and Large Stretchability for Strain Sensors[20]

Integration of 2D membranes into 3D macroscopic structures is essential to (1) __________ the intrinsically low stretchability of graphene for the applications in flexible and wearable electronics. Herein, the synthesis of 3D graphene films (3D-GFs) using chemical vapor deposition (CVD) is (2) __________, in which a porous copper foil (PCF) is chosen as a template in the atmospheric-pressure CVD preparation. When the 3D-GF prepared at 1,000 °C (noted as 3D-GF-1000) is transferred onto a polydimethylsiloxane (PDMS) membrane, the obtained 3D-GF-1000/PDMS hybrid film (3) __________ an electrical conductivity of 11.6 S cm^{-1} with good flexibility, (4) __________ by small relative resistance changes ($\Delta R/R_0$) of 2.67 and 0.36 under a tensile strain of 50% and a bending radius of 1.6 mm, respectively. When the CVD temperature is (5) __________ to 900 °C (generating a sample noted as 3D-GF-900), the 3D-GF-900/PDMS hybrid film (6) __________ an excellent strain-sensing performance with a workable strain range of up to 187% and simultaneously a gauge factor of up to ≈1,500. The 3D-GF-900/PDMS also (7) __________ a remarkable durability in resistance in repeated 5,000 stretching-releasing cycles. Kinetics studies (8) __________ that the response of $\Delta R/R_0$ upon strain is related to the graphitization and conductivity of 3D-GF which are sensitive to the CVD preparation temperature.

Ⅳ

High temperature deformation and microstructural evolution of core-shell structured titanium alloy[21]

The hot deformation behavior and microstructural evolution of a core-shell (soft coarse-grained Ti cores and hard Ti-N solid solution shells) structured titanium alloy were (1) ___________ under 800–950 °C at a strain rate range of 0.001–1 s^{-1}. The core-shell structured titanium alloys (2) ___________ a higher flow stress (195 MPa deformed at 800 °C/0.1 s^{-1}) (3) ___________ with that of pure Ti (48 MPa deformed at 800 °C/0.1 s^{-1}), (4) ___________ that core-shell structure promotes a strengthening effect of titanium alloys. The nitrogen solution (5) ___________ hard shells squash along compression direction playing a roll of skeleton and the soft cores accommodate the deformation, (6) ___________ to inhomogeneous deformation. Dynamic recrystallization is (7) ___________ mainly adjacent to the shells due to stress and strain concentration. The improvement compression stress of core-shell structured titanium alloys could be (8) ___________ to the nitrogen-induced solution strengthening and core-shell structure itself, together with the inhabited plastic deformation and dynamic recrystallization by core-shell structure.

V

Active pixel sensor matrix based on monolayer MoS_2 phototransistor array[22]

In-sensor processing, which can (1) ___________ the energy and hardware burden for many machine vision applications, is currently lacking in state-of-the-art active pixel sensor (APS) technology. Photosensitive and semiconducting two-dimensional (2D) materials can (2) ___________ this technology gap by (3) ___________ image capture (sense) and image processing (compute) capabilities in a single device. Here, we (4) ___________ a 2D APS technology based on a monolayer MoS_2 phototransistor array, where each pixel uses a single programmable phototransistor, (5) ___________ to a substantial reduction in footprint (900 pixels in ~0.09 cm^2) and energy consumption (100s of fJ per pixel). By (6) ___________ gate-tunable persistent photoconductivity, we (7) ___________ a responsivity of $\sim 3.6 \times 10^7$ A W^{-1}, specific detectivity of $\sim 5.6 \times 10^{13}$ Jones, spectral uniformity, a high dynamic range of ~80 dB and in-sensor de-noising capabilities. Further, we (8) ___________ near-ideal yield and uniformity in photoresponse across the 2D APS array.

Task 16: Write an abstract explicitly stating the background, purpose, methods, results, and discussion/conclusion of a recent study or experiment you have conducted. Present your abstract orally in groups and assess each other's performance.

Unit 2
Introduction Sections

Objectives

- Understand the multiple communicative purposes of research article introductions;
- Identify the generic structure and justification steps of research article introductions;
- Acquaint with quoting, paraphrasing, and summarizing strategies to incorporate previous literature into your writing;
- Familiarize with basic styles of in-text citations and reference lists at the end of research articles;
- Identify complex and compressed noun phrases conveying propositional contents;
- Identify commonly used linguistic features in reporting, synthesizing, and evaluating previous literature;
- Distinguish different authorial stances towards previous literature expressed through the use of reporting verbs, evaluative adjectives, and linking adverbials;
- Understand different rhetorical functions of common reporting verbs and the appropriate verb-noun collocations;
- Raise awareness of lexical and syntactic complexity in research article introductions.

Task 1: Discuss with each other which of the following communicative purposes a typical research article introduction is expected to fulfill.

Communicative Purposes	Yes/No?
(1) Claiming importance or value of a research area	

continued

Communicative Purposes	Yes/No?
(2) Emphasizing theoretical/methodological/practical necessity or urgency of addressing central issue(s) in a research area	
(3) Reviewing previous research clearly relevant to the present one	
(4) Referring to the importance or contributions of specific studies	
(5) Discussing negative outcomes of previous studies	
(6) Making generalizations or syntheses of recent studies in a research area	
(7) Pointing out research gap unaddressed by previous literature	
(8) Announcing research purposes of present study	
(9) Creating a theoretical framework for present study	
(10) Presenting research questions or hypotheses to be investigated in present work	
(11) Introducing test materials (or samples) of present study	
(12) Elaborating on research design (or method) of present study	
(13) Justifying theoretical or analytical perspective of present study	
(14) Rationalizing present research focus and design	
(15) Highlighting primary findings of present study	
(16) Stating the significance of present study	
(17) Discussing implications of present research	
(18) Outlining the structure of the research paper	
(19) Acknowledging the limitations of the present study	
(20) Making recommendations for future studies	

Task 2: Identify the main idea of each paragraph in Text Ⅰ with a focus on the topic sentence and general-specific and specific-general structure.

Text Ⅰ

Architected Cellular Piezoelectric Metamaterials: Thermo-Electro-Mechanical Properties[1]

INTRODUCTION

1 By tailoring their periodically repeated microarchitectures, metamaterials deliver many exotic properties beyond those found in natural materials or chemically synthesized substances. After the pioneering theoretical works on metamaterials over a decade ago [1], experiments first demonstrated the existence of negative permeability and/or negative permittivity in electromagnetic metamaterials [2,3]. Later, optical metamaterials [4,5], acoustic metamaterials [6,7], mechanical metamaterials [8], [9], [10], [11], [12] and thermal metamaterials [13,14] triggered a significant amount of research interest in the fields of engineering, physics, material science and chemistry mainly due to their unprecedented and unusual multifunctional properties (e.g. negative refractive index [4,5], negative effective density [6,7], negative incremental stiffness [9], negative Poisson's ratio [10], shape reconfigurability [11], ultra-low density [12], and negative thermal conductivity [14]). If these metamaterials are made of smart materials (e.g. piezoelectric, piezomagnetic, and magnetostrictive materials) in the form of optimized cellular architectures, they can open a new venue for designing lightweight advanced multifunctional materials responsive to arbitrary multiphysical stimuli. Cellular smart metamaterials promise optimized multiphysical properties not readily achievable by existing smart materials. Tailoring the microarchitecture of metamaterials is an effective method to develop optimized multifunctional materials. Rather than merely changing chemical constituents, deliberately optimizing the microarchitecture of smart materials can obtain not inherently achievable multiphysical properties for piezoelectric metamaterials [15,16].

2 With wide application in ultrasonic imaging devices, sensors, energy harvesting, and transducers, piezoelectric materials have been rapidly developing in recent years. Advanced 3D printing technology [17] and nanotechnology [18] provided reliable routes for manufacturing piezoelectric metamaterials with complex geometries in micro and nano scales. Typically, apart from modifying material constituents, there are two commonly used methods to modify the properties of piezoelectric materials: (1) adding a second phase (piezocomposites) [19], [20], [21], [22], [23], [24], [25], [26], [27], [28], [29] and/or (2) introducing pores (cellular piezoelectric materials) [30], [31], [32], [33], [34], [35], [36], [37], [38], [39], [40]. Based on the interconnection of the inclusions/pores, there are four types of composite/cellular piezoelectric metamaterials: (1) 0–3 type (inclusions/pores are enclosed by the matrix), (2)

1–3 type (inclusions/pores connect in one direction and matrix connects in all three directions), (3) 2-2 type (both inclusions/pores and matrix exhibit connectivity in two dimensions) and (4) 3-3 type (both inclusions/pores and matrix exhibit connectivity in all three dimensions).

3 Compared to cellular piezoelectric materials, encapsulating brittle piezoelectric ceramics by soft polymers provides a piezocomposite with excellent electro-mechanical performance. For polymer-based piezocomposites, a piezoelectric matrix (such as Polyvinylidene fluoride (PVDF) [19], [20], [21]) showed better figures of merit and piezoelectric properties than a non-piezoelectric matrix (such as epoxy [22], [23], [24], [25], [26], [27], [28], [29]). The ceramic-based piezocomposite (such as PZT-7A/$BaTiO_3$ [19]) was explored to modify the electro-mechanical properties of piezocomposites. Regarding the dimension of inclusion connectivity, 0–3, 1–3 and 2-2 types of piezocomposites with different inclusion shapes and unit cell arrangements were modelled and compared in Ref. [20]. The 0–3 piezocomposites with short piezoelectric fibers (cylinder) [21] and particulate inclusion (sphere) [22] have exhibited similar properties for different volume fractions of inclusion. However, due to their extensive use in biomedical and naval applications, 1–3 piezocomposites attracted more efforts than the other three types of piezocomposites. One method to modify the properties of 1–3 piezocomposites is to engineer the inclusion topology (prismatic inclusion (octagon and square) and non-prismatic inclusion (circle and ellipse) [20]) and unit cell position arrangement (hexagonal and square position arrangement [23]), especially for transversely polarized 1–3 piezocomposite (with polarization direction perpendicular to the inclusion). Another category of 1–3 piezocomposites, such as four-step braided cylindrical inclusions [24], randomly distributed cylindrical inclusions [25], and parallelogram periodic cells [26], have also been investigated to seek the possibility of improving electro-mechanical properties; it was found that changing the position arrangement of unit cells had obvious influence on effective electro-mechanical properties. In addition, functionally graded piezocomposites have been introduced by tailoring polarization direction and material gradation within the smart composites [27]. Compared to 1–3 piezocomposites with circular inclusions, one advantage of 2-2 piezocomposites was its straightforward processing method and avoiding direct connection between electrodes and piezoelectric phase [28]. One noticeable 2-2 piezocomposites was a Macro Fiber Composite (MFC) with interdigitated electrodes, which has been used as actuators and sensors by a company named as *Smart Material*. In general, 1–3 piezocomposites showed better figures of merit for energy harvesting (higher piezoelectric charge coefficients and coupling constants), while 0–3 piezocomposite revealed the lowest acoustic impedance for biomedical imaging devices. However, the separation of the interface between inclusion and matrix might deteriorate piezocomposites [29].

4 Introducing pores makes piezoelectric materials brittle; however, cellular piezoelectric metamaterials have shown remarkable improvement in figures of merit, especially for

hydrophone applications. Cellular piezoelectric metamaterials can be considered as piezocomposites with empty inclusions, periodically distributed in a piezoelectric matrix [30]. Many studies have recently been conducted to optimize electro-mechanical properties of cellular piezoelectric materials by tailoring their microarchitectures [30], [31], [32], [33], [34], [35], [36], [37], [38], [39], [40]. For 0–3 type of cellular piezoelectric metamaterials, alternative pores, e.g. flat-cuboidal, spherical and short-cylindrical, showed different effective electro-mechanical properties; flat-cuboidal pores resulted in better hydrostatic charge and hydrostatic voltage constants [31]. Similar to piezocomposite, electro-mechanical properties of 1–3 cellular piezoelectric materials can be improved by alternative parameters, e.g. pore shape [30], pore aspect ratio [32], and polarization direction [32], [33], [34], [35]. For longitudinally polarized porous piezoelectric materials, electro-mechanical properties were insensitive to pore shapes and aspect ratio; opposite behavior has been found for transversely polarized porous piezoelectric materials [30, [33], [34], [35], [36], [37]. In addition, acoustic impedance and hydrostatic charge constant can be modified by tailoring pore position arrangement [33], [34]. As for prismatic pores, auxetic [35], honeycomb [35], [36], [37], tetragonal [36], [37] and triangular microarchitectures [38, 39] have shown different deformation modes and electro-mechanical properties. It was found that along polarization direction, bending dominant deformation mode enhanced the piezoelectric figures of merit. Compared to 1–3 type of piezoelectric cellular materials, 3-3 type of piezoelectric cellular materials have shown great improvement in piezoelectric figures of merit at the cost of deteriorating mechanical performance [38]. Apart from the piezoelectric effect, some pyroelectric figures of merit with cellular microarchitectures also showed obvious improvement compared to that of dense materials [40], [41].

5 In order to obtain the effective properties of architected cellular materials with periodic microarchitectures, both analytical and numerical methods have been proposed. Based on the pioneering work conducted by Eshelby [42], several analytical mean field methods [43], [44], [45], e.g. dilute, self-consistent, and Mori–Tanaka, were proposed to predict effective electro-mechanical properties of piezocomposite with alternative inclusion topologies. Among these micromechanical models, Mori–Tanaka micromechanical scheme has provided the most accurate results compared to experimental data. However, local fluctuations of field quantities were not taken into account in these analytical micromechanical models. This restriction can be solved by combination of analytical models with numerical methods, e.g. finite element method. Recently, asymptotic homogenization (analytical method) [23], [24], [26], [27], [28], [30], [46], [47] and finite element method (numerical method) [20], [22], [25], [27], [31], [32], [33], [34], [35], [36], [37], [48] have been widely used to predict the effective properties of cellular materials with periodic microstructures. Although finite element method is straightforward to understand, asymptotic homogenization method has a robust mathematical basis, which provides closed-form expressions for accurate theoretical predictions of effective multiphysical properties of advanced materials with periodic microstructures.

6 Different 0–3, 1–3, 2-2, 3-3 types of piezoelectric cellular materials have been investigated by analytical (Mori-Tanaka [19], [29], [33] and asymptotic homogenization [23], [24], [26], [30], [31]) and numerical (finite element method [20], [21], [22], [25], [27], [32], [33], [34], [36], [37], [38], [48]) methods. Employing these methods, some researches have provided the relationship between effective properties and relative density or solid volume fraction. However, the design chart for piezoelectric properties (similar to material selection charts for mechanical properties) is yet to be developed to provide a broad view on feasible piezoelectric materials through devising advanced smart materials. In addition, despite of many structural and multifunctional advantages offered by piezocomposites and high figures of merit of lightweight cellular solids, studies on the combination of piezocomposites and cellular solids, i.e. cellular piezocomposite metamaterials, are limited. For example, for cellular piezocomposites made of alternative matrices, it was found that ceramic matrix-based cellular piezocomposites were more sensitive to pores than polymer matrix materials [21]. Based on micromechanical modelling, the introduction of pores to matrix was found to increase the hydrostatic performance of cellular piezocomposites [22]. However, both of the aforementioned researches did not consider the influence of polarization directions and pore shapes on the electromechanical properties. Furthermore, since most of commercial finite element software, e.g. ANSYS and ABAQUS, does not have an appropriate element for pyroelectric constants, investigations on the pyroelectric properties of piezoelectric cellular materials were mostly limited to experimental studies [40], [41]. To the best of authors' knowledge, there is no numerical study in the literature on exploring the effect of pore shapes and polarization directions on the effective pyroelectric properties of piezoelectric cellular metamaterials. Therefore, objectives of the present study are:

(a) To develop an asymptotic homogenization model for characterizing the thermo-electro-mechanical properties (e.g. effective elastic properties, piezoelectric properties, dielectric properties, pyroelectric properties and thermal conductivity) and evaluating the piezoelectric and pyroelectric figures of merit of architected cellular piezoelectric metamaterials.

(b) To design cellular piezoelectric metamaterials of 1–3 type with alternative pore shapes, polarization directions and non-piezoelectric or piezoelectric phases to optimize the thermo-electro-mechanical properties of cellular smart metamaterials.

(c) To establish a relation between microarchitectural features and multiphysical properties of piezoelectric metamaterials and to provide a design chart for optimizing their thermo-electro-mechanical properties by tailoring their microarchitecture or material composition.

(d) To elucidate the rationale for dissimilar thermo-electro-mechanical properties found in longitudinally and transversely polarized cellular metamaterials with and without non-piezoelectric phases.

7 The present paper is organized as follows: Four types of architected cellular piezoelectric metamaterials are introduced in Section 2. The constitutive and governing equations for uncoupled linear thermo-piezoelectricity are given in Section 3. Section 4 provides details on the implementation of asymptotic homogenization for obtaining the effective thermo-electro-mechanical properties of cellular piezoelectric metamaterials. The role of microarchitectural features of cellular metamaterials on their effective properties and figures of merit are presented in Section 5 compared to conventional honeycomb piezoelectric cellular materials and piezocomposites with solid circular inclusion. To better understand the effect of microarchitecture on thermo-electro-mechanical properties of piezoelectric metamaterials, a detailed discussion is provided in Section 6 on the thermo-electro-mechanical performance of four types of cellular piezoelectric metamaterials. Section 7 highlights the main conclusions extracted from the present study.

References

[1] V. G. Veselago, The electrodynamics of substances with simultaneously negative values of ε and μ, Sov. Phys. Usp. 10 (4) (1968) 509.

[2] A. N. Lagarkov, V. N. Semenenko, V. A. Chistyaev, D. E. Ryabov, S. A. Tretyakov, C. R. Simovski, Resonance properties of bi-helix media at microwaves, Electromagnetics 17 (3) (1997) 213-237.

[3] D. R. Smith, J. B. Pendry, M. C. K. Wiltshire, Metamaterials and negative refractive index, Science 305 (5685) (2004) 788-792.

[4] V. M. Shalaev, W. Cai, U.K. Chettiar, K. Yuan, A. Sarychev, V. Drachev, A. Kildishev, Negative index of refraction in optical metamaterials, Optic Lett. 30 (24) (2005) 3356-3358.

[5] V. M. Shalaev, Optical negative-index metamaterials, Nat. Photon. 1 (1) (2007) 41-48.

[6] J. Li, C.T. Chan, Double-negative acoustic metamaterial, Phys. Rev. 70 (2) (2004), 055602.

[7] H. Chen, C.T. Chan, Acoustic cloaking in three dimensions using acoustic metamaterials, Appl. Phys. Lett. 91 (18) (2007), 183518.

[8] A. Rafsanjani, A. H. Akbarzadeh, D. Pasini, Snapping mechanical metamaterials under tension, Adv. Mater. 27 (39) (2015) 5931-5935.

[9] J. T. B. Overvelde, K. Bertoldi, Relating pore shape to the non-linear response of periodic elastomeric structures, J. Mech. Phys. Solid. 64 (1) (2014) 351-366.

[10] S. Babaee, J. Shim, J. C. Weaver, E. Chen, N. Patel, K. Bertoldi, 3D soft metamaterials with negative Poisson's ratio, Adv. Mater. 25 (36) (2013) 5044-5049.

[11] J. T. B. Overvelde, T. A. D. Jong, Y. Shevchenko, S.A. Becerra, J. Whitesides, J. Weaver, C. Hoberman, K. Bertoldi, A three-dimensional actuated origami-inspired transformable metamaterial with multiple degrees of freedom, Nat. Commun. 7 (2016), 10929.

[12] M. G. Lee, J. W. Lee, S. C. Han, K. Kang, Mechanical analyses of "Shellular", an ultralow-density material, Acta Mater. 103 (8) (2016) 595-607.

[13] M. Maldovan, Sound and heat revolutions in phononics, Nature 503 (7475) (2013) 209-217.

[14] Ş. Narayana, Y. Sato, Heat flux manipulation with engineered thermal materials, Phys. Rev. Lett. 108 (21) (2012), 214303.

[15] T. Shimada, V. L. Le, K. Nagano, J. Wang, T. Kitamura, Hierarchical ferroelectric and ferrotoroidic polarizations coexistent in nano-metamaterials, Science Report 5 (2015), 14653.

[16] V. L. Le, T. Shimada, S. Sepideh, J. Wang, T. Kitamura, Polar and toroidal electromechanical properties designed by ferroelectric nano-metamaterials, Acta Mater. 113 (2016) 81-89.

[17] S. Bodkhe, G. Turcot, F. P. Gosselin, D. Therriault, One-step solvent evaporation-assisted 3D printing of piezoelectric PVDF nanocomposite structures, ACS Appl. Mater. Interfaces 9 (24) (2017) 20833.

[18] J. Briscoe, S. Dunn, Piezoelectric nanogenerators e a review of nanostructured piezoelectric energy harvesters, Nanomater. Energy 14 (2015) 15-29.

[19] R. Kargupta, C. Marcheselli, T. A. Venkatesh, Electromechanical response of 1e3 piezoelectric composites: effect of fiber shape, J. Appl. Phys. 104 (2) (2008), 024105.

[20] R. Kar-Gupta, T. A. Venkatesh, Electromechanical response of piezoelectric composites: effects of geometric connectivity and grain size, Acta Mater. 56 (15) (2008) 3810-3823.

[21] C. Marcheselli, T. A. Venkatesh, Electromechanical response of 1-3 piezoelectric composites with hollow fibers, Appl. Phys. Lett. 93 (2) (2008), 322903.

[22] S. Iyer, T. A. Venkatesh, Electromechanical response of (3e0, 3e1) particulate, fibrous, and porous piezoelectric composites with anisotropic constituents: a model based on the homogenization method, Int. J. Solid Struct. 51 (6) (2014) 1221-1234.

[23] R. D. Medeiros, R. Rodríguez-Ramos, R. Guinovart-Díaz, J. Bravo-Castillero, J. Otero, V. Tita, Numerical and analytical analyses for active fiber composite piezoelectric composite materials, J. Intell. Mater. Syst. Struct. 26 (1) (2015) 101-118.

[24] M. L. Feng, C. C. Wu, A study of three-dimensional four-step braided piezoceramic composites by the homogenization method, Compos. Sci. Technol. 61 (13) (2001) 1889-1898.

[25] S. Kari, H. Berger, R. Rodriguez-Ramos, U. Gabbert, Numerical Evaluation of effective material properties of transversely randomly distributed unidirectional piezoelectric fiber composites, J. Intell. Mater. Syst. Struct. 18 (4) (2007) 361-372.

[26] R. Guinovart-Díaz, P. Yan, R. Rodríguez-Ramos, J. C. Lopez-Realpozo, C. P. Jiang, J. Bravo-Castillero, F.J. Sabina, Effective properties of piezoelectric composites with parallelogram periodic cells, Int. J. Eng. Sci. 53 (4) (2012) 58-66.

[27] S. L. Vatanabe, G. H. Paulino, E. C. N. Silva, Design of functionally graded piezocomposites using topology optimization and homogenization - toward effective energy harvesting materials, Comput. Methods Appl. Mech. Eng. 266 (11) (2013) 205-218.

[28] A. Deraemaeker, H. Nasser, Numerical evaluation of the equivalent properties of Macro Fiber Composite (MFC) transducers using periodic homogenization, Int. J. Solid Struct. 47 (24) (2010) 3272-3285.

[29] R. Haj-Ali, H. Zemer, R. El-Hajjar, J. Aboudi, Piezoresistive fiber-reinforced composites: a coupled nonlinear micromechanical-microelectrical modeling approach, Int. J. Solid Struct. 51 (2) (2014) 491-503.

[30] C.N. Della, D. Shu, The performance of 1-3 piezoelectric composites with a porous non-piezoelectric matrix, Acta Mater. 56 (4) (2008) 754-761.

…

Task 3: Discuss in groups which paragraphs and sentences in Text Ⅰ convey each communicative purpose (main idea) in the following steps of research justification (text structure) with the help of the underlined phrases and sentences.

Steps of research justification

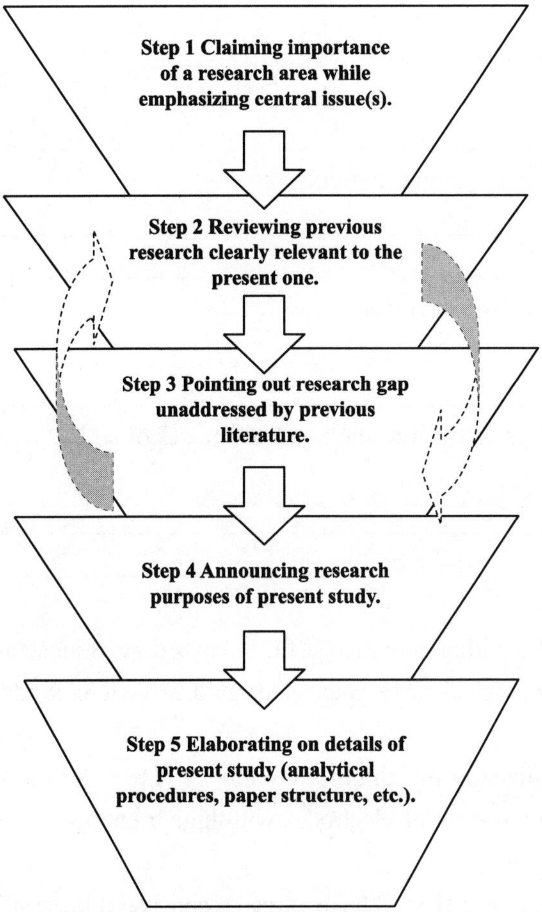

1) Importance of metamaterials: exotic properties
2) Cellular smart metamaterials: a new venue
3) How to deliberately optimizing the microarchitecture of smart piezoelectric materials

1) Two property modification methods and four types of piezoelectric materials:
  - Adding a second phase (piezocomposites) (e.g. encapsulating brittle ceramics)
  - Introducing pores (cellular materials)
2) Analytical and numerical methods

1) Undeveloped design chart for piezoelectric properties
2) Limited studies on the combination of piezocomposites and cellular solids
3) Mostly limited to experimental studies
4) No numerical study concerning pore shapes and polarization

Objectives of present study: a) -d)

Paper structure: Section 2-7

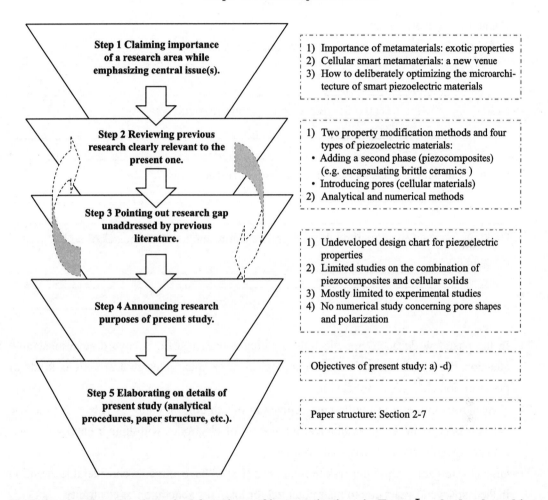

Task 4: Analyze the rhetorical functions of in-text citations in Text Ⅰ, whether launching an argument based on cited work(s), serving as a typical example for writers' statement, attributing certain methods/findings to the literature, or making evaluations with reliance on source text(s).

Task 5: Summarize orally how the authors of Text Ⅰ clearly justified their research of particular significance based on relevant literature in a specific research field.

Task 6: Identify the communicative purposes of the following statements which usually open research article introductions by clearly defining a key technical term, describing growing research interest, pointing out underexplored issues in an important research area, or highlighting its theoretical/methodological/practical contributions/advantages.

Communicative purposes:

- Defining a key technical term in an important research area:

  ____

  ____

- Describing growing research interest in an important research area

  ____

  ____

- Pointing out underexplored issues in an important research area

  ____

  ____

- Highlighting theoretical/methodological/practical contributions/advantages of an important research area

  ____

  ____

(1) In the past few decades, organic light-emitting diodes (OLEDs) have drawn substantial attention from both scientific and technological researchers interested in various kinds of practical optoelectronic applications.[1-3]

(2) Tremendous interests in the commercialization of lithium-ion batteries (LIBs) for electric vehicles necessitate the technological innovation of electrodes with higher energy density, greater power, and durability at a lower cost.[1-2]

(3) Organic-inorganic hybrid perovskite solar cells (PSCs) have attracted considerable attention in the last several years due to their potential for low-cost generation of electricity.

(4) Hydrogen evolution reaction (HER) in the electrolysis of water, as one of the promising approaches to produce hydrogen efficiently, has attracted much attention worldwide.[1]

(5) In the last decade, organic light-emitting diodes (OLEDs) have attracted intensive research interest for the displays and solid-state lighting applications.[1]

(6) In the past two-decades, highly ordered 3D architectures at nano-and microscales have been intensively explored for developing advanced material systems with unprecedented properties such as photonic and phononic bandgaps, negative Poisson's ratio, etc.[1-4]

(7) In the past few decades, UV photodetectors have been widely used in air and water sterilization, environmental monitoring, optical imaging, flame sensing and fire detection, military, and space, etc.[1-3]

(8) Porous polymer thin films with controlled network topologies and defined surface functionalization have attracted wide interest for applications in biotechnology and nanomedicine in the past decade.[1]

(9) Naked clusters of metals produced by laser desorption, sputtering, and thermal evaporation have been studied intensely in the past several years. [1-7]

(10) In the Internet of Things (IoT) era, a wide and increasing number of electronic devices are being connected to the Internet, which results in dense networks.

(11) Lithium-ion batteries (LIBs) have attracted abundant attention in the scientific and industrial fields during the past decades, driven by the ever-growing requirements for various portable electronic devices.[1-3]

(12) In the past decade, the proliferation of electron microscopy and scanning probe microscopy techniques have generated massive amounts of data on local chemical structure and atomic transformation.[1-5]

(13) Diamond grown by chemical vapor deposition (CVD) has recently drawn considerable attention due to its properties such as the extraordinarily high thermal conductivity, breakdown voltage, and stiffness, which open perspectives for the applications of this material in the next-generation electronics.[1]

(14) In the latest decades, wearable electronics have become the new-generation electronic devices in our lives due to their remarkable potential applications[1], including intelligent electronics, health monitoring, implantable medical devices, and smart robots.

(15) The development of soft robots which can mimic the motion strategies of natural organisms including crawling, swimming, jumping, and flying has been a longtime pursuit.[1-3]

(16) Fluorescence imaging has been proven to be a highly sensitive and noninvasive technology that offers researchers a very useful tool for analytical sensing and optical imaging.[1]

(17) Due to increasing energy and environmental challenges, there has been much attention focused on developing clean and sustainable energy sources.[1-6]

(18) Over the past decade, much research has been dedicated to the development of roll-to-roll, low-power consumption, and cost-effective organic photovoltaics (OPVs) and perovskite solar cells (PSCs) as promising alternatives to traditional photovoltaic (PV) applications.[1]

(19) The widespread use of magnesium and its alloys in lightweight structural elements for the transportation industry has been limited because of their poor formability at room temperature as compared to other ductile metals[1-3].

(20) High-pressure die casting (HPDC) of Mg alloys and Al alloys has been widely utilized in the automotive and electrical apparatus fields as a lightweight component.

(21) Since solar cells based on organometal lead halide perovskite were reported for the first time in 2009[1], hybrid perovskite materials with excellent optoelectronic and photovoltaic properties have attracted plenty of attention in scientific and industrial applications.[2-9]

(22) Flexible,[1-3] wearable,[4] and biocompatible soft physical sensors,[5-11] featured with reliable biocompatibility, high sensitivity,[12-13] lightweight, flexibility, and robust stretchability,[14-18] have attracted a myriad of research interest in the recent years due to their potential applications for human-machine interfaces,[19] human activity monitoring,[20-21] and personal healthcare diagnosis[22-24] in the near future.

(23) Surface-enhanced Raman spectroscopy (SERS) probes the significantly amplified Raman scattering signals from analytes attached to, or in the very close vicinity of metallic nanostructures.[1-3]

(24) In the past tens of years, a significant amount of research on infrared (IR) photodetectors has been conducted for their wide applications in military, commercial, public, and academic domains.[1-6]

(25) Fabrication of 3D cellular structures with controlled architectures is of increasing importance in the field of tissue engineering and pharmacy.[1]

(26) Recent advances in the chemical vapour deposition (CVD) of two-dimensional (2D) transition metal dichalcogenides (TMDs) on various substrate surfaces[1-2] have opened up prospects for the exploration of their fundamental physical properties[3] and practical device implementation schemes[4].

(27) The impact of nanomaterials in the discovery of new phenomena and the development of applications that exploit them has led to intense research efforts to expand the library of synthesized materials.

(28) Polyhydroxyalkanoates (PHA) are a group of polyesters that are promising alternatives to conventional plastics due to their biodegradability and capability of being produced from renewable resources (Braunegg et al., 1998).

(29) Failure of materials and structures is strongly dependent on the presence of stress risers, which can either be due to intrinsic heterogeneity of the material (for example, because of the presence of particles or defects) or artificially introduced due to application requirements (such as geometrical discontinuities and cut-outs).

(30) Titanium (Ti) and its alloys are widely used in almost every domain of industry such as machinery, electronics, metallurgy, petrochemical, and aerospace due to their excellent comprehensive properties such as high specific strength, excellent corrosion resistance, and good fatigue performance.

(31) Macroporous ceramics find applications in a wide variety of areas due to a combination of favorable properties like low thermal conductivity, low density, high porosity, and high chemical tolerance.

(32) Polymeric composite materials are used in a wide range of applications in our society due to their high specific strength and modulus.

(33) In recent years, biodegradable polymers have received an increasing attention due to solid waste pollution caused by the use of conventional petroleum-based polymers.

(34) Due to the intensification of white pollution and the shortage of global petroleum resources, biodegradable plastics, such as polylactic acid (PLA), poly (butyleneadipate-co-terephthalate) (PBAT), are increasingly used in packaging, daily necessities, and medical applications[1,2].

(35) The electrospinning technique has received much attention in recent years due to its versatility, and also due to the possibility of producing electrospun mats from many natural and synthetic polymers[1].

(36) Al—Zn—Mg—Cu alloys are widely employed in many light-weight structural applications due to their high specific strength, good corrosion resistance and workability, and relatively low production cost.

(37) Porous graphene and carbon nanotubes (CNTs) are promising materials for membrane separation[1−4] due to their atomically smooth carbon walls, their tunable pore size, and because they can be chemically functionalized.

(38) Surface adsorption is a ubiquitous phenomenon vital to many physical, chemical, and biological fields, and is widely used in industries, such as purification, catalysis, sensors, chromatography, and cell growth.

(39) The self-assembly of multicomponent nanomaterials by thermal equilibration is a scalable device fabrication technique that produces complex nanoscale architectures capable of having unique properties and narrow size distributions.

(40) In nanoplasmonics, the spaser (surface plasmon amplification by stimulated emission of radiation) is a promising quantum generator that can overcome fundamental problems brought by miniaturization beyond the diffraction limit imposed on electromagnetic waves.[1-2]

(41) The removal of extreme heat fluxes is a very important problem for electronics in applications including high-frequency radar,[1] renewable power generation and electric vehicles,[2] and high-performance computing.[3]

(42) Oil-water separation is a worldwide subject because of the increasing demands in numerous applications[1], such as oil-spill accidents, wastewater treatments, and chemical manufacturing.

(43) Photoelectric conversion is a phenomenon of great importance in nature since been initially revealed in the early 20th century.[1]

(44) The rapid development of information storage requires new concepts for fast and nonvolatile random-access memory with high density and low-power consumption, which is a significant and challenging task.

(45) Using two or more materials in organic blends is a strategy for engendering synergistic or even emergent properties; for instance, when blending an electron-donating and an electron-

accepting material to achieve charge separation after photoexcitation in organic solar cells (OSCs)[1].

(46) Electrolysis of water to produce H_2 and O_2 is a promising pathway for the storage of solar or other renewable energy in the form of chemical fuels.[1-2]

(47) Atomic layer deposition (ALD) is a highly useful technique to synthesize thin film materials, such as metal oxides, nitrides, sulfides, and metals themselves.[1−4]

(48) Hydrogen evolution reaction (HER) is a crucial step in electrochemical water splitting which is utilized as a clean and sustainable approach for hydrogen production.[1-2]

(49) Spark plasma sintering (SPS) is a widely applied advanced sintering technique for the fabrication of dense ceramics[1-2].

(50) In recent years, selective laser melting (SLM), which is a category of additive manufacturing technology, has been exploited to manufacture metallic components with complex surface geometries (Bremen et al., 2012).

(51) Precision surface texturing over larger surfaces is a powerful tool to improve surface functional properties, such as improved hydrophobicity, decreased reflectivity, reduced friction and wear, and others.

(52) Metal spinning is a very flexible incremental forming process by which a sheet blank or tube is rotated at a high speed and formed into a hollow body by rollers which progressively force the blank onto the mandrel (Xia et al., 2014).

(53) As a fundamental property of a solid surface, wettability plays a key role in addressing the problems related to energy, environment, resources, and health.[1-2]

(54) Hydrogen, an alternative and environmentally-friendly energy carrier, has attracted broad attention in recent years as a potential solution to global energy demands and environmental pollution.[1]

(55) Graphene, a kind of 2D carbon material, owns excellent mechanical, electrical, and thermal properties,[1] and can be utilized as a basic building block for constructing 3D materials with extremely low density for varied applications.[2]

(56) With the rapid development of society, antibacterial materials have been regarded to reduce the risk of infection and further enhance the healing quality [1].

(57) With the increasing development of additive manufacturing techniques, recently, the built quality has become a main barrier for its wider adoption, especially in the aerospace and medical industry.

(58) With the increasing demand for efficient energy usage, energy-saving buildings and automobiles using smart windows have received intense attention.

(59) The use of biodegradable polymers has increased considerably with the rapid increase in environmental awareness.

(60) With the development of various electronic devices in the trend of higher power, miniaturization and higher integration, efficient thermal dissipation has become important for their normal operation [1-2].

(61) Perovskite solar cells have emerged as one of the most promising photovoltaics by virtue of their highly improved power conversion efficiency (PCE), up to 22% within a few years.[1-3]

(62) Composite materials generally are the most advanced and adaptable engineering materials known to man, and the growth in the use of polymers and fibre/polymer composites is due to their outstanding mechanical properties, unique flexibility in design capabilities, and ease of fabrication.

(63) Epoxy resin has become one of the most widely used industrial thermoset resin owing to its excellent mechanical strength, great heat resistance, chemical resistance, minimal curing shrinkage, and excellent adhesion [1].

(64) Selective laser melting (SLM), inspired by the stereolithography technique, is considered as one of the most promising additive manufacturing (AM) techniques [1], [3].

(65) Polyarylene ether nitriles (PENs) are intensively researched in recent years because of their outstanding performances[1-7].

(66) Because of its fascinating properties, graphene is emerging as a highly promising plasmonic material for implementation in devices aimed at applications such as chemical and molecular sensing, ultrafast optical modulation, nonlinear optics, photodetection, light sources, and quantum optics.[1-6]

(67) Graphene-based photodetectors have attracted significant attention recently due to their ultrafast and broadband response, features that are important for next-generation optoelectronic applications such as imaging, high-speed communication, spectroscopy and displays [1-23].

(68) Ultrathin 2D semiconductors comprising transition metal chalcogenides (TMCs),[1-6] hybrid perovskites,[7-8] or group III–VIA compounds[9-11] have attracted intense attention as key building blocks for electronics and optoelectronics due to their large specific surface area, tunable optical properties, and quantum confinement effect, which is a result of their ultrathin form.[12-14]

(69) Superhydrophobic surfaces, which are extremely water repellent, with water contact angles greater than 150° and sliding angles less than 10°[1-7], have attracted significant attention because of their potential industrial applications such as self-cleaning[3], [8], drag reduction[9], corrosion-resistant[10], [11], and antiicing[7], [12] films.

(70) Since their discovery in the early nineties, single-wall carbon nanotubes (SWNTs) have attracted great attention due to their potential for nanoscale electronics[1], sensors[2], photodetectors[3-5], solar cells[6-7], LEDs[8-9], and, more recently, photonics.[10-12]

(71) Titanium (Ti) and its alloys are widely used in almost every domain of industry such as machinery, electronics, metallurgy, petrochemical , and aerospace due to their excellent

comprehensive properties such as high specific strength, excellent corrosion resistance, and good fatigue performance [1-4].

(72) Shape memory polymers (SMPs), as a sort of smart materials, can be deformed into arbitrary temporary shapes and when necessary, recover to their permanent shapes upon an external stimulus, such as heat, light, electric or magnetic field, PH, or moisture [1-5].

(73) Shape Memory Alloys (SMAs) are capable of recovering large, apparently permanent strains through a reversible transformation between two solid phases, austenite and martensite, that can be triggered by appropriate mechanical and/or thermal inputs.

(74) With the increasing industrial demand for micro-products, advanced manufacturing technologies are being developed continuously (Rajurkar et al., 2006).

(75) The increasing demand for portable electronic devices and electric vehicles favors the development of electrochemical energy storage (EES) devices with high capacity, enhanced power performance, as well as low cost, long cycle life, and improved safety.[1]

(76) The rapid growth of energy consumption and the corresponding environmental concerns have intensified the demand for renewable and clean energy sources.

(77) The ever-increasing demand for electric vehicles (EVs), stationary energy storage systems (ESSs), and portable smart electronics inspires the relentless pursuit of advanced rechargeable power sources with reliable/sustainable electrochemical performance and safety tolerance.[1-3]

(78) Emergingenergy[1-5], transportation[6-7], military[7-8], biomedica[l], [9] electronic[5], [10-12], and other technologies[13-14] create constantly increasing demand for materials with high mechanical strain/stress, electrical conductivity, thermal stability, toxicology, novel optical,[15] and other properties.[6], [16-17]

(79) With the fast growing demand for portable and wearable electronic devices, flexible light-weight high-performance energy storage devices are urgently needed.[1]

Task 7: Skim through Text Ⅱ, a research article introduction, and fill in the missing information in the outline below showing the text structure along with communicative purposes.

Text Ⅱ

Eddy Current Induced Dynamic Deformation Behaviors of Aluminum Alloy During EMF: Modeling and Quantitative Characterization[2]

INTRODUCTION

1 Electromagnetic forming (EMF) is a high-energy-rate sheet metal forming technique in which material deforms at high strain rates under time-varying eddy current field (Psyk et al., 2011). From the work by Balanethiram et al. (1994), EMF is proved to possess the potential to break through the conventional forming limit of materials. As a result, EMF has gained more and more attentions in the manufacture of hard-to-deform components. For example, Oliveira et al. (2005) investigated free-forming and two configurations of cavity fill operations of AA5754 and AA5182 aluminum alloy sheets in EMF.

2 With increasing demands of large-scale and large height-to-diameter ratio thin-walled components in aeronautic and astronautic fields, investigators continuously improve the EMF technique by further augmenting the discharging energy and utilizing hybrid processes. Luo et al. (2014) designed a novel multi-layer flat spiral coil to realize hole flanging for large and thick sheets. Lai et al. (2015) united multi-layer driving coil and side driving coil to form large-scale deep drawing part with large drawing ratio. Moreover, a uniform pressure coil was designed by Thibaudeau and Kinsey (2015) with increased forming efficiency and repeatability. For hybrid EMF processes, Cui et al. (2016a) combined stretch forming and EMF for manufacturing large-size and thin-walled ellipsoidal parts. Cui et al. (2016b) and Fang et al. (2016) proposed a new technique called incremental electromagnetic-assisted stamping (IEMAS) with radial magnetic pressure, by which the forming depth could be increased by 31%. The increase of discharging energy leads to the higher induced eddy current in the metal blank during EMF, e.g. peak current magnitude of 80 kA reported by Cao et al. (2014) and around 280 kA reported by Thibaudeau and Kinsey (2015). Hence, the sheet metal undergoes a current-carrying dynamic deformation during EMF, in which the influence of high-density current becomes non-ignorable.

3 Early studies by Troitskii (1969) and Conrad and Sprecher (1989) have shown that the flow stress of metallic materials can be reduced by high-density current charging during plastic deformation, which is defined as electroplasticity (EP) effect. Moreover, the intensity of the EP effect was proved to be positively correlated with the electric current density by Conrad (2002) and Kopanev (1991). Recently, Roh et al. (2014) used a newly defined parameter, electric energy density, to reflect the intensity of the EP effect. By actively applying current

to the metal blank during or after plastic deformation, EP effect is successfully utilized in electrically-assisted manufacture to reduce forming load (Salandro et al., 2010; Nguyen-Tran et al., 2015), enhance material plasticity (Li et al., 2012) and ameliorate microstructure (Xie et al., 2015; Sánchez Egea et al., 2016). Moreover, in recent decade, a renewed understanding of EP effect by Satapathy and Landen (2006) and Landen et al. (2007) indicates that high-density electric current leads to flow softening of metals during plastic deformation, even in high strain rate regimes and under short current durations.

4 However, the effect of eddy current on the deformation behaviors of metals has been not yet considered in current numerical simulation of EMF, which may result in an under-prediction of the forming depth within the current-acting zone. In order to realize the precision prediction of the EMF processes, quantitatively characterizing the effect of electric parameters (e.g. current density, charging time) on the stress responses of sheet metals becomes a hinge issue. However, the real-time measurement of deformation information in EMF process is still challenging in present state-of-art due to the high forming velocity and the time-varying electromagnetic field, which makes the quantification of the effect of eddy current a tough task.

5 Provided with the superiorities of simple loading mode and convenience in velocity and current measurement, electromagnetic ring expansion (EMRE), as a uniaxial stress EMF process, has become an important method to quantitatively investigate the current-carrying dynamic deformation behaviors in EMF (Grady and Benson, 1983). Earlier investigations by Niordson (1965) focused on the enhanced plasticity with a description of multiple necking and fragmentations. In the works of Altynova et al. (1996) and Hu and Daehn (1996) the enhanced plasticity/ductility of metals was attributed to the inertial effect on the necking and fracture behavior. In the more recent work by Zhang and Ravi-Chandar (2006, 2008), diffuse necking strain in EMRE was predicted to be around the Considère strain and the steady state deformation of EMRE was believed to be responsible for the enhanced plasticity. In addition, Altynova et al. (1996) pointed out that the temperature rise and microstructure change caused by combined action of high-density current and wide-range strain rate variation might result in the change of the deformation behaviors of materials in EMRE.

6 With the aid of advanced experimental technologies, researchers exerted efforts to characterize the effect of induced eddy current on the deformation behaviors of metals from material aspect. With digital high-speed camera and micro-hardness tests, Janiszewski (2012) found that micro-hardness increased slightly for cold-rolled copper and 7075 aluminum alloys but decreased for steel and tungsten alloy with the increase of expansion velocity in EMRE. Hence, it is suggested that apart from inertial effects, the change of deformation behaviors of materials under induced eddy current was also responsible. Huang et al. (2014) found an increase of micro-hardness and a decrease of grain size with the increase of the discharging current. Ma et al. (2014a) reported that micro-hardness of EMRE specimen was

lower than the one of split Hopkinson tensile bar (SHTB) specimen at comparable strain rates. Additionally, the dimples observed in EMRE specimens were larger and more homogeneously distributed. Apart from EMRE process, significant stress drop (Li et al., 2012; Nguyen-Tran et al., 2015) and microstructure changes such as recrystallization (Jiang et al., 2012), retarded martensitic transformation (Liu et al., 2013), local grain boundary melting (Fan et al., 2013) and detwinning (Xie et al., 2015; Sánchez Egea et al., 2016) were also found in some electric assisted forming processes, indicating the significant effect of electricity charging. These investigations indicate that EMF is not simply a high strain rate deformation process, viz. the eddy current has influences on the material constitutive behaviors possibly by inducing microstructure transformation, pure EP effect or Joule heat softening.

7 However, the detailed effect rules of eddy current in EMRE are still lack of quantitative characterization, thus the interaction of strain rate hardening and EP induced softening is still unclear. To this end, an accurate description of the material constitutive behaviors under the combined action of high strain rate loading and current charging is urgently demanded. Analytical method is capable of describing the relations of different variables efficiently and concisely, and makes it convenient to analyze the combined effects with analytical expressions in dealing with problems with simple boundary conditions. Thus, it becomes an appropriate strategy to investigate the effect of induced eddy current in EMF.

8 Gourdin (1989) established an analytical model of EMRE by combining a set of electromagnetic, kinematics and constitutive equations to describe the deformation history of ring specimens. The constitutive model was rate-independent and derived from the measured stress-strain curves during the deceleration of ring specimens. However, since the intensity of eddy current and strain rate both change with time, the measured constitutive model is only suitable for the deceleration stage under certain discharging condition. Altynova et al. (1996) followed the model framework of Gourdin and studied the applicability of rate-independent Hollomon model and rate-dependent physically based models by Zerilli and Armstrong (1987) and Follansbee and Kocks (1988). A discrepancy of 7% was found in predicted history of ring radius between the two models.

9 Huang et al. (2014) studied the EMRE of A5083 aluminum alloy with experiments and analytical modeling. In their works, quasi-static constitutive model was used in the analytical calculations. In view of significant strain rate hardening of 5xxx series aluminum alloys which is investigated by Ma et al. (2014b) and Yan et al. (2014), the assumption of rate-independent constitutive relation apparently lacks physical basis. However, acceptable predictive precision was obtained, indicating the strain rate hardening was attenuated by electric current in EMRE process. In order to explain this phenomenon and clarify the effect rules of electric current, Han et al. (2014) and Huang (2014) utilized uniaxial tensile tests with pulsed current charging to preliminarily investigate the stress responses of aluminum

alloy in cutirrent-carrying plastic deformation. The results demonstrated that periodical high density direct current (D.C.) charging led to the periodical stress softening of A5083 aluminum alloy, which is also observed in 5754 alloy (Roth et al., 2008), 5052-H32 alloy (Salandro et al.,2010; Kim et al., 2014; Roh et al., 2014). More specifically speaking, the current-induced softening was believed to offset certain part of strain rate hardening (Huang, 2014). However, the effects of strain rate variation and electric current charging were separately considered in above investigations, without a consideration of their coupling effect.

10 Henchi et al. (2008) implemented a Johnson-Cook typed constitutive model (Johnson and Cook, 1983) into the FE model with initial model parameters determined by SHTB tests. By comparison between simulated displacement histories of ring specimen and experimental obtained ones in EMRE, the predetermined model parameters were modified. Resultantly, the modified initial yield stress and hardening exponent were found to be lower than the preset ones. To predict the stress responses in EMRE, Satapathy and Landen (2006) established a Johnson-Cook typed constitutive model with a consideration of thermal expansion and the attenuation of hardening exponent with temperature by adiabatic heating. The predicted results were compared with experimental ones which were obtained by an inverse analytical solution based on the measured data of current intensity, velocity, and displacement. Consequently, predicted stress-strain profiles of 11,000 copper 6061-T6 and 7075-T6 aluminum alloys exhibited less significant softening than the experimental ones. Gallo et al. (2012) built a viscoplastic model of Al 6061-T6 and Cu-102 to explain the combined effect of mechanical load and high-intensity electric current. In this work, the importance of considering thermal expansion effect was also stressed.

11 To the knowledge of the authors, no existing constitutive model takes into account the combined effect of high strain rate loading and electric current charging on the deformation behavior (or stress responses) in EMF processes. Some models of EP induced stress drop in quasi-static deformation have already been established and applied to electrically-assisted forming (EAF) processes. For example, Li et al. (2013) proposed a theoretical model for EP induced stress drop on the basis of quantum mechanics and thermally-activated theory. Tskhondiya (2009) and Tskhondiya and Beklemishev (2012) considered the enhanced plasticity by EP effect and modified the Cockroft-Latham damage model. With a consideration of the theory of dislocation dynamics, Hariharan et al. (2017) established a new constitutive model for electroplastic deformation in which the mechanical behaviors under continuous and pulsed electric current were well captured. However, the possible components of thermal expansion, pure EP effect and Joule heat softening were not definitely distinguished and coupling modelled. Moreover, the effect of strain rate on the EP effect was not considered so that the extension of existing models to high strain rate regimes was limited.

12 In this study, the EP-induced stress drop and high-strain-rate-induced hardening were characterized respectively, on the basis of the results of tensile tests with pulsed current

charging (Han et al., 2014) and the results of SHTB experiments (Yan et al., 2014). Then, grounded on the classical theories of electro-dynamics and incremental theory of plasticity, an analytical model was established by implementing the EP and thermal expansion effect into the high-strain-rate constitutive mode (Yan et al., 2016), to study the histories of current, temperature, strain rate and stress responses of aluminum alloy during EMRE. Meanwhile, EMRE experiments of 5A06 aluminum alloy were done using specimens with three kinds of section geometries under a discharging voltage of 7 kV ~11.6 kV to optimize the undetermined model parameters and further verify the proposed model. Finally, the effect rules of eddy current on the deformation behaviors of 5A06 aluminum alloy within a wide strain rate range during EMRE were quantified with the aid of the proposed model.

...

References

[1] Altynova, M., Hu, X.Y., Daehn, G.S., 1996. Increased ductility in high velocity electromagnetic ring expansion. Metall. Mater. Trans. A. 27, 1837–1844.

[2] Balanethiram, V.S., Hu, X.Y., Altynova, M., Daehn, G.S., 1994. Hyperplasticity: enhanced formability at high rates. J. Mater. Process. Technol. 45, 595–600.

[3] Cao, Q., Li, L., Lai, Z., Zhou, Z., Xiong, Q., Zhang, X., Han, X., 2014. Dynamic analysis of electromagnetic sheet metal forming process using finite element method. Int. J. Adv. Manuf. Technol. 74, 361–368.

[4] Conrad, H., 2002. Thermally activated plastic flow of metals and ceramics with an electric field or current. Mater. Sci. Eng. A 322, 100–107.

[5] Conrad, H., Sprecher, A.F., 1989. In: Nabarro, F.R.N. (Ed.), Dislocations in Solids. Elsevier Science, New York, pp. 497 ch. 43.

[6] Conrad, H., Sprecher, A.F., Cao, W.D., Lu, X.P., 1990. Electroplasticity-the effect of electricity on the mechanical properties of metals. JOM. 42 (9), 28–33.

[7] Cui, X., Mo, J., Li, J., Xiao, X., Zhou, B., Fang, J., 2016a. Large-scale sheet deformation process by electromagnetic incremental forming combined with stretch forming. J. Mater. Process. Technol. 237, 139–154.

[8] Cui, X., Li, J., Mo, J., Fang, J., Zhou, B., Xiao, X., Feng, F., 2016b. Incremental electromagnetic-assisted stamping (IEMAS) with radial magnetic pressure: a novel deep drawing method for forming aluminum alloy sheets. J. Mater. Process. Technol. 233, 79–88.

[9] Dibartolo, B., 1991. Classic Theory of Electromagnetism. Prentice-Hall, Englewood Cliffs, New Jersey.

[10] Fan, R., Magargee, J., Hu, P., Cao, J., 2013. Influence of grain size and grain boundaries on the thermal and mechanical behavior of 70/30 brass under electrically-assisted deformation. Mater. Sci. Eng. A 574, 218–225.

[11] Fang, J., Mo, J., Cui, X., Li, J., Zhou, B., 2016. Electromagnetic pulse-assisted incremental deep drawing of aluminum cylindrical cup. J. Mater. Process. Technol. 238, 395–408.

[12] Follansbee, P.S., Kocks, U.F., 1988. A constitutive description of the deformation of copper based on the use of mechanical threshold stress as an internal state variable. Acta Metall. 36, 81–93.

[13] Gallo, F., Satapathy, S., Ravi-Chandar, K., 2012. Plastic deformation in electrical conductors subjected to short-duration current pulses. Mech. Mater. 55, 146–162.

[14] Gao, C.Y., Zhang, L.C., 2012. Constitutive modeling of plasticity of fcc metals under extremely high strain rates. Int. J. Plast. 32–33, 121–133.

[15] Gourdin, W.H., 1989. Analysis and assessment of electromagnetic ring expansion as a high-strain-rate test. J. Appl. Phys. 65, 411–422.

[16] Grady, D.E., Benson, D.A., 1983. Fragmentation of metal rings by electromagnetic loading. Exp. Mech. 23 (4), 393–400.

[17] Han, X.T., Huang, L.T., Shi, J.T., Ni, B.B., Duan, X.Y., Chen, Q., 2014. Development of mechanical measurement system applied for electroplastic effect research. 2014 IEEE International Instrumentation and Measurement Technology Conferece 945–948.

[18] Hariharan, K., Kim, M.J., Hong, S., Kim, D., Song, J., Lee, M., Han, H.N., 2017. Electroplastic behaviour in an aluminum alloy and dislocation density based modelling. Mater. Design. 124, 131–142.

[19] Henchi, I., L'eplattenier, P., Daehn, G.S., Zhang, Y., Vivek, A., Stander, N., 2008. Material constitutive parameter identification using an electromagnetic ring expansion experiment coupled with LS-DYNA® and LS-OPT®. 10th International LS-DYNA User Conference. Optimization 14 (2), 1–9.

[20] Huang, L.T., 2014. Effect of Induced Current and High-strain Rate on the Mechanical Property of Aluminum Alloy in Electromagnetic Forming. Doctoral Dissertation. Huangzhong University of Science and Technology, Wuhan, China (in Chinese).

…

Steps of research justification
1. Claiming importance of a research area while emphasizing central issue(s) · Importance of Electromagnetic forming (EMF)
2. Reviewing previous research clearly relevant to the present one 1) Improvement of EMF technique: · Method 1: augmenting discharging energy (e.g. ______________________) · Method 2: ______________________ (e.g. ______________________) · Problems encountered during EMF: ______________________ 2) EP effect · Definition: ______________________ · Applications: ______________________ · Factor less considered in EMF numerical simulation: ______________________ 3) Quantitative characterization of the effect of electric parameters · Problems existing: ______________________ 4) EMRE · Deformation behaviors investigated: ______________________ · Potential influencing factors: ______________________ 5) Further efforts to characterize induced eddy current effect on metals deformation behaviors: · Typical investigations and their findings: ______________________ · Important implications: ______________________ · Underexplored issues: ______________________ 6) Analytical methods/models · Typical examples and their features and limitations: ______________________

continued

3. Pointing out research gap unaddressed by previous literature
• ______________   • ______________   • ______________
4. Announcing research purposes of present study   • Characterize the EP-induced stress drop and high-strain-rate-induced hardening   • Establish an analytical model   • Verify the proposed model   • Quantify the effect rules of eddy current on the deformation behaviors
5. Elaborating on details of present study (analytical procedures, paper structure, etc.)   • ______________   • ______________   • ______________   • ______________

Task 8: Identify the rhetorical functions of in-text citations in Text Ⅱ, whether launching an argument based on cited work(s), serving as a typical example for writers' statement, attributing certain methods/findings to the literature, or making evaluations with reliance on source text(s).

Task 9: Compare the basic styles of in-text citations and reference lists in the following two segments of research article introductions, and work in groups to identify the potential differences.

Note:

In terms of how the source texts are used, in-text citations could be classified into five main categories, namely 1) a direct quotation usually placed in quotation markers or italicized, 2) a block quotation for direct quotes of no less than 40 words from the source text; 3) a paraphrase that states information from a source text in your own words and sentence structures, 4) a summary to concisely present the main ideas rather than the details found in a source text in your own words, and 5) a generalization to synthesize and incorporate the commonalities of multiple source materials in your text.

In terms of whether a citation is included within the grammar of a sentence, in-text citations could be categorized as integral or non-integral. Integral citations incorporate the cited author(s) or literature within the grammar of the sentence, emphasizing the author(s) or the source of the ideas, opinions, and assertions (Example sentences 1-3); on the contrary, non-integral citations refer to the forms such as family name(s) of the cited author(s) and publication year in brackets (Example sentence 4), a number in round or squared brackets (Example sentences 5-6), or superscript references (Example sentence 7), highlighting the reported message rather than the author(s).

For the in-text author-date citations as a whole, whether integral or non-integral, the detailed end-of-text references are documented in alphabetical order of the family name(s) of the cited author(s). For the in-text numbering citations (i.e., parenthetical or superscript citations), whether integral or non-integral, the end-of-text references are listed in the same order as the source literature is cited consecutively in the text.

The formats of in-text citations and end-of-text reference documentation, whether integral vs. non-integral or author-date vs. numbering, are primarily subject to journal requirements. No matter what format is used, precise and consistent citation practices in line with disciplinary conventions are highly valued.

Example sentences:

(1) Venkateswaran and Reynolds (2012) found that increasing tool rotation to extend the interpenetration and promote mechanical interlocking between metallic phases could result in reasonable transverse strength in FSW of 6063 and AZ31 alloys.

(2) Sarazin et al. [11] used glycerin as a plasticizer for PLA/TPS composites and found that the effect of plasticization was significant.

(3) In this paper the equivalent modelling strategy proposed in [35] was applied on structures containing multiple clinched joints.

(4) Micro-ultrasonic machining (MUSM) is an effective way to process microstructures made of hard-brittle materials (Hu et al., 2006; Kandaa et al., 2006).

(5) Diamond field-effect transistors and microelectro-mechanical radio frequency switches have been demonstrated to be superior candidates for high-power high-speed devices, [2-3] while diamond-GaN composite devices offer significant performance-enhancing potential for active and passive future high-power radio frequency and microwave communications devices. [4]

(6) Conductive additives, such as, carbon nanotubes [10], graphene [11], intrinsically conducting polymer (ICPs), such as polypyrrole (PPy) [12] and polyaniline (PANI) [13], may be dispersed in polymer solutions and then electrospun into mats. Among these fillers, PPy displays interesting electrical and optical properties and appears very promising from the electrical perspective [9], [14].

(7) The use of polymeric hydrophobic barriers, just on top of the semiconductor layer, such as polystyrene[17-18] and carbon materials,[19] has been suggested to relax the constraints of the encapsulating materials.

Segment (1)

Experimental validation of an equivalent modelling strategy for clinch configurations[3]

1 Modern mechanical design focuses on increasing performance whilst simultaneously providing environmentally friendly structural solutions. This has led in recent years to increased attention for lightweight structures. This sparked a more rapid development of novel and adapting mechanical joining techniques such as self-piercing rivets (SPR), riveting and clinched joints. Clinching (also referred to as clinch joining or press joining) is a joining technique that makes use of a die and a punch to create a severe local plastic deformation of the sheet material, creating a permanent mechanical interlock.

2 Clinch joining has already been used for more than 35 years as a successful joining technique. Besides the fact that the joining technique itself does not add weight to the assembly, its possibility to join similar [1-3] as well as dissimilar materials [4] makes this an interesting technique for lightweight constructions. This has led to an increasing interest for the joining technique in recent years [5]. The suitability of using clinching as a joining technique for high strength steels has been investigated by Varis et al. and Abe et al. [6-8]. For brittle materials, however, the formability represents a limitation for the clinch process. As the process induces large deformations, cracks potentially occur during the clinching of brittle aluminium alloys [3], [9], which might have an effect on the structural integrity of the joint. Hence, the increased attention with respect to fracture and fatigue life of clinched joints [3], [10], [11]. To improve the formability during the clinch process, heat assisted clinching was proposed [9], [12], [13], [14] in order to improve the ductility of the material during the clinching process. Clinching is mainly used in non-critically loaded constructions, because of its limited axial load or pull-out strength. In order to increase this limited strength, clinched joints are combined with adhesives [15-16] to create hybrid joints with increased strength, new possibilities and applications.

3 Although the forming process, the influence of the tool geometry and the mechanical performance of a single clinched joint have been extensively investigated by finite element simulations [1], [17-22], the knowledge on the multi-joint mechanical behaviour of clinched joints is currently limited in literature. The mechanical behaviour of clinched groups in comparison with conventional joining techniques has been reported by Pedreschi et al. [23-24] and Davies et al. [25-26] for construction applications. For many spot joining technologies, up to thousands of joints can be present in a structure. In order to predict the mechanical behaviour of these constructions, a full scale model of the joint is not computationally feasible. Therefore, a simplified model was introduced for several joining techniques such as bolts [27], spot welds [28], rivets [29-30] and SPR [31-34]. An equivalent modelling strategy for a single clinched joint was proposed by Breda et al. [35] enabling to reproduce the mechanical behaviour of a single clinched joint under several loading modes. The field of interest of these

equivalent models, however, lies in applications were a large number of clinched joints are combined to assemble a complete structure or product. Applying multiple joints in a configuration can result in synergistic effects such as strengthening and stiffening of the structure. The key point under investigation in this paper is whether an equivalent model for a single clinched joint can be used to predict the behaviour of groups of clinched joints. Until today, limited information is available on the mechanical behaviour of clinched joints in configurations containing multiple joints, and the application of the equivalent modelling strategy for multiple joints.

4 In this paper the equivalent modelling strategy proposed in [35] was applied on structures containing multiple clinched joints. The accuracy and performance of the equivalent modelling strategy for a single clinched joint were validated to predict the mechanical behaviour of structures containing multiple joints. This was done using lab scale tests on DC01 steel sheet material under different controlled load cases (pull-out+peel, shear and mixed mode loads). Finally, a structural analysis of a galvanised steel feed intake boot was performed using the FEA along with the proposed equivalent modelling strategy. This FE model is validated using the force displacement behaviour captured by the tensile machine.

…

References:

[1] S. Coppieters, Experimental and Numerical Study of Clinched Connections, Ph.D. thesis. KU Leuven. 2012.

[2] X. Baoying, H. Xiaocong, W. Yuqi, Y. Huiyan, D. Chengjiang, Study of mechanical properties for copper alloy H62 sheets joined by self-piercing riveting and clinching, J. Mater. Process. Technol. 216 (2015) 28–36.

[3] S. Coppieters, H. Zhang, F. Xu, N. Vandermeiren, A. Breda, D. Debruyne, Process-induced bottom defects in clinch forming: simulation and effect on the structural integrity of single shear lap specimens, Mater. Des. 130 (2017) 336–348.

[4] M.M. Eshtayeh, M. Hrairi, A.K.M. Mohiuddin, Clinching process for joining dissimilar materials: state of the art, Int. J. Adv. Manuf. Technol. 82 (1) (2016) 179–195.

[5] R. Hörhold, M. Müller, M. Merklein, G. Meschut, Mechanical properties of an innovative shear-clinching technology for ultra-high-strength steel and aluminium in lightweight car body structures, Weld. World 60 (3) (2016) 613–620.

[6] J. Varis, The suitability of clinching as a joining method for high-strength structural steel, J. Mater. Process. Technol. 132 (1) (2003) 242–249.

[7] Y. Abe, K. Mori, T. Kato, Joining of high strength steel and aluminium alloy sheets by mechanical clinching with dies for control of metal flow, J. Mater. Process. Technol. 212 (4) (2012) 884–889.

[8] Y. Abe, T. Kato, K. ichiro Mori, S. Nishino, Mechanical clinching of ultra-high strength steel sheets and strength of joints, J. Mater. Process. Technol. 214 (10) (2014) 2112–2118.

[9] F. Lambiase, A. Di Ilio, A. Paoletti, Joining aluminium alloys with reduced ductility by mechanical clinching, Int. J. Adv. Manuf. Technol. 77 (2015) 1295–1304.

[10] K. Mori, Y. Abe, T. Kato, Mechanism of superiority of fatigue strength for aluminium alloy sheets joined by mechanical clinching and self-pierce riveting, J. Mater. Process. Technol. 212 (9) (2012) 1900–1905.

…

Segment (2)

Experiments and simulations of micro-hole manufacturing by electrophoresis-assisted micro-ultrasonic machining[4]

1 With the increasing industrial demand for micro-products, advanced manufacturing technologies are being developed continuously (Rajurkar et al., 2006). The use of precision parts in the manufacturing of micro-products has created new micro-manufacturing process requirements, such as those in terms of the microstructure (Zhao et al., 2008). Hard and brittle materials are commonly used in microstructure manufacturing, but their physical characteristics can make them difficult to process; in particular, edge chipping (EC) is inevitable during processing. Micro-ultrasonic machining (MUSM) is an effective way to process microstructures made of hard-brittle materials (Hu et al., 2006; Kandaa et al., 2006). Kumar and Singh (2018) used rotary ultrasonic machining (RUM) to drill holes in BK-7 optical glass and then scrutinized the spindle speed, feed rate, and ultrasonic power to evaluate the machining proficiency in terms of the surface roughness and the material removal rate (MRR). Wang et al. (2018) used RUM combined with diamond grinding with small-amplitude tool vibration to improve the machining of hard and brittle materials. This technique was applied successfully to the machining of a number of brittle materials, including optical glasses, advanced ceramics, and ceramic matrix composites.

2 However, traditional ultrasonic machining has some inherent limitations (Tateishi et al., 2009 a,b,c), one being the low utilization ratio of abrasive particles (Tateishi et al., 2009 a,b,c). MUSM results from the comprehensive mechanical impact and polishing effect of the abrasive particles under the actions of ultrasonic vibration and cavitation, the main factor being the impact of the free abrasive particles. Under the impact of a large number of free abrasive particles, the surface of workpiece starts to form microstructure. The impacting effect of the free abrasive particles is shown schematically in Figure 1.

3 Agarwal (2015) investigated the mechanism of the MRR for glass when using ultrasonic machining. That analysis showed that material is removed mainly by micro-brittle fracture on the workpiece surface. For this fracture mode, a relationship with abrasive particles and manufacturing quality was established for the MRR based on simple fracture mechanics and considering the abrasive grains impacting the workpiece directly. Tateishi et al. (2009 a,b,c) discovered that ultrasonic vibration of the micro-tool or workpiece and rotation of the micro-tool expelled abrasive particles from the machining region, leaving none therein in the worst case. This increases the likelihood of the tool touching the workpiece directly, leading to unwanted EC and low MRR.

4 Simulation modeling is an effective method for studying particle motion. The motion of abrasive particles in electrophoresis-assisted ultrasonic machining is yet to be reported, but some scholars have studied particle tracking in other area. Zheng et al. (2018) studied the

electrical characteristics of particle transport by considering the electrohydrodynamics and the effect of particle space charge through numerical simulation. The influence of particle space charge is considered in particle motion and is thus worthy of reference in practical application. Guha et al. (2007) used non-invasive computer-automated radioactive particle tracking to study the solids flow field in dense solid–liquid suspensions. That method provides a Lagrangian description of the solids flow field, which is then used to obtain time-averaged velocity fields and turbulent quantities.

5 Because of the electrophoretic characteristics of the ultrafine abrasive particles, electrophoresis-assisted micro-ultrasonic machining (EPAMUSM) can attract the abrasive particles to the processing area by means of an electric field during processing. Herein, the manufacture of high-quality micro-holes by EPAMUSM during micron-level processing is studied. The trajectories of the abrasive particles are predicted by modeling and simulation. The proposed technique raises the MRR and lowers the edge chipping rate (ECR) significantly in micro-hole manufacturing. The effects of the EPAMUSM parameters on the machining quality based on the electrophoretic features of the ultrafine particles are studied, whereupon the parameters are optimized to make high-quality micro-holes in silicon.

…

References:

[1] Agarwal, S., 2015. On the mechanism and mechanics of material removal in ultrasonic machining. Int. J. Mach. Tools Manuf. 96, 1–14.

[2] Anupam, V., Tao, L., Yogesh, G., 2014. High resolution micro ultrasonic machining for trimming 3D microstructures. J. Micromech. Microeng. 24, 494–497.

[3] Cai, P., 2008. The advancement and application on electrophoretic technique. Life Sci. Instrum. 8, 3–7.

[4] Feng, H., Jing, Z., Fang, Y., Mo, S., 2009. Two-dimensional electrophoresis and its application. Biotechnol. Bull. 1, 59–63.

[5] Hu, X., Yu, Z., Rajurkar, K.P., 2006. State-of-the-art review of micro ultrasonic machining. International Conference on Manufacturing Science and Engineering. pp. 1017–1024 2006.

[6] Kandaa, T., Makinoa, A., Onoa, T., Suzumoria, K., Moritab, T., Kurosawa, M.K., 2006. A micro ultrasonic motor using a micro-machined cylindrical bulk PZT transducer. Sens. Actuators A Phys. 127, 131–138.

[7] Kumar, V., Singh, H., 2018. Machining optimization in rotary ultrasonic drilling of BK-7 through response surface methodology using desirability approach. J. Braz. Soc. Mech. Sci. Eng. 40, 83–97.

[8] Lian, H.S., Guo, Z.N., Liu, J.W., Huang, Z.G., He, J.F., 2016. Experimental study of electrophoretically assisted micro-ultrasonic machining. Int. J. Adv. Manuf. Technol. 85, 2115–2124.

[9] Peng, W., Xu, X., He, X., Chen, Z., 1999. An electrophoresis grinding process and its application. China Mech. Eng. 10, 85–88.

[10] Rajurkar, K.P., Levy, G., Malshe, A., Sundaram, M.M., Mcgeough, J., Hu, X., Resnick, R., DeSilva, A., 2006. Micro and nano machining by electro-physical and chemical processes. CIRP Annal.: Manuf. Technol. 55, 643–666.

[11] Tateishi, T., Shimada, K., Yoshihara, N., Yan, J.W., Kuriyagawa, T., 2009 a. Control of the behavior of abrasive grains by the effect of electrorheological fluid assistance-study on electrorheological fluid-assisted micro ultrasonic machining. Adv. Mater. Res. 76, 696–701.

[12] Tateishi, T., Shimada, K., Yoshihara, N., Yan, J.W., Kuriyagawa, T., 2009b. Effect of electrorheological fluid assistance on micro ultrasonic machining. Adv. Mater. Res. 69, 148–152.

[13] Tateishi, T., Yoshihara, N., Yan, J., Kuriyagawa, T., 2009c. Study on electrorheological fluid-assisted micro-ultrasonic machining. Int. J. Abrasive Technol. 2, 70–82.

[14] Wang, J., Zhang, J., Fenga, P., Guo, P., 2018. Damage formation and suppression in rotary ultrasonic machining of hard and brittle materials: a critical review. Ceram. Int. 44, 1227–1239.

[15] Zarepour, H., Yeo, S.H., 2012. Predictive modeling of material removal modes in microultrasonic machining. Int. J. Mach. Tools Manuf. 62, 13–23.

Task 10: Identity the typical expressions used to point out a central issue/problem in a disciplinary field that deserves more scholarly attention, and then necessitates a review of relevant existing literature.

(1) In all the numerous fundamental and applied investigations of vdW materials, the weak interlayer interactions play a central role in the formation, intercalation, exfoliation and layer-by-layer building of the materials as well as being decisive for their unique properties.

(2) For low-grade waste-heat (<100 °C) thermoelectrics are inherently limited by their low ZT, a dimensionless figure of merit (FoM) that requires high electrical and low thermal conductivity; a combination that is difficult to achieve[4].

(3) To combat infectious and chronic diseases, safe and effective adjuvant platforms for prophylactic and therapeutic vaccinations[1-2] are needed.

(4) Fast human genome sequencing has initiated radical changes in clinical diagnosis, personalized medicine and the study of genetic diseases[2-3]. However, in addition to low cost, for many of these benefits to be realized, sequencing needs to be significantly faster.

(5) An ideal material for a white light emitter requires the simultaneous emission of the three primary colors (red, green, and blue) with a nearly identical distribution of intensities covering the entire visible spectrum.

(6) One of the key challenges in SNT self-assembly is tuning their diameter at the nanoscale(4, 5) which is essential to enable accommodation of various molecular cargos.

(7) However, the room-temperature brittleness, resulting from the localized deformation in shear bands (SBs), limits the widespread application of most BMGs [4-6]. Introducing a crystalline secondary phase to fabricate the BMG matrix composites (BMGMCs) is proved to be an efficient way to improve the ductility of BMGs [7-9].

(8) Due to the increasing demand in individualizing products, additive manufacturing techniques have strongly emerged over the past decade. Even though there is a broad variety of additive processing techniques available, only a few of them fulfill the requirements for producing small lot sizes with desired properties and therefore possessing the ability to adapt to dynamic and complex markets [2].

(9) Wherever there is a joint, bolted or bonded, it will most likely be the site where failure initiates under operational conditions, unless damage is imparted by foreign objects, such as impact. Bolts require a hole or cut-out which acts as a stress raiser within a structure. Because of this, the need to size the bolted joint becomes of utmost concern in a structural design.

(10) One of the top challenges being faced right now is how to produce uniform and highly oriented ML-MoS_2 on a large scale, at low cost, and in a reproducible manner.

(11) Nowadays, the demand to get more efficient and miniaturized energy storage microcapacitors for portable electronic devices, such as cellphones, tablets, watches, or implantable medical devices, has given rise to the development of advanced capacitor designs.[1-6] Moreover, the capacitors are one of the primary building blocks of many types of electrical circuits, from microprocessors to power supplies.[5-7]

(12) As the latest carbon allotrope, graphdiyne (GDY) has recently emerged and sparked tremendous interest in its potential applications in energy storage, electrocatalytic activity, and rapid heterogeneous electron transfer. In view of its remarkable electronic, mechanical, and thermal properties, GDY is hoped to offer a better alternative. Nonetheless, further realizing the applications of graphene and GDY in many fields still requires many fundamental research studies.

(13) Preparing and engineering the single atoms on the proper sites of the substrates is essential to maximize the catalyst's activity and stability.

(14) Dielectric ceramics is becoming increasingly important in both high energy-storage density (W_d) and energy-storage efficiency (η) for most power electronics applications due to their nanosecond grad charge/discharge speed and excellent temperature stability [1-2]. Compared to ferroelectric ceramics (FE), relaxor ferroelectric (RFE) possess higher W_D and η, since they have large polarization in ferroelectric phase and a small remnant polarization (P_r) in weak polar phase [3]. However, most of the RFE contain lead, which is harmful to human health and ecological environment. Therefore, the research for the development of new relaxor systems is a growing trend [4-9].

(15) Scanning probe microscopy is useful to image complex molecular structures with submolecular resolutions;[10-15] however, acquiring detailed molecular structures inside β-carotene/chlorophyll-a complexes is hindered by a high degree of complexity in their conformations.

(16) A magnetostriction phenomenon relies on a change of either size or volume of magnetic materials placed in the magnetic field for lots of applications, such as in ultrasound generators, vibrators, 'artificial muscles', sensors, actuators and so on [1-2]. Especially in the latter applications, the sufficiently high value of saturation magnetostriction is very essential [3]. Unfortunately, the saturation magnetostriction of the currently known magnetic materials ranges between wide limits from 10^{-9} for amorphic and non-crystalline substances to 10^{-2}

for intermetallic compounds of rare-earth elements such as dysprosium or terbium, Ni-Mn-Ga alloys and so on [4-8].

(17) InGaN-based light emitting diodes (LEDs) have attracted intense research interest due to their long lifetime, energy savings, and environmental friendliness [1-3]. Conventional semiconductor solid-state light (SSL) sources are obtained by combining ultraviolet/purple/blue LEDs with down-conversion phosphors but of unavoidable Stokes shift energy losses and low phosphor reliability [4-7]. The SSL sources may also be realized by a red-green-blue (RGB) approach via color mixing of multiple primary color LEDs [8]. The RGB approach is favorable for applications in smart lighting, high color rendering index display, and high-speed light-fidelity communication. However, the growth of high-quality high indium content InGaN for high-performance long-wavelength LEDs is still a big challenge. With the increase of indium content in InGaN for long-wavelength emission, the InGaN alloy phenomena become severe, leading to poor crystalline quality and an increase of nonradiative recombination rate [9-12]. In addition, strong piezoelectric polarization fields (PFs) built in high indium content InGaN grown on polar GaN lead to a significant reduction in carrier recombination rate [13-15].

(18) To join aluminum alloys to steel materials, the application of coatings poses new challenges.

(19) Since developed, micropatterning technology has been widely used in various fields. Recent applications of micropatterns in biomedical research enable the manipulation of cytoskeletal structures by restraining cells in a controlled size and geometry, which plays a crucial role in regulating cell spatial and mechanical functions.[1] Micropatterned surfaces with various geometrical features can provide controllable and reproducible cell morphology with relatively stable cytoskeletal structure which cannot be fulfilled with conventional uniform cell culture substrates.[2] The quantitative description of the influence of subcellular structures can also be achieved on micropatterns without the influence caused by the diversity of cell morphology.[3] Therefore, a systematical investigation of cell nanomechanics on micropatterned surfaces is required to elucidate the relationship between nanomechanics and subcellular structures.

(20) The development of new protocols to prepare nanoparticles (NPs) is an exciting task of nanoscience. The perfect mastering of the synthesis stages is a prerequisite for the obtention of NPs exhibiting reproducible properties.

(21) The casting process is indispensable in the manufacture of ingots and hollow articles with complex shapes; however, cast products sometimes have crack defects owing to solidification shrinkage and thermal contraction. Recently, cracks have become a more serious problem owing to the diversification of alloy composition and the increasing size of ingots. Against this background, crack prediction using thermal stress analysis (Monroe and Beckermann, 2005) is worthy of note. Consequently, the mechanical properties of the solidifying alloys are imperative and essential for the analysis.

(22) In such a low-aspect-ratio nanopore structure, the temporal resolution of the cross-membrane ion transport measurements is of key importance to track the small rapid changes in the ionic blockade by fast translocation of an object.[10]

(23) Serpentine layouts in the interconnects,[20-27] typically of periodic arcs and straight segments, can offer levels of stretchability that meet requirements for many applications. In many of these designs, the serpentines fully bond to the elastomeric substrate to prevent damage from external physical contact.[17], [28-32] Additionally, across the wide range of examples of this design, the substrate material type and thickness can vary significantly.[33-36] As a result, the interplay between substrate properties and deformation of the interconnects is critical. In particular, mechanical properties such as tensile stiffness and bending stiffness depend on the thickness and elastic modulus of the substrate, and these properties can strongly influence the behavior of the interconnects.

(24) The ubiquitous deployment of QDs in biology has been constrained due to a lack of methodologies for permanently and homogenously monofunctionalizing them[1]. Traditional approaches to achieve this involve coupling QDs and ligands in fixed ratios in solution[2-7]. This has met with limited success due to stochastic control over the ligand stoichiometry, leading to sample heterogeneity.

(25) Although decorating functional groups in covalent or noncovalent frameworks has been developed in a few works,[7-8] the 2D polymeric layers incorporated with the charge functionality are not yet achieved.

(26) The casting of brass alloys traditionally leads to coarse and heterogeneous microstructures, even when the die casting process is used, requiring suitable melt treatment to refine and modify the microstructure morphology to promote an increase in the castings sanity and mechanical behaviour.

(27) Grain refinement is crucial to achieving high mechanical properties, being imperative to find viable alternative routes, more efficient and clean, which may lead to the development of new melt processing technologies.

(28) However, it has remained to be challenging to fabricate well-defined nanostructures with a large density of catalytic active sites. Therefore, it is necessary to develop new strategies to construct a well-defined nanostructure of Mo_2C featuring high porosity and numerous exposed active sites.[27-29]

(29) Considerable interest has been given to exploring new phases with non-trivial, real-space order parameter topologies and emergent continuous symmetry that can occur at the mesoscale (for example, complex spin textures such as vortices and skyrmions[17-19].

(30) The periodic bricks-and-mortar structure has been mimicked frequently to acquire outstanding mechanical properties, especially for high tensile strength, Young's modulus, and work to fracture.[7-10] Nevertheless, high fragility and poor flexibility always exist, making

these biomineral materials far from real applications, and up to now, very limited flexible free-standing $CaCO_3$ mineral films have been prepared successfully.

(31) Supramolecular nanotubes (SNTs) represent an important class of noncovalent arrays[1-7], whose nanoporosity enables the capture, storage, transport, and release of different molecular and macromolecular cargos.[6, 7]

(32) The ultrasonic treatment, successfully used with aluminium and magnesium alloys, can be a promising alternative to promote brass grain refinement through the cavitation effect, as stated by Qian et al. (2009).

(33) The water transport velocity on the *Sarracenia* trichome is about three orders of magnitude faster than that on cactus spine. Therefore, it is very important to understand the underlying mechanism of ultrafast water harvesting and transport on the *Sarracenia* trichome.

Task 11: Identify the commonly used language features to initiate a literature review in research article introductions, which quotes, summarizes, or synthesizes the most important literature related to the central issue(s)/problems in a specific disciplinary field. Meanwhile, pay attention to the format and function of in-text citations, integral or non-integral with the purpose of launching an argument based on cited work(s), serving as a typical example for writers' statement, attributing certain methods/findings to the literature, or making evaluations with reliance on source text(s).

(1) A number of strategies have been developed to create white light emission, including 1) fabrication of a multilayer structure with different dyes that emit the three primary colors—red, green, and blue[3-6] and 2) generation of a single layer structure that contains all of them.[7, 8]

(2) Therefore, the development of a white light emitter based on a single component has been widely attempted[9-10] to improve the uniformity, and reproducibility and to simplify the fabrication process.

(3) Many researches have shown that the BMGMCs prepared by *in-situ* precipitation method always show better mechanical properties than those by *ex-situ* direct adding method [10-12]. In addition, the refractory metal reinforced BMGMCs have been studied a lot [10-12].

(4) Considerable literature is available which have identified the parameters that influence the failure of bolted joints when the parts to be joined are fiber-reinforced laminates.

(5) Recently, much effort has been devoted to the wafer-scale production of MoS_2 films. These approaches include MoO_3 sulfurization, pulsed laser deposition, atomic layer deposition, and metal-organic chemical vapor deposition. The as-produced films are polycrystalline with the existence of many randomly orientated domains and domain boundaries.[10-14]

(6) In the last decade, the attention of the scientific community has been focused on 2D materials[8], such as graphene[9-16] and transition metal dichalcogenides (TMDs)[17-24], for obtaining efficient

supercapacitors. Particularly, layered metal sulfides such as MoS_2[17], [19], [20], WS_2[18], [23] , and VS_2[21] were reported as electrochemically active materials for microsupercapacitor applications.

(7) Coprecipitation and impregnation methods are the main stream in synthesizing single-atom catalysts.

(8) Intensive work has focused on tailoring the composition by linear additives to achieve superior energy-storage performance.

(9) Memristors have been extensively studied for both digital memory and analog logic circuit applications[3-7]. At the device level, memristors have been shown to be able to emulate synaptic functions by storing the analog synaptic weights and implementing synaptic learning rules[8-12]. When constructed into a crossbar form, memristor networks offer the desired density and connectivity that are required for hardware implementation of neuromorphic computing systems[13-15]. Recently, memristor arrays and phase-change memory devices have been used as artificial neural networks to perform pattern classification tasks[16-18]. Other studies have shown memristors can be used in recurrent artificial neural networks for applications such as analog-to-digital convertors[19].

Memristor-based architectures have also been proposed and analysed for tasks such as sparse coding and dictionary learning. The ability to sparsely encode data is believed to be a key mechanism by which biological neural systems can efficiently process large amounts of complex sensory data[22-24] and can enable the implementation of efficient bio-inspired neuromorphic systems for data representation and analysis[25-28].

(10) Platforms in which (signatures of) Majorana modes have been observed are semiconductors with Rashba spin-orbit interaction[9-11], ferromagnetic atom chains[12] , and topological insulators[13-14], all in combination with superconductors.

(11) Recently, magnetic elastomers have attracted lots of attention for application in magnetostriction due to be relatively large magnetostriction of the order 10^{-2}, easily manufactured and formed into a myriad of shapes.

(12) Single-chip phosphor-free SSL sources have been developed to overcome these problems. Multicolor emission is obtained either by growth of cascade or lateral InGaN/GaN quantum wells (QWs) with different indium contents in various InGaN QWs [16-17], or by growth of InGaN active layers on GaN islands, nanorods, or nanowires [18-24].

(13) Degradation of polymers due to ultraviolet (UV) radiation is a well-known phenomenon that induces physical processes and chemical reactions in polymers. It has been recognized that photooxidative degradation results in breaking of polymer chains, producing free radicals and, subsequently, oxidation products, and reducing molecular weight [1-4], although radical-induced crosslinking can also increase the molecular weight. Many attempts to identify the degradative products of polyurethane-based materials have been presented [5-13].

(14) Among the commonly used joining techniques, resistance spot welding is desirable due to its robustness, high throughput, low cost, and flexibility. Qiu et al. indicated that resistance spot welding (RSW) of aluminum alloys to steel poses unique challenges due to the large difference in physical and thermal properties between aluminum alloys and steel.

(15) Atomic force microscopy (AFM) has been widely used for the measurement of cell nanomechanics.[4] It has high spatial resolution and a large force range. Both cell elastic properties and adhesion information can be acquired simultaneously. It was proved to be one of the least invasive techniques for nanomechanical measurement of cells.

(16) Metal oxides possess quite unique and peculiar properties in optics, electronic, catalysis, etc. This has stimulated the development of numerous synthesis pathways toward NPs with well-controlled sizes and shapes.

(17) To reach this goal, numerous parameters have to be controlled, especially when chemical syntheses are conducted. For example, studies of growth mechanisms of colloidal nanocrystals especially focus their works on the effect of temperature,[1-3] the solvent,[4] the nature of the precursor,[5-6] the effects of ligands,[7-10] their relative concentrations,[11] the surface species,[12-15] the activation reagents,[16-17] and even side reactions;[18] more recently addition rates have also been considered.[19]

(18) Several studies using this approach for semi-solid alloys have been published. Sharifi and Larouche (2015) obtained the instantaneous Young's moduli of Al-Cu alloy during tensile analyses. They did not mention the inelastic properties. Phillion et al. (2008) and Sistaninia et al. (2012, 2013) compared the numerical and experimental stress-strain curves of aluminum alloys. Although the stress-strain curves were compared, derivation of the properties and their validation were not obtained.

(19) Driven by the prospect of applying the simple sensing mechanism for genome sequencing, [5-8] recent efforts have been devoted to enhancing the spatial sensitivity by shrinking the channel length through employing atomically thin membranes so as to make the ion exclusion effects more sensitive to the local shape of analytes. [9-13]

(20) Tremendous effort has thus been devoted to slow-down the motions of electrophoretically drawn analytes inside the channel to better reflect their fine topology on the ionic current signatures. [15-17] Progress has also been made in development of amplifier-integrated nanopore chips that enabled high current sampling rates exceeding 1 MHz. [18-19]

(21) Many researches [8-15] referred to the strengthening below 800 °C by carbon solid solution and microstructure refinement in conventional TiAl alloys. However, the various solidify path and the microstructure in high niobium containing TiAl alloys would cause different strengthening or toughening mechanism above 800 °C. In addition, carbon solid solution is less efficient for strengthening the alloys than different types of carbides [16]. Tian [17] et al. reported that the fine carbides could strengthen the TiAl alloys while the Ti_3AlC(P-type) and

Ti_2AlC(H-type) contributed to the strength. There are also less references about the carbides effect on further property improvement of the novel high niobium containing TiAl alloys.

(22) Structural DNA nanotechnology has yielded diverse, well-defined, nanoscale polyhedra[8-9]. These polyhedra enclose an internal void that can house nanoscale cargo and possess a well-defined surface for molecular display with both stoichiometric and spatial precision[8], [10-17].

(23) To date, several studies have been conducted that focused on manipulating peptide self-assembly into the desired nanoarchitectures[9-10].

(24) To date, extensive magnetotransport measurements have been performed on TIs, and intriguing physical phenomena have been identified, such as weak antilocalization, [10-11] Shubnikov–de Haas (SdH) oscillations,[12-14] universal conductance fluctuations,[15] quantum Hall effect,[16-17] and anisotropic magnetoresistance (MR).[18-22] Recently, as triggered by the discovery of topological semimetals,[23-24] the negative MR in nonmagnetic systems has attracted renewed interest.

(25) In the case of ferroelectrics, several theoretical studies[20-24] have proposed complex polarization topologies (reminiscent of rotational spin topologies) in low-dimensional structures. The models have predicted the formation of polarization waves, vortices and so on that can be characterized by an emergent order parameter, a so-called electric toroidal moment.

(26) Most tubular assemblies reported so far have diameters ranging from tens of nm to several micrometers, while SNTs having diameters of several nm are rare. [5], [8-9]

(27) Various nanostructures of molybdenum carbides, including nanowires, nanosheets, nanospheres, and nanoparticles, have been explored to improve the electrocatalytic performance.[22-26] However, it has remained to be challenging to fabricate well-defined nanostructures with a large density of catalytic active sites.

(28) Under the inspiration of these water harvesting and transport species, a variety of fog collector[12-13], underwater bubble collector[14] and oil–water separation devices[15] are proposed in diverse fields. Many attempts to achieve high-speed liquid spreading or transport have been conducted with the use of a high-intensity femtosecond laser structuring technique[16] or Leidenfrost effect[17].

Task 12: Discuss in groups what authorial stance towards the cited literature, whether positive, neutral, negative, or mixed, is conveyed by each underlined reporting verb in the sentences below.

(1) A high plasmon to hot electron conversion efficiency 35% was <u>reported</u> when the scanning probe technique was combined to the detection.

(2) Cui and co-workers reported on a mid-infrared transparent nanoporous polyethylene for efficient human body cooling.

(3) Clauer reported the mechanism of laser peening using a nanosecond laser.

(4) An interesting transition from a low dark current photodiode to a high gain photoconductor is observed upon UV light exposure.

(5) Moreover, Raman signal is enhanced when the emission energy is in resonance with neutral exciton emission, but such a phenomenon has not been observed in trions.

(6) Sunoqrot et al. also observed that methoxypoly(ethylene glycol) (mPEG)–polycaprolactone nanoparticles became mucoadhesive following PDA coating.

(7) Here, interestingly, we observed a significantly different cellular response under anisotropic contact by applying the hNSCs to the novel high-resolution nanogroove (HRN) structure with an extremely limited contact width of 15 nm.

(8) Several previous studies already suggested that the red-shift in the CDs' emission can be caused by nitrogen.

(9) On the other hand, recent theoretical and experimental studies have suggested the presence of ferroelectricity in MAPbI3 perovskite materials.

(10) Our findings suggest a carefully designed Marangoni effect can be an effective tool to regulate the growth mechanism of the organic crystalline thin films.

(11) Our results suggest that GPS can be used as a nanodevice for extracting molecular information from the microenvironment.

(12) Subsequently, a number of experiments have confirmed the existence of this elusive tetragonal phase.

(13) Also, work by Al-Saidi and Umrigar using a diffusion Monte Carlo method confirmed diradical structure 3a.

(14) However, for a specific functional part, the dimensional accuracy of AM parts can be negatively affected by thermal post-processing, as illustrated by Patterson et al.

(15) The formation of this hybrid nanosystem and the proposed mechanism of action after intravenous administration are illustrated in Figure 1.

(16) Dynamically controlling surface microstructure is believed to be an effective way to realize the switch between the isotropic and anisotropic wetting.

(17) Previous reports suggested that Bi_2S_3 NPs can be an ideal nanotheranostic platform.

(18) Experimental evidence suggested a strong link between *N* and the structure of platelets, initiating the idea that *N* possibly plays an important role in the formation and structure of platelets.

(19) They also reviewed the wetting behavior of different micro-nano scale surface modification approaches and concluded that the fabrication approach significantly affects the surface characteristics which in turn affect the wetting behavior.

(20) The authors concluded that the double patch supported higher maximum loads.
(21) This result suggests ML-based approaches to materials selection may provide significant acceleration over the guess-and-check research paradigm of the past.
(22) By tuning the reaction stoichiometry and time, and thus roughness, a favorable correlation between surface area and membrane flux has been proposed, suggesting a powerful route to manufacture thin film membranes with high permeance.
(23) The SCOSC system does not rely on the donor-acceptor blend ratio, suggesting that a proper molecular design can lead to a stable phase separation for OSCs.
(24) Computed tomography (CT), magnetic resonance imaging (MRI), positron emission tomography (PET), and ultrasound imaging have been widely implemented for theranostic imaging.
(25) We note that the disclination line is distorted in the images of the axial droplets shown in (C).
(26) Several studies confirmed the grafting of methanol onto the aluminol function of the interlayer space of methoxykaolinite.
(27) AMPs are thought to primarily act via mechanically destroying the cell membrane, formation of pores or ion channels, or a multihit strategy including additional intracellular targets.
(28) Encouraged by these inspiring results and simplified procedures, we believe this TMSI-based method will become more universal in obtaining highly bright and stable CPI PQDs.
(29) We confirmed CSR of the multilayer islands using the statistical test by Clark-Evans.
(30) The results from these experiments ultimately confirmed their broad spectrum of potential applications ranging from general industries to medical fields.
(31) Structural water is thought to play an important role in controlling the lifetime and phase transformation rates of amorphous carbonates.
(32) Previous research used XCT to observe structural changes and void evolution in cement paste during the hydration of cement.
(33) This is a pioneering milestone, but we believe this is not the limit.
(34) A recent work of Hu et al. provides a meaningful experimental demonstration of this concept, implementing carbon nanotube (CNT) random bits by leveraging the randomness of a CNT assembly process.
(35) Previous studies showed that FSP could be applied to the Al-cast alloys to achieve considerable microstructure refinement, generating a fine equiaxed grain structure by implementing a severe plastic deformation, as reported by Su et al.
(36) At the device level, memristors have been shown to be able to emulate synaptic functions by storing the analog synaptic weights and implementing synaptic learning rules.
(37) Conventionally, Newton-Raphson (NR) method is employed to obtain solutions implementing the least square technique.

(38) Previous work within our group has developed PolyHIPE-based porous woodpile structures for bone tissue engineering applications.

(39) On the other hand, MRE-based isolators have also been developed by implementing various control strategies such as fuzzy logic control and clipped optimal control.

(40) For example, Gregory and Phillips conducted an evaluation of four key methods for monitoring household waste prevention campaigns in the UK, and quantified the direct waste tonnage impacts of implementing a targeted household waste campaign.

(41) Soares and Martins identified socio-political-economic barriers to the process implementing alternative and complementary technologies for generating electricity from MSW in So Paulo, Brazil, using LCA.

(42) In the present work, we propose a cost-effective and easy-to-implement preparation technique that provides Si-based anode materials with different capacities.

(43) The aim of the present work is to implement a material point method on the molecular dynamics program LAMMPS developed by Plimpton and apply the implementation to 3D micro-milling.

(44) Low-temperature study reports that this energy difference strongly increases with MoS_2 Fermi energy due to the recoil effect.

(45) A recent study reports that a negatively charged DPP-based probe can improve its water solubility and has the potential to label biomacromolecules.

(46) Removing the binder after film deposition is necessary but implies a great challenge at low temperatures.

(47) Moreover, the chemical control is typically nonvolatile in nature; this implies that the altered magnetic states can be retained without an electrical power supply.

(48) More importantly, the surface of this material is highly reducible, which implies a strong reactivity of the intercalated H^+ toward surrounding chemicals with an appropriate redox potential.

(49) Recently, Guo et al. and Feng et al. pointed out the conditions and limitations of underwater wet laser welding and cladding of selected metals.

(50) Previous studies pointed out that the buckling of individual Fe ion in these compounds can help the diffusion of each Na ion.

(51) Bailey et al. evaluated the influence of end tab materials and clamping force on the fatigue test results for glass fibre reinforced epoxy composites and pointed out the high risk of failure in or near the gripping region.

(52) It was theoretically demonstrated that dislocations in STO do not accelerate oxygen diffusion.

(53) From these studies it was concluded that the roughened surface is covered with nanoscale islands of (at most) a few atoms high.

(54) Some studies have concluded that amorphous metal oxides have lower capacity compared to their crystalline counterparts, contrary observations have been shown in other amorphous systems.

(55) As concluded by Wang et al. and Feng et al., microstructures and mechanical properties of materials subjected to SPD with grain refining show improvement in strengths, hardness, fatigue life, corrosion, wear resistances, etc.

(56) However, a point worth noting is that the power capability of the LIC devices is generally limited by the Li-ion battery electrode side.

(57) Recent experiments demonstrated that some metal oxides can be deposited on the desired sites of seed materials by selective ALD.

(58) It is worth noting that fracture toughness may be insensitive to loading rates within a certain range.

(59) It is worth noting that the above study did not investigate the influence of the hydraulic pressure on the mechanical properties of FRPs.

(60) Wan et al. claimed that Fe/Fe_3C nanoparticles encapsulated in graphitic layers could boost the activity of Fe-N_x, and Mukerjee et al. proved that Fe/Fe_3C nanoparticles could activate the carbon-nitrogen structure in the outer skin, while Kramm et al. pointed out that the removal of metal-based particles from the catalyst would lead to a tremendous increase in the ORR activity.

(61) After confirming our structural model by comparing the experimental INS data with our calculations, we analyze the role and aspects of entropy and hydrogen bonding interactions in determining the thermodynamics.

(62) In a recent work the nucleation of this material has been studied, confirming the formation of high amounts of (intersecting) stacking faults and, therefore, numerous-nuclei by rapid quenching.

(63) Strong electric field could however bring about a cubic-like to rhombohedral distortion confirming that the appearance of the different global structure cubic-like/monoclinic in the unpoled state is not related to structural change on the scale of unit cell, but in the assemblage of the nanometer sized domains, as suggested by Levin and Reaney.

(64) The first works reporting the application of NHts in magnetic hyperthermia (MH) date back to the 1950s, and more recently, numerous studies showing the success of proofs of concept and clinical trials have boosted the research interest into them.

(65) While it may be argued that changes in the average grain size and grain size distribution with time are relatively well understood, we know much less about the behavior of individual grains in this process and what factors control whether a specific grain shrinks or grows.

(66) It has been argued that the 10% substitution of Bi ions by Er^{3+} ones enhances only the ferroelectric property.

(67) We had <u>argued</u> that this two-step process could lead to incomplete reactions and mixed-phase formation of MoS_2 and MoO_2.

(68) It must be <u>noticed</u> that the glide of b_1 and b_3 on the twin plane results in the opposite shear, corresponding to compression or tension twinning.

(69) Further, it was also <u>noticed</u> that in using lateral stress gauges to determine the shear strength, rise times were shorter in the T0 state compared to T6, explained in terms of the precipitates inhibiting dislocation motion.

(70) Considering the changes in surface roughness, the authors have <u>noticed</u> that there is a low surface modification for processing time lower than 60 s/mm.

(71) Similarly, Ringsberg et al. <u>studied</u> the RCF behaviour of a Co-Cr alloy layer cladded on the pearlitic UIC 900A (R260) rail steel and <u>reported</u> that excellent agreement between experimental and numerical results of residual stresses was noticed.

(72) We <u>observe</u> a large positive MR of 225% in the out-of-plane magnetic field at 2 K.

(73) It is well <u>acknowledged</u> that nanoparticles can provide distinguishing characteristics to improve electrochemical performance of anode materials based on alloying/dealloying lithium storage mechanism.

(74) Several authors have <u>conducted</u> numerical studies of friction stir processes, with the objectives of understanding the process mechanisms, and reducing the number of empirical trials needed to optimise processing conditions, as <u>emphasized</u> by Threadgill et al.

(75) Eselun et al. <u>noted</u> the influences of resin micro cracks, generated by thermal cycling, on fatigue life, tensile strength, and interlaminar shear strength.

(76) They have <u>claimed</u> that this approach could successfully represent the initial high stiffness, broaden cyclic deformation, and Payne effect that usually occurs in carbon black filled compounds.

(77) Until now, FSSW has been <u>employed</u> to the assembling of dissimilar alloys, such as aluminum to steel and aluminum to magnesium, and some attempts have <u>claimed</u> a success in joining these dissimilar alloys.

(78) Others <u>argue</u>, however, that voids might have a strengthening effect or even enhance the tensile ductility due to dislocation-void interactions.

(79) We <u>reasoned</u> that elasticity and cell interactions could be achieved if tropoelastin could dominate a novel material, blended with silk fibroin that serves a secondary role to impart strength and durability.

(80) Meanwhile, this paper <u>emphasized</u> the enhancement of filler material content and curing agent ratio on rheological behaviors, and the potential of MRE as a multifunctional strain sensor.

(81) Others <u>claim</u> such lock-ins and competition as myths, and stress the need for incinerating non-recyclable waste and the fact that plants can be converted to other fuels if needed (Swedish Waste Management Association, 2016).

(82) Nandan et al. agreed that friction coefficient can not be determined from fundamental principles or by simple experiments relevant to FSW conditions.

(83) Yang and coworkers thought that the addition of nanoparticles generally enhanced the molecular interactions within the epoxy matrix due to strong interactions around nanofillers.

(84) On the other hand, the use of weak-base ammonia solution can be advantageous that forms stable complexes with major surface-coated metals (Cu and Ni) even at the storage temperature of ammonia (~20 °C), rejecting any possibilities for the degradation of ABS polymers.

(85) Murr et al. commented that there was some evidence for cracking within columnar grains in EBM-built Rene 142 although they did not show any crack features.

(86) Likewise, most other studies also have insisted that anisotropic geometry-induced cellular differentiations initiate from enhanced FA formation or cytoskeletal organization.

(87) Furthermore, the postulations of the "modified perturbation theory" have been disputed owing to a lack of physical correspondence and incoherent elucidation of the pinch-off events.

(88) This concurs with the coarse-grained computational modeling that we have recently performed.

(89) Several authors assert that WtE can be an integral part of sustainable waste management without violating waste hierarchy principles or compromising reuse and recycling (Ng et al., 2014, Brunner and Rechberger, 2015, Cucchiella et al., 2017).

(90) The $Li_{0.5}FePO_4$ phase forms diagonal interfaces, not parallel to any primary axis, in both LFP and FP, qualitatively agreeing with observations by synchrotron X-ray diffraction (XRD) and transmission electron microscopy (TEM).

Task 13: Choose the most appropriate statement to fill in each blank in the review article introduction below. Then discuss in groups whether similar justification steps in research article introductions are utilized here, namely 1) claiming the importance of a research area while emphasizing central issue(s), 2) reviewing previous research clearly relevant to the present one, 3) pointing out research gap unaddressed by previous literature, 4) announcing research purposes of the present study, and 5) elaborating on details of the present study.

Toward the Scale-Up of Solid-State Lithium Metal Batteries: The Gaps between Lab-Level Cells and Practical Large-Format Batteries[5]

INTRODUCTION

1 Lithium-ion batteries (LIBs) have achieved triumph since their commercialization in the 1990s due to the wide range of applications including mobile electronic devices, medical devices, satellites, and military electronic equipment.[1] However, commercial LIBs provide a normal energy density of around 260 Wh kg^{-1} (Figure 1), which gradually falls behind the

market requirements.[2] In addition, LIBs still suffer from high cost, limited durability, and poor safety.[3] As a consequence, an alternative rechargeable battery system is crucial to cope with the growing demands of electric vehicles.[4]

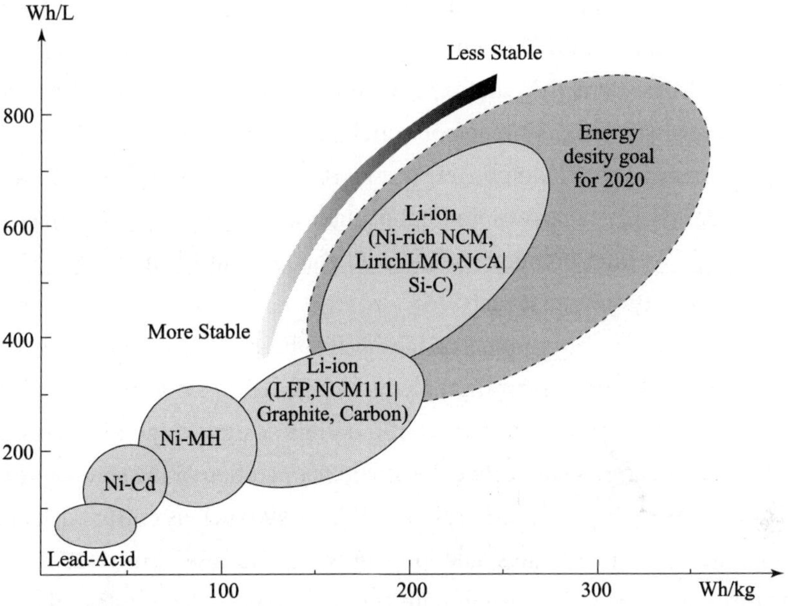

Figure 1 Energy density level of industrial products[5] and goals for 2020

2 ___________ (1) ___________. The conventional power batteries suffered from frequent combustion accidents. The ascending voltage of cathodes will lead to stability concerns. Most of the thermal runaway is triggered by the reaction derived from electrodes.[6] The decomposition of cathode materials releases heat and oxygen and will trigger the combustion of flammable organic electrolytes. However, the high-energy cathode is essential for an energy-dense battery. Hence the substitution of conventional organic electrolytes is feasible to both enhance the energy density and battery safety.[5], [7]

3 The replacement of organic liquid electrolytes (OLEs) with solid-state electrolytes (SSEs) provides a promising future for the large-scale application of lithium metal batteries (LMBs). SSEs with wide electrochemical stability windows and high modulus possess the potential to enable high-capacity electrode materials and prevent Li dendritic deposition.[8] ___________ (2) ___________. The introduction of SSEs allows a simplified battery design and minimization of inactive materials, consequently increasing energy density at the cell-level. Moreover, the solid-state electrolytes enable the employment of Li metal anodes, which is considered as the most promising anode for next-generation rechargeable batteries due to its

ultrahigh theoretical specific capacity of 3860 $mA \cdot h \cdot g^{-1}$ and lowest negative electrochemical potential (-3.04 V vs the standard hydrogen electrode).[9] In conventional organic electrolytes, lithium metal suffers from the unstable solid-state interphase, dendrite penetration, and pulverization issues.[1c], [10] The rapid deterioration of LMBs is attributed to the high reactivity of lithium metal.[11] Virtually most conventional OLEs can be reduced at the Li surface, forming unstable solid electrolyte interphase (SEI). During repeated charge–discharge cycles, Li tends to generate dendritic morphology because of nonuniform current/ion distribution.[12] Lithium dendrites normally lead to the fracture and regeneration of SEI, further aggravating the dendrite growth.[12a], [13] The vicious circle at Li anode surface during cycling is known as the main cause of low Coulombic efficiency (CE) and even explosion hazards of LMBs.[14] Among these issues caused by the application of Li metal anode, considering the safety concern, SSEs are appropriate candidates to couple with Li metal anode, improving safety and energy density simultaneously.[15]

4 __________ (3) __________. The lab-level solid-state lithium metal batteries (SSLMBs) aim at dealing with the universal problems in SSLMBs such as poor interfacial contact, low ionic conductivity, rational battery configuration design, electrode structure optimization, etc.[16] These researches can realize the effective operation of the solid-state batteries, but the operation is based on precise control toward internal (such as current densities) and external factors (such as temperatures and pressures) in lab-level coin cells or mold batteries. Comparing with batteries using conventional liquid electrolytes, the scale-up toward SSLMBs needs to consider the preparation of scale-up electrodes, SSE pellets, and fully consider their compatibility. Hence the SSLMBs of large dimensions are hard to realize satisfying quality control and comparative performances. The assembly and operation between lab-level and scale-up SSLMBs remain a huge gap. The lab-level strategies should be rationally integrated with battery manufactory technology to realize the practical applications.

5 In this review, we mainly aim at highlighting the gaps and challenging between the practical level batteries and lab level batteries. We firstly present a succinct introduction to SSEs and their applications in SSLMBs. __________ (4) __________. We then deduce the key parameters for batteries scale-up and focus on the gaps between lab-level cells and commercial SSLMBs in electrodes and electrolytes construction. Moreover, suitable practical technologies toward battery commercialization are proposed. A conceptual schematic diagram of the gaps and promising strategies in the scale-up process of solid-state lithium batteries is presented in Figure 2.

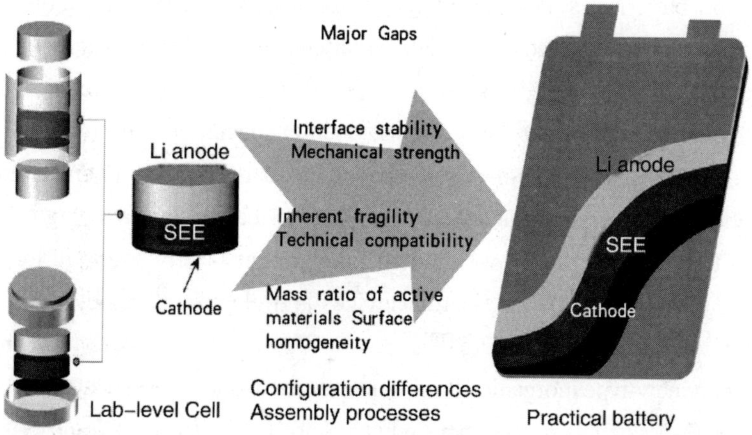

Figure 2 A schematic illustration of the gaps in the process of the scale-up of solid-state lithium batteries

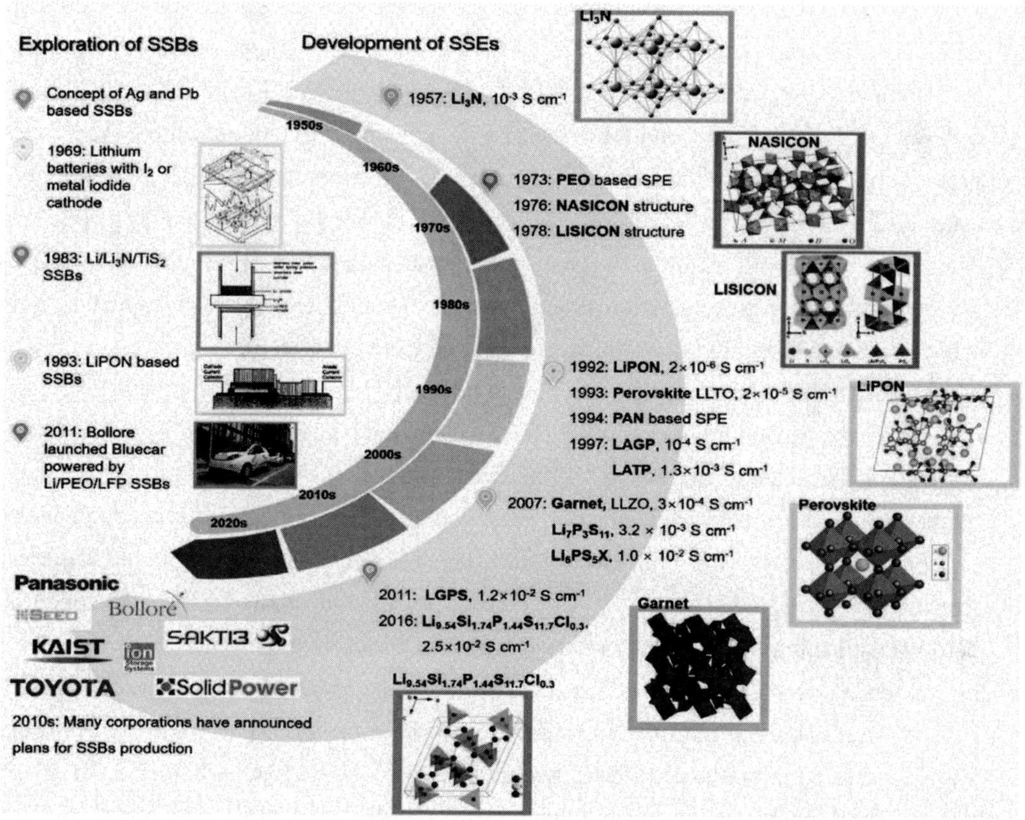

Figure 3 A chronology of the development of typical SSEs and the process of SSBs[26], [28]

2. General Remarks on SSLMBs

6 Long before the coming out of lithium insertion materials which are the core components of

lithium-ion batteries, solid-state battery technologies had been promoted since solid ionic conductors were discovered. Solid-state batteries with metal anodes and metal salts electrolytes were emerged since 1950s, such as Ag/AgI/V_2O_5, Ag/AgBr/$CuBr_2$, Ag/AgCl/$KICl_4$,[17] Li/LiI/AgI, and Li/LiI/I_2.[18] They are still used in special devices as primary batteries. In the 1960s, a solid-state sodium–sulfur secondary battery was invented by Ford with the application of Na ion conductor β"-Al_2O_3. Na–S batteries with a high theoretical specific energy of 760 Wh kg^{-1} were well-suited for stationary energy storage for the electrical grid but not suitable for portable electric devices and electronic vehicles (EVs) because of the elevated operating temperature.[19] ___________ (5) ___________. In the 1970s, the discovery of new-type inorganic lithium superionic conductor (LISICON) and sodium superionic conductor (NASICON) opened new windows for SSEs. LISICON and NASICON as typical structures inspired many derivations and new design of SSEs, such as $Li_{1+x}Al_xGe_{2-x}(PO_4)_3$ (LAGP)[20] and $Li_{1+x}Al_xTi_{2-x}(PO_4)_3$ (LATP).[21] Other structures such as LiPON[22] and perovskites[23] were proved the ability for ionic conduction in 1992 and 1993, respectively. After 2000, the discovery of solid-state ionic conductor with high performance exhibited a blowout such as garnet-based SSEs (e.g., $Li_7La_3Zr_2O_{12}$, LLZO[24]) with high stability against Li metal and sulfide SSEs (e.g., argyrodite Li_6PS_5Cl[25] and thio-LISICON $Li_{10}GeP_2S_{12}$[26]) with extremely high ionic conductivity. SSEs are generally divided into three types: solid-state inorganic ionic conductor, solid polymer electrolytes (SPEs), and composite materials.[27] In 2019, the Nobel Prize was awarded to the lithium-ion battery, which will guide the further development of the energy storage area. And the solid-state battery will lead the revolution of next-generation energy storage technology. Nowadays, many battery enterprises have begun to invest the layout on solid-state batteries.

7 The Li-ion conduction is generally accepted to be related to the concentration and distribution of defect sites in inorganic SSEs. The vacancy-mediated mechanism and relatively nonvacancy- mediated mechanism are the two main branches of ion diffusion mechanism based on Schottky and Frenkel defects.[29] The conduction of lithium-ion is accomplished by the conductive channels in bulk electrolytes, which is decided by the sublattices in crystal structures. ___________ (6) ___________. These processes need to overcome the conductive barriers, reflecting as the activation energy.[30] The activation energy represents the difficulties on ionic transition, which is determined by the crystal structure, type of conductive pathway (e.g., 1D, 2D, 3D), the size of the conductive pathway, etc.[31] To the best of our knowledge, inorganic solid electrolyte with the highest ionic conductivity of 2.5×10^{-2} S cm^{-1} was demonstrated by Kanno and co-workers in 2016. Li ion conductor $Li_{9.54}Si_{1.74}P_{1.44}S_{11.7}Cl_{0.3}$[28d] with LGPS-type crystal structural enabled all-solid-state $Li_4Ti_5O_{12}$–$LiCoO_2$ cell a stable cycle performance with a high current density of 18 C, which imports new opportunities for the development of solid-state batteries for fast charging. At present, sulfide SSEs possess high ionic conductivity and high stability against Li metal have

been discovered, e.g., $2Li_3PS_4LiX$(X = Br, I),[32] $Li_{5.5}PS_{4.5}Cl_{1.5}$,[33] $Li_{10}P_3S_{12}I$.[34] The discovery of these new sulfide electrolytes avoids using expensive Ge elements, and fulfilling the high ionic conductivity of >6 mS cm^{-1} or even >10 mS cm^{-1}. __________ (7) __________.

8 Polymeric SSEs were initially proposed by Fenton and coworkers in 1973.[35] Poly(ethylene oxide) (PEO)-based complexes via evaporating solvent from the PEO and alkali metal salts were found to be ionic conductive. In 1993, Lightfoot et al. demonstrated the crystal structure of the archetypical polymer electrolyte PEO:$LiCF_3SO_3$[36] and detected Li-ion migration in the crystalline complexes by nuclear magnet resonance (NMR) and AC impedance. Above the glass transition temperature (*Tg*), the ionic conduction in polymeric SSEs is accomplished by continuous complexation and dissociation between Li-ion and negatively charged atoms in the amorphous phase.[37] The high flexibility of polymeric SSEs enables the compatibility toward roll-to-roll technique. __________ (8) __________. Therefore, a variety of dimensionally stable solid polymer electrolytes were reported afterward, including polyacrylonitrile (PAN)[38], poly(vinyl pyrrolidinone) (PVP), poly(vinyl chloride) (PVC), polyvinylidene fluoride (PVDF)[39], and polymethyl methacrylate.[40] The ionic conductivity at room temperature (RT) varied between 10^{-6} and 10^{-3} S cm^{-1} for those polymer-based Li-ion conductors. However, the high crystallinity of SPEs at RT leads to unsatisfied ionic conductivity for daily usage. It is essential to further develop the SPEs due to their irreplaceable merits. Incorporating inorganic fillers into polymer-based electrolytes brings new opportunities to the development of composite electrolytes. Inert fillers (ZrO_2,[41] TiO_2,[42] SiO_2,[42] and Al_2O_3[43]), and active fillers (LLZO,[44] LAGP,[45] LLTO,[46] and LPS[47]) were introduced into polymer complexes matrices, and help to improve the ionic conductivity.

9 __________ (9) __________. The advantage of high-voltage cathodes such as $LiCoO_2$, $LiNi_{0.5}Mn_{1.5}O_4$, and $LiNi_{1/3}Co_{1/3}Mn_{1/3}O_2$ can be fully released after coupling with Li metal anode.[26] Those intercalation materials are recognized as high-rate-capacity materials with rapid Li^+ deintercalation reaction and stable cycling performance, shedding a light on constructing high-energy-density SSBs with fast charging performance and long lifespan.[48] Besides, using high-specific-capacity active cathodes is significantly effective to improve the overall energy density. The transferring from OLEs to SSEs can guide considerable breakthrough toward Li–S batteries, which is regarded as one of the most promising next-generation battery systems with ultra-high theoretical energy density (2,600 Wh kg^{-1}).[49] The "shuttle effect" of polysulfides hinders the application of conventional Li–S batteries with liquid electrolytes.[50] Importing rigid solid-state single-ion conductors can impede the diffusion of polysulfide.[51] Li metal anodes are protected from the corrosion of shuttling polysulfide at the same time.[52] In all-solid-state mode, the mechanism of sulfur is fully changed, sulfur will be directly transferred into Li_2S, completely avoiding the production of the soluble intermediates.

10 Benefit from the high safety and high energy density, SSLMBs hold a great possibility to be fully applied in varying life scenes. Long-mileage EVs and energy storage power stations equipped with SSLMBs would be more endurable to withstand an emergency. However, the lab-level SSLMBs have not reached the comparable level of commercial lithium-ion batteries in cycle life and stability. Moreover, the energy density decreases sharply once transfer those achievements to practical pouch cells. ___________ (10) ___________. Therefore, a comprehensive understanding of those gaps between lab-level cells and large-scale practical batteries is of great significance for follow-up research.[53] Subsequently, the gaps between practical level batteries to lab level batteries are investigated from the views of SSE preparation, cathode manufacturing, anode construction, and the battery assembly.

…

(The end of the passage)

A) Li-ion transport is accomplished by hopping from one site to adjacent equivalent sites in the crystalline framework of inorganic electrolytes.

B) Considering the requirement of high-energy-density battery systems, research on lithium metal anode has been revived to be the hotspot in recent years.

C) The prior demand for a power battery is security.

D) The reduced cost of SSE ingredients helps to further improve the battery scale-up.

E) Challenges and opportunities have also been fully discussed both for lab-level researches and scale-up development.

F) As shown in Figure 3, the development of SSEs is nearly synchronous to the development of the batteries.

G) There remain challenges in solving fundamental issues of SSLMBs and still huge gaps between lab-level cells and large-scale practical batteries.

H) Besides, the SSEs also possess good thermal stability, which enables a wide operation temperature range and avoid the combustion risk.

I) However, the applications of solid-state batteries are still developed at initial stages.

J) Hence the polymeric SSEs have attracted great attention because of the industrial potential.

Task 14: Answer the following questions based on Text Ⅲ, and then discuss in groups how the proposed research is justified.

(1) What is the purpose of this research article?

____

____

(2) What is considered a critical factor in the design and development of innovative materials?

____

(3) For what reasons have the full-field measurement methods been mostly used to study the mechanical behavior of innovative materials?

(4) What strengths and weaknesses of interferometric techniques and digital image correlation (DIC) methods are described here?

(5) What alternative imaging methods have been developed to capture materials structure at a smaller scale?

(6) What typical functions could be performed by different types of profilometers?

(7) What is the operating principle of the scanning probe microscopy (SPM) method?

(8) Compared to stylus-based profilometers, what advantages do light-based devices have?

(9) What problems still exist in the combined use of various full-field measurement techniques?

(10) What advantages and disadvantages does the combined use of OCT and DIC techniques have?

(11) What advantages and disadvantages of the combined use of profilometry and DIC are discussed?

(12) What limitations of the Integrated Digital Image Correlation (I-DIC) method in the AFM and DIC combination are pointed out?

(13) To what fields has the combination of DIC and CM analyses been applied?

(14) Compared to the combined use of different methods in previous literature, why does this paper attempt to combine the CM method and the commercial DIC algorithm mentioned in paragraphs 12-13?

(15) What two case studies were carried out utilizing the proposed procedure of CM and DIC analyses?

(16) What purpose does each citation in paragraph 2 serve? In what order are the citations presented?

(17) What purposes do the citations in paragraph 11 serve?

(18) What purposes do the citations in paragraph 12 fulfill?

Text Ⅲ

Full-field Measurement with Nanometric Accuracy of 3D Superficial Displacements by Digital Profile Correlation: A Powerful Tool for Mechanics of Materials[6]

INTRODUCTION

1 The wide-reaching technological advances made over the last decades have allowed for the development and diffusion of a number of new materials [1], [2], [3]. This process of innovation is not an end in itself, but is being driven and accelerated by different needs, coming from several fields.

2 Engineers are perpetually trying to improve the performance of the mechanical properties with the aim of using them in increasingly critical and extreme applications – e.g. sports competitions, aerospace, energy, defense and military [4]. Applications in electronics, computer science, and tailored sensors demand materials that accomplish specific tasks that are nonetheless contained to the greatest extent in weight and dimension [5]. Biomedical and

biomechanical studies and research show the advantages of new materials used for their biocompatibility [6], capability to act as cure carriers [7], or capacities for inspecting [8]. In this framework, mechanical properties often represent a crucial aspect of design and development of these innovations, even if the primary aim has not been to satisfy specific stiffness and/or strength constraints.

3 Nowadays, the tools available to carry out experimental stress analysis are applied in several fields of mechanics of materials, covering a wide spectrum of analyses and serving numerous ends. It would be, as such, a daunting task to undertake a thorough classification of them all. It would result difficult, in fact, to merely list those that are most used or most popular. This lies beyond the scope of the present paper, in which, instead, the author intends to limit the focus on those techniques that are most typically used for measuring displacement field (i.e. mechanical strain) or material structure (i.e. microscopic imaging and profile). Among the methods suitable for studying the mechanical behavior of innovative materials, the most popular are those capable of performing full-field measurements, i.e. to extract a large amount of information on a significant portion of the material under investigation. This distinctive feature can play a key role when a material behaves in a non-conventional way, due to its heterogeneity and/or its complex structure.

4 Speckle interferometry [9] and digital holography [10] are two interferometric methods able to measure, in their basic configurations, a single displacement component with interferometric accuracy. The strengths of these techniques reside in their high sensitivity (few nanometers) and the ease of use of their recording media (CCD/CMOS cameras). On the other hand, the experimental setups are complex to assemble and manage, and their sensitivity to external disturbances is significantly high. On the contrary, digital image correlation (DIC) methods [11], sharing some of the strengths of the interferometric techniques (full-field capability, information recorded by cameras), allow for the acquisition of two (by the conventional 2D-DIC) or even all three (by the 3D-DIC arrangement) displacement components through much simpler experimental configuration and with a much smaller disturbance sensitivity. In contrast, the displacement sensitivity of DIC is much smaller (typically a few micrometers) and depends on displacement gradients and on the quality of the imaging process.

5 Instead of cameras, which are extremely useful for the aforementioned methods for measuring deformation, it is possible to use alternative imaging techniques that are capable of capturing the structure of the materials at a smaller scale, such as: Scanning Electron Microscopy (SEM) for reflection imaging of any opaque material, Transmission Electron Microscopy (TEM) for transmission imaging of thin film or transparent materials, Optical Coherence Tomography (OCT) for retrieving the structure of transparent tissues, and Electron BackScatter Diffraction (EBSD) for identification of crystal orientation, phases and defect.

6 Today, different types of profilometers are used to characterize the outer surface structure of a material, thanks to the wide applicability of this specific machine to characterize thickness, texture, roughness, thin films, coatings, semiconductor, wear, corrosion, and perform biomedical and biological analyses. The profilometers can be classified in two categories: stylus-based and light-based. The devices based on the stylus require a probe that follows the profile of the surface under test, according to different operating principles. In its most basic configuration, the interaction is purely mechanical. According to this concept, the first profilometers were built to measure roughness along a line, and even today they are profitably used in different applications to this end. Nevertheless, at present, significantly higher accuracy is attained by applying the scanning probe microscopy (SPM) method, which obtains the outer surface geometry by exploiting the interaction between the stylus and the surface under investigation at atomic level. Among the possible methods of implementation, the two most popular are the Scanning Tunneling Microscopy (STM) [12] and the Atomic Force Microscopy (AFM) [13]. On the other hand, the light-based devices have the advantage of preventing any contact between the measurement system and the surface under investigation. Among the several light-based implementation methods, the most successful are those based on White Light Interferometry (WLI) [14] and Confocal Microscopy (CM). The effectiveness of this type of analysis in material science [15-20] is demonstrated today by the numerous commercial apparatuses on the market that integrate these methods, available at increasingly reasonable prices.

7 The combined use of all of the aforementioned techniques can be found in a number of scholarly papers. Particularly common is the application of correlation algorithms to the various imaging system ways different form the conventional one – i.e. focusing an object on a sensor by way of an optical lens. This approach requires the acquisition of two configurations of the same portion of the surface under investigation, before and after the application of a deformation field. One of the most popular implementation techniques consists of applying the 2D-DIC on the images obtained by SEM, which measured and monitored deformations occurring, at a very low scale, in different type of materials [21-27]. Similar results can be attained through TEM [28] and EBSD [29]. In all these cases, the spatial resolution is much higher than that of conventional imaging systems. On the other hand, this equipment works in a vacuum, they are expensive, and they do not allow for the acquisition of the out-of-plane component of the displacement vector. This is due to the lack of information along the direction orthogonal to the plane containing the image. In addition, local contrast is often slightly different in the two loading configurations, decreasing the accuracy and reliability of the correlation algorithms. Consequently, bigger subsets have to be used for properly running DIC procedures, and displacement fields characterized by high gradients are retrieved with lower accuracy. In fact, it is worth noting that when correlation

algorithms are successfully applied to the images acquired by these techniques, the subset size is hardly smaller than 40 ÷ 50 pixels, and the displacement fields are considerably regular.

8 The specific case of the combined use of OCT and DIC [30-31] deserves different consideration. In this case, spatial resolution is comparable to those techniques based on a conventional imaging system, and when different configurations are considered, the decay of image quality is even worse than that occurring in electron scanning techniques. Nevertheless, it works for biological materials, and has acted as a useful investigation tool in different medical fields. In addition, thanks in large part to its capacity to penetrate through tissues of the light sources used by OCT machines, the images can be analyzed by complex correlation algorithms that are able to retrieve volume deformations, rather than simply planar ones – i.e. Digital Volume Correlation (DVC) [32].

9 The SPM methods display similar spatial resolution to that of electron scanning techniques, and suffer less local contrast loss, thanks to the fact that the imaging operation is not performed by lens, but consists of a digital representation of the physical topography of the surface under investigation, shown at nanometric scale. In addition, SPM methods can retrieve the out-of-plane component, thanks to the topographic information provided by the measurement method. Vendroux et al. in 1998 [33-35] published a series of three papers by which they proposed the combined use of profilometry and DIC to retrieve displacement vectors, with particular emphasis on the STM; in their research, the authors proposed the idea, carried out an extensive analysis of DIC procedures, and reported some preliminary experimental results.

10 A combination of AFM and DIC was used by a research group of University of South Carolina for some applications: in-plane mechanical characterization of polymeric thin film [36], compensation of AFM in-plane distortion due to the translation stage of the device [37], and the mapping of nanoscale wear occurring on a gold coating [38]. Other interesting applications of this method in the field of mechanics of materials include residual stress measurement [39], elastic characterization of MEMS [40], and identification of fracture mechanics parameters of a silicate glass at nanoscale [41]. In all these cases, the nanometric resolution of the AFM device allows for the in-plane displacement components to be determined by correlation algorithms with sub-nanometric resolution. However, the out-of-plane component was usually neglected. In particular, in the case of wear evaluation [38], the consumption of the materials – i.e. the lowering of the peak – was evaluated by analyzing the correlation coefficient and comparing profiles along the same segment of the gold coating. Instead, when applying this approach to an analysis of displacement field around the crack tip of a silicate glass [41], all three displacement components were retrieved. In this application, a specific analytic model was assumed according to the elastic solution for linear-elastic materials, which required the estimation of a limited number parameters by means of an optimization process. This approach is known as the Integrated Digital Image Correlation (I-

DIC) method, the assumptions of which do not necessarily involve kinematic variables – i.e. displacement field. Additional and alternative parameters can be included in analytical models in order to retrieve the displacement components. These can include the stress intensity factor and the residual stress state or any mechanical property (e.g. Young's modulus, Poisson's ratio, yield strength, hardening coefficient), all of which significantly impact the mechanical response.

11 Correlation procedures, and in particular the Global Digital Image Correlation method (GDIC) [42], were also applied to the profiles retrieved by CM. The GDIC approach is based on the definition of a mesh on the surface of the specimen and an algorithm applied iteratively to identify displacement components. The Eindhoven research group of TUE, together with other colleagues, developed a series of outstanding applications obtained through the combined use of GDIC and CM profile data. Van Beeck et al. [43-45] studied 3D deformations of polymer-coated metals induced by metal sheet stretching. The procedure consisted of non-standardized dog bone samples analyzed by a micro-tensile machine; displacement component maps and profiles were obtained at different points on the specimens. Bertin et al. [46] performed a micro-mechanical characterization of a two-grains, low-carbon interstitial free steel specimen subjected to a tensile test. Kleinendorst et al. [47] demonstrated how to study complex geometries through a procedure based on adaptive mesh by which they successfully retrieved the displacement field on an S-shaped stretchable electronic part. Through a similar experimental approach, the local curvature of a bulged membrane [48] was measured, as was a micro-beam under bending loads [49] on parts designed for MEMS applications. In the last application, the specimen's profile was measured by Dual Wavelength Digital Holographic Microscopy (DWDHM) and the entire procedure showed a curvature resolution of $1.5 \times 10^{-6}\ \mu m^{-1}$. Another novel application obtained through the combination of DIC and CM analyses is reported in [50], in which the authors carry out an extensive study of dislocation mechanisms in austenitic stainless steel due to stress corrosion cracking.

12 In the present paper, the author demonstrates how to profitably combine CM and DIC in the field of mechanics of materials. It is done with the aim of exploiting the strength of this approach, while proposing a method capable of measuring the whole 3D displacement field with an accuracy that is satisfactory for several applications. More specifically, the use of CM provides nanometric accuracy and eliminates the need to work in a controlled environment, as is required by SEM, TEM and EBSD. CM can retrieve out-of-plane information, as shown in [43], [46-49], that electron-scanning techniques are not able to obtain. In addition, such electron scanning techniques are also prone to defocusing concerns. In contrast with SPM techniques, with CM there is no required contact between a probe and the surface, which could damage or alter a weak specimen.

13 In addition to the benefits of CM, the author shows how a commercial DIC algorithm can be successfully applied to process images without the need of any assumption on the displacement field, as required by I-DIC, or by a FEM-like mesh definition, as in G-DIC. The fact that nothing should be assumed in order to retrieve all three displacement components offers the flexibility to manage any displacement field occurring on the surface of the specimen. Hence, a non-kinematically compatible deformation field can also be measured without knowing a priori any information about its spatial distribution – e.g. displacement around a crack tip. Thanks to the high sensitivity of the profilometer (about 10 nm), the spatial carrier necessary to make the DIC algorithm work – i.e. the surface roughness can show an extremely low contrast, and the subset size can be chosen small enough ($15 \div 25$ pixels) to retrieve displacement fields characterized by steep gradients. Through this procedure, it was possible to obtain an accuracy of approximately 10 nm for all three displacement components, even if a standard confocal microscope was used.

14 The proposed procedure was tested in two case studies. The first case study consists of the measurement of the displacement field around the crack tip of a Ni-alloy specimen under uniaxial tensile load. In this case, the correlation algorithm is able to identify displacements on the entire sample area, given that essentially the entire surface area is subject to elastic deformations that do not significantly alter the roughness. As such, almost every portion of the area is able to correlate after the deformations occur. In the second case, the 3D displacement field around the indentation obtained by a standard Brinell test on an AISI 1040 steel specimen was measured. This test emphasized how high plastic strains occurring on the area where the indentation is applied cause the DIC algorithm to fail, while the same algorithm worked soundly just beyond the indentation area, where the displacement gradients (for all the three components) were clearly retrieved by the correlation algorithms.

…

References

[1] J. Rasson, O. Poncelet, S.R. Mouchet, O. Deparis, L.A. Francis, Vapor sensing using a bio-inspired porous silicon photonic crystal, Mater. Today Proc. 4 (2017) 5006–5012, https://doi.org/10.1016/j.matpr.2017.04.107.

[2] Y. Han, S. Li, F. Chen, T. Zhao, Multi-scale alignment construction for strong and conductive carbon nanotube/carbon composites, Mater. Today Commun. 6 (2016) 56–68, https://doi.org/10.1016/j.mtcomm.2015.12.002.

[3] S. Brundavanam, R.K. Brundavanam, G.E.J. Poinern, D. Fawcett, Flower-like Brushite structures on mg, Mater. Today 20 (2017) 92–93, https://doi.org/10.1016/j.mattod. 2017.01.014.

[4] P. Parandoush, D. Lin, A review on additive manufacturing of polymer-fiber composites, Compos. Struct. 182 (2017) 36–53, https://doi.org/10.1016/j.compstruct.2017. 08.088.

[5] G. Dehm, B.N. Jaya, R. Raghavan, C. Kirchlechner, Overview on micro- and nanomechanical testing: new insights in interface plasticity and fracture at small length scales, Acta Mater. 142 (2018) 248–282, https://doi.org/10.1016/j.actamat.2017. 06.019.

[6] W.S.W. Harun, R.I.M. Asri, J. Alias, F.H. Zulkifli, K. Kadirgama, S.A.C. Ghani, J.H.M. Shariffuddin, A comprehensive review of hydroxyapatite-based coatings adhesion on metallic biomaterials, Ceram. Int. 44 (2018) 1250–1268, https://doi.org/10. 1016/j.ceramint.2017.10.162.

[7] C.H. Nguyễn, J.-L. Putaux, G. Santoni, S. Tfaili, S. Fourmentin, J.-B. Coty, L. Choisnard, A. Gèze, D. Wouessidjewe, G. Barratt, S. Lesieur, F.-X. Legrand, New nanoparticles obtained by co-assembly of amphiphilic cyclodextrins and nonlamellar singlechain lipids: preparation and characterization, Int. J. Pharm. 531 (2017) 444–456, https://doi.org/10.1016/j.ijpharm.2017.07.007.

[8] J. Lee, M. Morita, K. Takemura, E.Y. Park, A multi-functional gold/iron-oxide nanoparticle-CNT hybrid nanomaterial as virus DNA sensing platform, Biosens. Bioelectron. 102 (2018) 425–431, https://doi.org/10.1016/j.bios.2017.11.052.

[9] A.F. Doval, A systematic approach to TV holography, Meas. Sci. Technol. 11 (2000) R1–R36, https://doi.org /10. 1088/0957-0233/11/1/201.

[10] W. Osten, A. Faridian, P. Gao, K. Körner, D. Naik, G. Pedrini, A.K. Singh, M. Takeda, M. Wilke, Recent advances in digital holography [invited], Appl. Opt. 53 (2014) G44–G63, https://doi.org/10.1364/AO. 53.000G44.

[11] M.A. Sutton, J.-J. Orteu, H.W. Schreier, Image Correlation for Shape, Motion and Deformation Measurements, Springer-Verlag, New York, 2009https://doi.org/10.1007/ 978-0-387-78747-3.

[12] G. Binnig, H. Rohrer, Scanning tunneling microscopy, Surf. Sci. 126 (1983) 236–244, https://doi.org/10. 1016/0039-6028(83)90716-1.

[13] G. Binnig, C.F. Quate, C. Gerber, Atomic force microscope, Phys. Rev. Lett. 56 (1986) 930–933, https://doi.org/10.1103/PhysRevLett.56.930.

[14] L. Deck, P. De Groot, High-speed noncontact profiler based on scanning white-light interferometry, Appl. Opt. 33 (1994) 7334–7338, https://doi.org/10.1364/AO.33. 007334.

[15] B.S. Lee, T.C. Strand, Profilometry with a coherence scanning microscope, Appl. Opt. 29 (1990) 3784–3788, https://doi.org/10.1364/AO.29.003784.

…

Task 15: The following statements excerpted from research article introductions specifically function to highlight research gap(s) underexplored in previous literature, reflecting the problem-oriented nature of scientific research with an aim for viable solutions. Discuss in groups the use of negation, whether realized grammatically by "not", or achieved by a determiner like "less" or an adverb like "insufficiently", or constructed by the words/phrases with negative meanings like "the lack of", "has long been neglected".

(1) Previous work has focused on the size, shape, charge or compositions of the antigen carriers[3-5]. However, the fluidity of endocytosis targets has long been neglected in vaccine design[6].

(2) However, because the readouts from each nanopore must be addressed separately, it is difficult to see how this significant increase in scale might be achieved without sacrifices in device complexity, size and cost.

(3) However, a layer containing different light-emitting components has a color imbalance problem due to the different energy transfer rates among the dopants.

(4) However, the water harvesting and transport capability, especially the transport velocity, is severely limited to ~21 $\mu m\, s^{-1}$ for bio-inspired spider silk and ~3 $\mu m\, s^{-1}$ for bio-inspired cactus spine.

(5) Yet, synthetic schemes to control SNT structures are lacking due to complex assembly mechanisms that are only partially understood.[5], [14-15]

(6) Due to their specific anisotropic mechanical behavior as well as the high surface roughness, which is dependent on the orientation during the building process [8] and the inner structures like porosity [2], the transferability is questionable.

(7) With all the information available from prior literature, it is possible to get a fair idea of how bearing strength is influenced by the different parameters. However, the progression of failure is not clearly understood.

(8) The addition of the carbon-carbon triple bonds to GDY (the resulting product of graphene) may change physicochemical and biological properties of graphene to a certain degree. However, the speculation has not been proven yet.

(9) Coprecipitation and impregnation methods are the main stream in synthesizing single-atom catalysts. However, these traditional methods make it difficult to precisely control the doping level and anchoring sites. The atomic layer deposition (ALD) method and physical vapor deposition (PVD) were also used to deposit isolated single Pt atoms on supporting material through vapor transportation. Yet the expensive equipment and high vacuum environment are hurdles to its mass production. In both cases, the single-atom precursor needs to be controlled at extremely low concentrations.

(10) In previous work, we have certified a high P_m and slim P-E loops in $0.775Na_{0.5}Bi_{0.5}TiO_3$-$0.225BaSnO_3$ ceramics. However, the electrical breakdown strength (E_b) and η are relatively low.

(11) While most work is focused on achieving higher PCEs, the fundamental understanding of the physics of PSCs is still limited. Hysteresis has been one major issue that has impeded progress on the achievement of higher PCEs under steady state conditions. While significant hysteresis has been regularly observed in the current-voltage (I-V) characteristics of PSCs, the origin of this behavior has not been definitively explained.

(12) The first signatures of superconductivity in DSMs have been reported (for example, by applying pressure[15] or by using point contacts[16-17], but topological aspects of DSM superconductivity have not been studied.

(13) Although, these materials could generate field-induced strains, yet the field-induced strain was also relatively small (less than 1.0%) [9-14]. Furthermore, the magnetic elastomers showed large remanent magnetostriction and poor stability. These problems have hampered their potential application in "artificial muscles" and as sensors and actuators in robotics.

(14) Extensive theoretical work, particularly that using first-principles density functional theory (DFT) calculations, has shown progress in predicting the spectral distribution, but is limited by computational costs. Experimentally, inelastic neutron scattering is a classical technique that provides a powerful tool to study phonon spectra and lifetimes, but this approach is limited to the characterization of single-crystal materials and requires facilities that are not widely available.

(15) Generally, these structural changes on the molecular level accumulate and ultimately lead to the deterioration of macroscopic physical and mechanical properties of polymers. The effects on the friction and wear behavior have, however, rarely been reported.

(16) In that context, a lack of reproducibility that may appear at first glance to be due to an unreliable synthesis or misfortune should rather suggest that key parameters have still not been identified.

(17) Several studies using this approach for semi-solid alloys have been published. Sharifi and Larouche (2015) obtained the instantaneous Young's moduli of Al-Cu alloy during tensile analyses. They did not mention the inelastic properties. Phillion et al. (2008) and Sistaninia et al. (2012, 2013) compared the numerical and experimental stress-strain curves of aluminum alloys. Although the stress-strain curves were compared, derivation of the properties and their validation were not obtained.

(18) Meanwhile, to what extent the ionic current can respond to the actual in-pore ion exclusion events together with contributions of nanopore materials therein has remained to be a fundamental yet crucial question for realizing the single-particle tomography via resistive pulse analyses. In fact, unlike bionanopores formed in a low-permeability lipid membrane,[20] the ultrathin solid dielectrics inevitably bring large capacitance that would degrade the temporal resolution of the ionic current measurements through posing longer charging/discharging time compared to the time-scale of the actual translocation events occurring in the nanosystem.[21]

(19) Many researches [8-15] referred to the strengthening below 800 °C by carbon solid solution and microstructure refinement in conventional TiAl alloys. However, the various solidify path and the microstructure in high niobium containing TiAl alloys would cause different strengthening or toughening mechanism above 800 °C. In addition, carbon solid solution is less efficient for strengthening the alloys than different types of carbides [16]. Tian [17] et al. reported that the fine carbides could strengthen the TiAl alloys while the Ti_3AlC(P-type) and Ti_2AlC(H-type) contributed to the strength. There are also less references about the carbides effect on further property improvement of the novel high niobium containing TiAl alloys.

(20) However, carbon nanotube normally shows a poor dispersion in elastomer matrix[7-8]. Thus, the challenge is to reduce the percolation threshold concentration of CB in conductive composite.

(21) Among the numerous coordination structures, Zn^{2+} (or Co^{2+}) ions can coordinate with organic imidazoles in tetrahedron units, making it feasible to form 2D silicate-like structures.[5], [15] However, a simple method to introduce functional groups to the structure without changing the 2D geometry is still not available.

(22) To the best of our knowledge, the luminescence properties of Ce^{3+}-, Eu^{2+}- and their co-doped $Ca_{10}M(PO_4)_7$ (M = Li, Na, K) for w-LEDs applications have been reported, but the site occupancy luminescence properties have not been investigated in detail so far.

(23) To date, several studies have been conducted that focused on manipulating peptide self-assembly into the desired nanoarchitectures[9-10]. Nonetheless, it remains unclear whether fluorescent peptide nanoparticles can be synthesized through self-assembly to change the intrinsic optical properties of peptide nanostructures.

(24) However, it has remained to be challenging to fabricate well-defined nanostructures with large density of catalytic active sites.

(25) Regarding intrinsic charge states, Coulomb blockade peaks were reported in ruthenium (Ru)-containing wires[18], but not confirmed in self-assembled monolayers[19].

(26) To the best of our knowledge, there has been only one successful attempt, i.e., free-standing calcite films were prepared with organic content less than 4%.[14] It should be noted that, however, such films are still easy to crack once exceeding a bending angle of about 50° and a tedious six-step preparation procedure is required.

(27) From these reports, the difference in the photovoltaic characteristics seems to be associated with the nature of A cations. However, there is a scarcity of reports in the literature discussing the role of A cations on the intrinsic and interfacial dynamics of PSCs.

(28) Most studies have focused on the effect of doping edge sites, with only a few reports on basal plane doping. These reports focused on the effect of dopant chemical composition. Little is known about the effect of doping concentration on HER activity due to the challenges associated with achieving controlled, variable dopant concentrations.

(29) However, a layer containing different light-emitting components has a color imbalance problem due to the different energy transfer rates among the dopants.

(30) However, how to align these individually nucleated domains precisely, and more importantly, form a continuous monolayer film at wafer scale by domain–domain stitching is a main hurdle to overcome.

(31) In summary, none of the researchers both derived numerical viscoelastic properties and experimentally validated the properties.

(32) As assessed by Sadayappan et al. (2002), the use of common chemical refiners exhibits a fading effect over time, resulting in a loss of its effectiveness, which is particularly important in situations where the time interval between the refinement operation and pouring is long. Moreover, chemical refinement usually leads to a significant amount of dross, which entails an important environmental impact, and the process is difficult to control.

(33) Although the ultrasonic treatment is well established in the treatment of the Al and Mg alloy melts, as demonstrated by Puga et al. (2011) and Eskin and Eskin (2017), there are no studies of this application in the treatment of liquid brass (Cu-Zn alloys).

(34) Various nanostructures of molybdenum carbides, including nanowires, nanosheets, nanospheres, and nanoparticles, have been explored to improve the electrocatalytic performance.[22-26] However, it has remained to be challenging to fabricate well-defined nanostructures with large density of catalytic active sites.

(35) The periodic bricks-and-mortar structure has been mimicked frequently to acquire outstanding mechanical properties, especially for high tensile strength, Young's modulus, and work to fracture.[7-10] Nevertheless, high fragility and poor flexibility always exist, making these biomineral materials far from real applications, and up to now, very limited flexible free-standing $CaCO_3$ mineral films have been prepared successfully.

(36) Recently, as triggered by the discovery of topological semimetals[23-24], the negative MR in nonmagnetic systems has attracted renewed interest. An intriguing phenomenon is the negative longitudinal MR observed in TIs when a magnetic field is applied along the current direction.[25-28] However, TPT-induced triaxial vector MR has never been reported.

Task 16: Work in groups to identify the typical sentence patterns and verb-noun collocations used for announcing research purposes/objectives, as exemplified by the first two sentences below.

(1) <u>Here we provide</u> the experimental <u>quantification</u> of interlayer interactions in the archetypal vdW material TiS_2 through modelling of structure factors extracted from accurate single-crystal synchrotron X-ray diffraction measurements.

(2) <u>Here we demonstrate</u> the <u>potential</u> of an approach based on pyroelectric materials.

(3) Here we show a new ultrafast water harvesting and transport mechanism found on the trichome surface of Sarracenia operculum.

(4) Inspired by this mechanism, we posited that exploitation of the pliability (mechanosensing deformation) and mobility of antigen delivery systems may confer an effective strategy to elicit robust prophylactic and therapeutic immune responses.

(5) Here we report a semifloating gate (SFG)-controlled field-effect transistor (FET) made of graphene, hexagonal boron nitride (h-BN) and WSe_2. We demonstrate that high-performance non-volatile programmable p–n junctions can be achieved and modulated in the 2D WSe_2 by applying voltage pulses to the control gate.

(6) Starting from commercially available silicon-based polymers, the luminescence properties are developed by means of a simple photoreaction in this study.

(7) Herein, we report a SNT self-assembly from symmetric bolaamphiphiles based on the same hydrophobic core, where precise structural control is enabled by a combination of specific molecular interactions and solvation.

(8) In the present study, we have successfully introduced the *in-situ* ductile Ta-rich particles into a $Zr_{55}Cu_{30}Ni_5Al_{10}$ (at.%) BMG by minor alloying with 7, 9 and 10 at.% Ta.

(9) In this study, in order to enhance the two dominant failure modes (kink and delamination), while keeping the through thickness load due to bolt constant, distance between bolt hole and longitudinal edge was changed.

(10) In this work, we report the growth of highly oriented continuous ML-MoS_2 films on 2 in. sapphire wafers with only two mirror-symmetric domain orientations present.

(11) Here, we present a completely new concept of on-demand reconfigurable self-propelled supercapacitor based on WS_2 nanoparticles-polyaniline/platinum microrobot design.

(12) In the present study, we did head-to-head comparisons between graphene oxide (GO) and GDY oxide (GDYO) in their physicochemical, biological, and mutagenesis properties with the hope that the well-controlled comparisons could provide clear information about the true advantages and disadvantages of the two unique carbon allotropes, define characters of alkenes and alkynes, and guide the future applications of the two carbon allotropes.

(13) In this article, an electroplating method was developed to deposit monodispersed noble metal atoms on a variety of 2D materials. The single-atomic dispersion can be precisely modulated with parameters, such as anode voltage and deposition time.

(14) In present work, $(1-x)(0.775Na_{0.5}Bi_{0.5}TiO_3\text{-}0.225BaSnO_3)\text{-}x\ BaZrO_3$ ceramics are prepared to further enhance its energy-storage performance.

(15) In this article, we present a new way of preparation of raspberry-like poly(styrene-co-2-vinylpyridine)/silica nanocomposite particles using an alcoholic dispersion polymerization which was chosen because of its convenience compared to the other polymerization systems available.

(16) Here we use photoluminescence (PL) measurements to investigate one aspect of the hysteresis phenomenon, namely, the transient variation in open-circuit voltage following a change in operating conditions.

(17) In this article, we experimentally demonstrate the implementation of a sparse coding algorithm in a memristor crossbar, and show that this network can be used to perform applications such as natural image analysis using learned dictionaries.

(18) Here, we report on the realization of proximity-induced superconductivity into an accidental DSM, and we reveal a significant contribution of 4π-periodic Andreev bound states to the supercurrent.

(19) In this work, we systematically investigate conformations of individual β-carotene and chlorophyll-a using molecular manipulation schemes with a low-temperature scanning tunneling microscope (STM), which enables the identification of their intimate structures in the molecular clusters.

(20) Here, a new class magnetic rubber foam with cellular structure was developed and prepared by one-step solution foam processing, in which the curing and the chemical foaming process were carried out simultaneously.

(21) In this study, without any *ex situ* lithography processes, we achieve self-organized aligned multifaceted GaN stripes by control of the initial growth conditions for semipolar GaN on m-sapphire substrates.

(22) In this article we demonstrate a desk-top approach to experimentally determine and construct the full spectral distribution of the thermal conductivity of a variety of materials by utilizing observations of quasi-ballistic thermal transport.

(23) In this paper, the effects of weathering exposure on unfilled and filled TPU materials are described as performed under different humidity conditions. The effects on friction are presented along with surface analyses to characterize the chemistry of the degradative process.

(24) In the present study, sheets of 0.8 mm thick X626 aluminum were welded to 0.9 mm thick low carbon steel (LCS) sheets using GM's patented welding process for welding aluminum alloys to steels.

(25) In this study, we prepared micropatterns with various sizes using UV photolithography. The nanomechanical properties of the major osteosarcoma microenvironment cells (NHOst, MSCs, and MG-63) were compared on micropatterned surfaces.

(26) Here, we report that a simple and previously overlooked experimental parameter, the delay between the mixing of reactants and the hydrolysis, is of tremendous importance in the preparation of metal oxide nanoparticles using this versatile organometallic approach.

(27) In the present study, therefore, we systematically studied the influence of cross-membrane resistance and capacitance on the resistive pulse waveforms by engineering the material and structure of nanopore membranes.

(28) This manuscript investigates the influence of substrate stiffness (modulus and thickness) on the mechanical performance of serpentine interconnects through combined finite element analyses (FEA), analytical modeling, and experiments.

(29) Therefore, the effect of carbon addition on high-temperature tensile properties and microstructure of Ti-43Al-6Nb-1Mo-1Cr alloys are studied in this paper.

(30) The main aim of this study was to prepare conductive CB filled ENR-50 composite with low percolation threshold concentration.

(31) In this paper, we investigate the correlation between crystalline, structural, surface morphology properties and optical parameter spectra of Bi_2Se_3 films prepared by thermal evaporation technique.

(32) Based on this process, we demonstrate here the growth of InP crystals atop group IV substrates.

(33) Here, we combine the photostability of QDs with the molecular programmability of the DNA polyhedra by encapsulating QDs inside DNA icosahedra that display a solitary biomolecular tag.

(34) Here, we demonstrate a simple, room-temperature, and scalable synthesis of free-standing and nanometer-thin zinc benzimidazolate coordination polymer (ZBCP) layers on water surfaces, where the desired hydroxyl (−OH) groups with charge functionality are stoichiometrically coordinated into the 2D structures, characterized as $Zn_2(Bim)_2(OH)_2$.

(35) In this paper, the luminescence properties of Ce^{3+}-doped $Ca_{10}Li(PO_4)_7$ (CLP), $Ca_{10}Na(PO_4)_7$ (CNP), $Ca_{10}K(PO_4)_7$ (CKP) phosphors under vacuum ultraviolet-ultraviolet (VUV-UV) excitation are reported. The assignment of site occupancies and thermal quenching of Ce^{3+} at two sites have been systematically studied.

(36) Inspired by organized assemblies in nature, we demonstrate the formation of tryptophan-phenylalanine (Trp-Phe) nanoparticles that can shift the peptide's intrinsic fluorescence signal from the ultraviolet to the visible range.

(37) This work describes the development and evaluation of the ultrasonic refinement technique applied to brass alloys, capable of significantly improving their mechanical properties when compared to those obtained through traditional chemical refinement techniques.

(38) Herein, we present an approach to synthesize Mo_2C porous nanostructures in controlled ways, including 2D nanosheets and 1D nanowires. The approach is through carburizing cobalt or zinc-based zeolite-type metal-organic framework (MOF) (ZIF-67 or ZIF-8) cladding MoO_3 nanosheets or nanowires under high temperature. The resulting nanostructures are periodically porous, largely preserving the precursor morphology. The porous Mo_2C materials may provide an exceptionally large number of active sites.

(39) Here, by intentional indium doping, we have driven the TPT in the nonmagnetic $(Bi_{1-x}In_x)_2Se_3$ series, as revealed by ARPES.

(40) In this work, we propose a fast and simple method to construct soft hydrogel bridges for the connection of $CaCO_3$ particles together.

(41) In this work, we analyzed the PSCs based on $FAPbBr_3$ and $CsPbBr_3$ to understand the recombination losses, capacitive characteristics, and charge transport and extraction within the absorber layer and across various interfaces.

(42) Herein, we overcome this challenge by first synthesizing doped transition metal oxides and then sulfurizing the oxides to sulfides.

(43) In the present study, an electron irradiation experiment of several sintered UO_2 disks was performed to attribute the Raman triplet defect bands to damages created either in the electronic or nuclear-stopping power regimes.

(44) Here we report the first measurements of electronic structure and comprehensive magnetic investigations carried out for this compound.

(45) Here, we systematically investigate the physical exfoliation of eight prototypical 2D materials with diverse molecular bonding forces at the same experimental conditions.

Task 17: Transform the following high-frequency verb-noun collocations used in research article introductions into passive voice. Determine what verb-noun collocations are typically used for each of the main communicative purposes, namely 1) claiming the importance of a research area while emphasizing central issue(s), 2) reviewing previous research clearly relevant to the present one, 3) pointing out research gap unaddressed by previous literature, 4) announcing research purposes of the present study, and 5) elaborating on details of the present study (analytical procedures, paper structure, etc.)

No.	(Phrasal) Verbs	Noun collocates	Examples
(1)	play	role part function	play a(n) important/crucial/key/significant role play multiple/critical/ functions play a crucial part
(2)	attract	attention interest effort	attract much/considerable/ significant/great attention attract great/considerable/significant interest attract tremendous/substantial research interest
(3)	use	method technique model material approach process microscopy technology system analysis strategy	use the finite element method use complex fabrication techniques use independent thermal and deformation models use various filler materials use wet chemistry synthesis approaches use proper process, use this new process use scanning/transmission electron microscopy use these three technologies use conventional laser-sintering systems use systematic experimental and numerical analyses use photochemical or photophysical strategies

continued

No.	(Phrasal) Verbs	Noun collocates	Examples
(4)	improve	property performance stability efficiency resistance strength conductivity activity quality toughness efficacy	improve the mechanical properties improve the electrochemical performance improve the thermal stability improve the power conversion efficiency improve the tracking and erosion resistance improve compressive strength improve the thermal/electrical conductivity improve the catalytic/electrochemical conductivity improve the surface/machining/film quality improve the fracture/impact/translaminar toughness improve the therapeutic/radiation efficacy
(5)	have	effect potential advantage property structure impact conductivity limitation	have no/little/a significant effect have great/high/significant potential(s) have many/numerous/distinct/unique advantage(s) have unique physical and chemical properties have a porous/layered/hexagonal structure have a strong/tremendous/significant impact have relatively low electrical conductivity have some/many/inherent/certain limitations
(6)	investigate	effect property influence behavior mechanism performance relationship	investigate the (microstructural/coupling) effect(s) investigate the mechanical/tensile properties investigate the influence investigate the compression/welding behavior investigate the formation mechanism investigate the electrochemical performance investigate the structure-property relationships
(7)	limit	application use performance efficiency study	limit their applications limit its practical use limit the overall performance limit their power conversion efficiency limit the study
(8)	receive	attention interest	receive considerable/much/increasing attention receive great/considerable/significant interest(s)

continued

No.	(Phrasal) Verbs	Noun collocates	Examples
(9)	study	effect property behavior influence mechanism interaction characteristic	study the synergistic effect study the physical and chemical properties study the mechanical behavior study structure influence study reaction mechanism study the interfacial interactions study the electrical characteristics
(10)	exhibit	activity behavior capability capacity conductivity effect efficiency performance property resistance response stability strength structure	exhibit much higher photocatalytic activity exhibit stress - stiffening behavior exhibit poor cycling capability exhibit a reversible capacity exhibit extremely low diffusive thermal conductivity exhibit nonlinear magnetoelectric effects exhibit a high luminous efficiency exhibit excellent rate performance exhibit robust mechanical properties exhibit an excellent sheet resistance exhibit fast and stable responses exhibit excellent long-term stability exhibit excellent tensile strength exhibit 3D nanoscale pore structure
(11)	remain	challenge issue problem task	remain a great/significant/big/major challenge remain a critical/unsolved/important/key issue remain a major/prominent problem remain a challenging/significant/difficult task
(12)	develop	approach method model device material system strategy technique process	develop a novel approach develop a software-based drift correction method develop data-driven models develop high-efficiency photovoltaic devices develop novel electrode materials develop a smart control system develop a simple, rapid, and robust strategy develop novel assemble techniques develop an innovative molding process

continued

No.	(Phrasal) Verbs	Noun collocates	Examples
(13)	make	effort progress attempt material process advance contribution	make significant/systematic/great efforts make great/remarkable/crucial/ great progress make a(n) valuable/useful attempt make the carbon materials make the powder coating process make important/significant advances make a great/significant/valuable contribution
(14)	enhance	activity conductivity efficiency performance property stability	enhance the water splitting activity enhance their electrical conductivity enhance charge separation efficiency enhance electrochemical energy storage performance enhance the electrical transport properties enhance the operational stability
(15)	exhibit	activity behavior capability capacity conductivity effect efficiency performance property resistance response stability strength structure	exhibit much higher photocatalytic activity exhibit stress-stiffening behavior exhibit excellent Li+ storage capability exhibit a reversible capacity exhibit extremely low diffusive thermal conductivity exhibit nonlinear magnetoelectric effects exhibit a high luminous efficiency exhibit excellent rate performance exhibit robust mechanical properties exhibit an excellent sheet resistance exhibit fast and stable responses exhibit excellent long-term stability exhibit excellent tensile strength exhibit 3D nanoscale pore structure
(16)	develop	approach process method model device material strategy	develop instrumental imaging approaches develop a low-pressure injection molding process develop a software-based drift correction method develop a general multiscale model develop high-efficiency photovoltaic devices develop novel electrode materials develop a simple, rapid, and robust strategy

continued

No.	(Phrasal) Verbs	Noun collocates	Examples
(17)	address	issue challenge problem limitation question	address the above mentioned issues address these challenges address problem address these performance-related problems address these critical questions
(18)	solve	problem issue equation	solve these multifaceted problems solve the wearable-electronics power-supply issue solve two-temperature governing equations
(19)	overcome	limitation problem drawback challenge issue disadvantage shortcoming obstacle	overcome the aforementioned limitations overcome this brittleness problem overcome the noted drawbacks overcome these technical challenges overcome the safety issues overcome their intrinsic disadvantages overcome the potential shortcomings overcome the multiple obstacles
(20)	meet	requirement demand need criterion	meet the necessary mechanical requirements meet the ever-increasing demand(s) meet the increasing energy need(s) meet the practica-economic criteria
(21)	show	activity behavior capacity conductivity decrease effect efficiency improvement increase performance potential property resistance stability trend	show remarkable high OER catalytic activity show strong rectifying behavior show much higher reversible capacity show a highest thermal conductivity show a resistivity decrease show a significant radio-sensitizing effect show a high efficiency show a significant improvement show a linear increase show unprecedented electrochemical performance show realistic potential show inferior photocatalytic properties show higher erosion resistance show enhanced operational stability show similar qualitative trends

continued

No.	(Phrasal) Verbs	Noun collocates	Examples
(22)	propose	method model process approach concept design solution mechanism strategy	propose the MD simulation method propose the nonlinear viscoelastic model propose an abrasive-free polishing process propose a chemical and structural approach propose an alternative concept propose a microstructure design propose an effective solution propose a reaction mechanism propose an alternating photothermal strategy
(23)	report	approach method strategy synthesis	report a template-free approach report a general post-annealing method report a tailored microwave-aided synthetic strategy report a new large-scale colloidal synthesis
(24)	conduct	study experiment test investigation analysis research simulation	conduct a longitudinal empirical study conduct multiple experiments conduct tensile and impact tests conduct a thorough investigation conduct a comprehensive sensitivity analysis conduct thorough research conduct a thermodynamic simulation
(25)	demonstrate	application approach feasibility performance potential strategy	demonstrate their potential applications demonstrate structural interface engineering approach demonstrate the feasibility demonstrate excellent antistatic performance demonstrate the great potential demonstrate a synchronous reduction strategy
(26)	hinder	application development use formation performance	hinder their practical applications hinder the future development hinder the high temperature use hinder the formation hinder the catalytic performance

Task 18: Arrange the sentences in each set logically and coherently into a research article introduction.

I

Phosphor-free white emission from InGaN quantum wells grown on in situ formed submicron-scale multifaceted GaN stripes[7]

(1) In addition, strong piezoelectric polarization fields (PFs) built in high indium content InGaN grown on polar GaN lead to a significant reduction in carrier recombination rate.

(2) Conventional semiconductor solid-state light (SSL) sources are obtained by combining ultraviolet/purple/blue LEDs with down-conversion phosphors but of unavoidable Stokes shift energy losses and low phosphor reliability.

(3) Single-chip phosphor-free SSL sources have been developed to overcome these problems. Multicolor emission is obtained either by growth of cascade or lateral InGaN/GaN quantum wells (QWs) with different indium contents in various InGaN QWs, or by growth of InGaN active layers on GaN islands, nanorods, or nanowires.

(4) In this study, without any *ex situ* lithography processes, we achieve self-organized aligned multifaceted GaN stripes by control of the initial growth conditions for semipolar GaN on m-sapphire substrates.

(5) With increase of indium content in InGaN for long-wavelength emission, the InGaN alloy phenomena become severe, leading to poor crystalline quality and an increase of nonradiative recombination rate.

(6) Cool-daylight white emission is observed for the InGaN/GaN QWs on the GaN stripe templates.

(7) The SSL sources may also be realized by a red-green-blue (RGB) approach via color mixing of multiple primary color LEDs.

(8) However, the growth of high-quality high indium content InGaN for high-performance long-wavelength LEDs is still a big challenge.

(9) These stripes have various smooth sidewall facets of submicron-scale height and width, which are ideal growth templates for multi-color-emitting InGaN/GaN QWs.

(10) InGaN-based light emitting diodes (LEDs) have attracted intense research interest due to their long lifetime, energy savings and environmental friendliness.

(11) The RGB approach is favorable for applications in smart lighting, high color rendering index display, and high-speed light-fidelity communication.

(12) Multicolor emission is obtained either by growth of cascade or lateral InGaN/GaN quantum wells (QWs) with different indium contents in various InGaN QWs, or by growth of InGaN active layers on GaN islands, nanorods, or nanowires.

Sequence of sentences: ______________________________

II

On the Origin of Hysteresis in Perovskite Solar Cells[8]

(1) PSCs also show great potential for tandem solar cells in combination with silicon, because of a suitable band gap.

(2) By measuring the PL emission from the active perovskite film of the cell and concurrently measuring the open circuit voltage of the same cell, the observed relationship between the two variables can be investigated and compared to behavior predicted by various models.

(3) While significant hysteresis has been regularly observed in the current–voltage (*I–V*) characteristics of PSCs, the origin of this behavior has not been definitively explained.

(4) The power conversion efficiency (PCE) of these devices has increased rapidly, and PCEs exceeding 20% have been reported.

(5) Hysteresis has been one major issue that has impeded progress on the achievement of higher PCEs under steady state conditions.

(6) While most work is focused on achieving higher PCEs, the fundamental understanding of the physics of PSCs is still limited.

(7) Here we use photoluminescence (PL) measurements to investigate one aspect of the hysteresis phenomenon, namely, the transient variation in open-circuit voltage following a change in operating conditions.

(8) On the basis of the results we suggest a plausible mechanism, whereby the voltage hysteresis is caused by a time-varying potential drop across at least one of the perovskite/transport layer interfaces (perovskite/TiO_2 or perovskite/spiro).

(9) Organic–inorganic hybrid perovskite solar cells (PSCs) have attracted considerable attention in the last several years due to their potential for low cost generation of electricity.

(10) Hysteresis has been linked to a variety of processes such as the ferroelectric effect, a combination of ionic displacement and carrier trapping at the interface between the perovskite and the electron transport layer (ETL) or hole transport layer (HTL), and charge accumulation at interfaces of the PSC.

Sequence of sentences: ______________________________

III

Influence of hole eccentricity on failure progression in a double shear bolted joint (DSBJ)[9]

(1) As with general application of laminated composites, bolt bearing strength is influenced by ply stack-up, hole size, ratio of hole diameter to width.

(2) With all the information available from prior literature, it is possible to get a fair idea of how bearing strength is influenced by the different parameters.

(3) However, wherever there is a joint, bolted or bonded, it will most likely be the site where failure initiates under operational conditions, unless damage is imparted by foreign objects, such as impact.
(4) In this study, in order to enhance the two dominant failure modes (kink and delamination), while keeping the through thickness load due to bolt constant, distance between bolt hole and longitudinal edge was changed.
(5) Interrupted test were conducted for different configurations to study the progression of failure through the complete loading process.
(6) Unlike bonding, bolted joints provide ease in maintainability, ease of connecting dissimilar material and less prone to failure due to manufacturing induced defects.
(7) However, the progression of failure is not clearly understood.
(8) The primary joining mechanism for most applications in the aerospace industry are bolted joints.
(9) Bolts require a hole or cut-out which act as a stress raiser within a structure. (10) Because of this, the need to size the bolted joint becomes of utmost concern in a structural design.
(10) In addition, bearing strength of bolted joints are significantly influenced by clamp up load, washer size, bolt configuration.
(11) Considerable literature is available which have identified the parameters that influence failure of bolted joints when the parts to be joined are fiber reinforced laminates.
(12) The sequence of failure is of importance when numerical models are being developed.

Sequence of sentences: ______________________________

Ⅳ

Laser-Assisted Chemical Modification of Monolayer Transition Metal Dichalcogenides[10]

(1) In particular, for layered and 2D materials, laser thinning of layered MoS_2, WS_2, and WSe_2 have proven to be an efficient way to produce on-demand monolayer films.
(2) Additionally, laser-induced site-specific doping of ultrathin MoS_2 and WSe_2 in a phosphine environment has been demonstrated, facilitating the localized modification of the intrinsic photoluminescence as well as the electrical response in a 2D device configuration.
(3) Here, we report the successful laser-induced chemical modification of suspended TMD monolayer films via local exchange of the chalcogen atoms: selenides to sulfides.
(4) The time constants associated with the different photochemical mechanisms involved in the conversion process were studied by in situ monitoring the Raman and photoluminescence spectra of the samples.
(5) Our results suggest that postgrowth laser-induced chemical modifications could be considered as an alternative route for the fabrication of spatially localized ternary alloys and in-plane 2D heterostructures in a controlled gas environment.

(6) Laser-assisted modifications have emerged as an alternative and reliable technique for local tailoring of the chemical, structural, optical, and electrical properties in a variety of nanomaterials.

(7) The success of the above-mentioned experiments suggests that postgrowth laser-assisted methods can become a reliable route to tailor the physicochemical properties of 2D materials.

(8) Local oxidation using a focused laser beam was also demonstrated to be effective in modifying the photoresponse of phosphorene-based 2D devices.

(9) With the proposed method, total or partial replacement of the chalcogen atoms was achieved, in both WSe_2 and $MoSe_2$ suspended films.

(10) Additionally, a laser patterning technique has been employed to fabricate micro-supercapacitors using paintable graphene7 and MoS_2 film.

(11) However, other than inducing evaporation, oxidation or doping, the potential of postgrowth laser-assisted method is yet to be verified for in situ changing of the chemical composition of transition-metal dichalcogenides (TMDs), such as creating localized ternary alloys, or even completely replacing the chalcogen atoms.

(12) Enhancement of the electrical conductivity of WSe_2 monolayer was observed during the laser-assisted oxygen passivation of chalcogen vacancies.

Sequence of sentences: ______________________________

V

Defect Creation in HKUST-1 via Molecular Imprinting: Attaining Anionic Framework Property and Mesoporosity for Cation Exchange Applications[11]

(1) For example, the physicochemical properties of HKUST-1 can be tuned by replacing the aqua ligands and by doping with different organic linkers or metal ions.

(2) For example, anionic MOFs, which consist of negative-charged frameworks and charge-balancing species grafted within the channels, can undergo convenient incorporation of extraneous cationic species via ion exchange.

(3) Among them, HKUST-1 ($[Cu_3(BTC)_2(H_2O)_3]_n$, BTC = benzene-1,3,5-tricarboxylate) has been widely studied and has been commercialized (e.g., Basolite C300 by BASF).

(4) For the first time, herein we demonstrate a facile approach to prepare HKUST-1 with an anionic framework and a mesoporous hierarchical structure.

(5) First, anionic HKUST-1 with hierarchical ringlike structure (HKUST-1-R) was synthesized through mixing of Cu^{2+}, hexadecyltrimethylammonium bromide (CTAB) and BTC^{3-} organic linker deprotonated by triethylamine (TEA).

(6) Modifications can be made to MOFs to enhance their performance for specific uses.

(7) More importantly, HKUST-1-R is anionic, thus a simple postsynthetic cation exchange in ethanolic solution allows us to obtain mesoporous metal-doped analogues (*M*/HKUST-1-R, where M = Ca, Cd, Ce, Co, Li, Mn, Na, Ni, or Zn).

(8) On the other hand, MOFs with charged frameworks have also gained significant attention due to the ease of functionalization.

(9) Over the past decades, they have received massive research attention, because of their superior porosity and chemical versatility, compared to conventional porous solids.

(10) Due to its high surface area and unsaturated copper sites, HKUST-1 has shown good potential for industrial applications, such as gas storage, separation, and catalysis.

(11) Much effort has also been devoted to creating MOFs with hierarchical structures and mesopores to improve mass transfer within the intrinsic microporous framework, though successful attempts are scarce.

(12) Metal–organic frameworks (MOFs), also known as porous coordination polymers, are a new class of hybrid materials formed through coordinative bonds of metal ions and molecular linkers.

(13) The ion exchange is an effective way to tailor MOFs for various applications that include removal of toxic ionic compounds, drug delivery, ion chromatography, gas adsorption, and catalysis.

(14) This synthesis process took place rapidly (30 min) in an aqueous medium at room temperature.

Sequence of sentences: ______________________________

Task 19: Choose the most appropriate statement to fill in each blank in the following text.

Sustainable Recycling Technology for Li-Ion Batteries and Beyond: Challenges and Future Prospects[12]

Introduction

1 Rapid development of industry and increasing wealth of modern society highlight the great importance of developing a sustainable and environmentally friendly economy.(1) The world fossil fuel consumption in 2018 was 11,743.6 million tons of oil equivalent (Mtoe), accounting for 84.7% of the world's energy consumption.(2) Furthermore, fossil fuel combustion is expected to dominate the world's energy consumption for the near future (Figure 1a). It is projected that total world energy demand will increase to 17,651 Mtoe by 2040 under the IEA's New Policies Scenario (NPS), of which renewable energy will only account for 20% (Figure 1b).(3) However, the use of fossil fuels worldwide has released large amounts of carbon dioxide and other greenhouse gases (GHG), which contribute to global warming. In 2018, global CO_2 emissions from fuel combustion were 33.1 Gt CO_2.(4) To

address the energy and environment issues associated with nonrenewable fossil fuel combustion, a new approach to energy and sustainable development, namely the sustainable development scenario (SDS), has been proposed to reduce energy-related CO_2 emissions and develop renewable energy sources (Figure 1c and 1d). In response, carbon-neutral electricity from renewable energy sources and electrification of the transportation sector are considered to be the most promising solutions. __________ (1) __________.

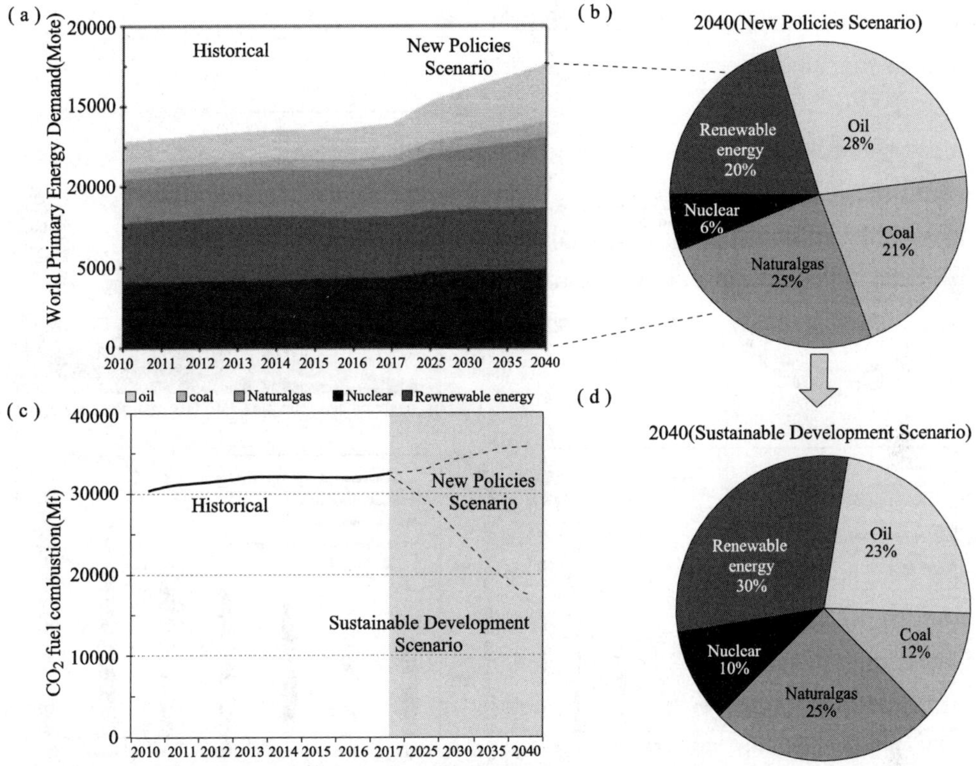

Figure 1 (a) World primary energy demand from 2010 to 2040; percent share of primary energy by source in 2040 under the NPS (b) and the SDS (c); (d) global CO_2 emissions from 2010 to 2040. Adapted with permission from ref (3). Copyright 2019 International Energy Agency. (NPS: Incorporates existing energy policies as well as an assessment of the results likely to stem from the implementation of announced policy intentions; SDS: an integrated approach to achieving internationally agreed upon objectives on climate change, air quality, and universal access to modern energy.)

2 Lithium (Li)-ion batteries (LIBs) show great promise for applications to electrical transportation and grid storage owing to their high energy efficiency, high power density, and environmental friendliness. Among electrochemical energy storage technologies, the cumulative installed capacity of LIBs in 2018 accounted for the largest proportion, exceeding 86% (Figure 2a).[5] In 2018, the global electric vehicle stock and newly registered electric vehicles, including battery-electric vehicles (BEVs) and plug-in hybrid electric vehicles (PHEVs), exceeded 5.12 million and 1.97 million, respectively, representing an expansion of

63% and 68%, compared with 2017.[6] By 2018, more than 2.3 million electric vehicles will be on the road in China, accounting for approximately 45% of the world's total, compared with approximately 24% for the European Union (EU) and 22% for United States of America (USA) (Figure 2b).[6] By 2030, the forecast number of EVs on the road will reach 253 million under the EV30@30Scenario (Figure 2c).[7] As the most popular technology for energy storage in applications ranging from consumer electronics to EV and grids, it is predicted that the global lithium battery market demand in 2025 will reach $99.98 billion, and the shipment volume will be 439.32 GWh (Figure 2d).[8] __________ (2) __________.

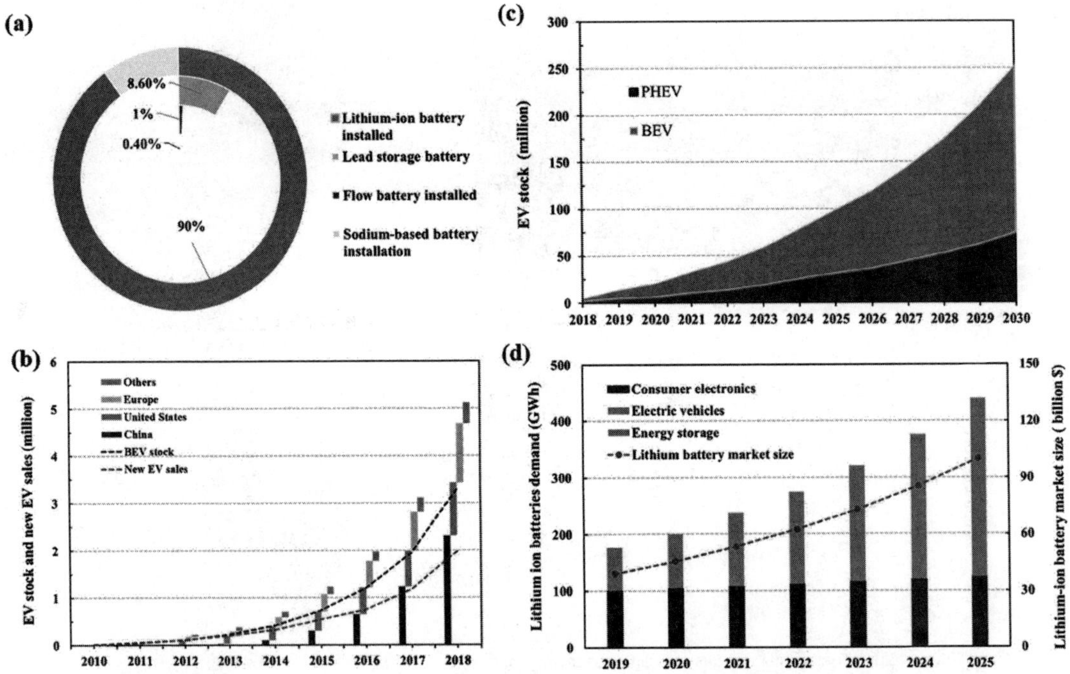

Figure 2 (a) Installed share of various electrochemical energy storage technologies in 2018. Data derived from ref (5). (b) Evolution of the global electric car (EV and PHEV) stock and EV sales by country from 2010 to 2018. Adapted with permission from ref (6). Copyright 2019 International Energy Agency. (c) Global EV stock in the EV30@30 scenarios from 2018 to 2030. Adapted with permission from ref (7). Copyright 2018 International Energy Agency. (d) Lithium ion batteries demand by application and its market size forecast from 2019 to 2025. Data derived from ref (8).

3 Limited resources and strong demand for high energy densities in EVs have motivated the development of next-generation rechargeable batteries to replace current LIBs. __________ (3) __________. It is predicted that by 2030 the global LIB recycling market will reach $23.72 billion.[9] Considering the need for effective use of these limited resources and environmental sustainability, spent LIBs should be properly handled and recycled. Spent LIBs contain heavy metal elements, such as nickel (Ni) and cobalt (Co), which are classified as carcinogenic and mutagenic materials, as well as toxic organic electrolytes, which

adversely affect human health and the environment. As essential raw materials for synthesis of the cathode materials of LIBs, Li and Co are in greater demand than other metals because of their low relative abundance and high price. According to the London Metal Exchange, the average price of Co in 2018 was $75991.27/t, which is more than five and ten times as high as the prices of Ni and Mn, respectively.(10, 11) In addition, the average export price of battery-grade Li_2CO_3 reached almost $12,514/t in March of 2019.(12) The high content of metal in spent LIBs represents an important metal resource, especially because global reserves are limited to approximately 62 million tons of Li and 145 million tons of Co.(13) Therefore, recycling of spent LIBs is highly worthwhile considering the need for sustainable use of these metals. ___________ (4) ___________. However, continuous development and replacement of rechargeable batteries poses challenges to recycling processes.

4 The outline of this review is illustrated in Figure 3. As the main source of spent batteries in the future, the development of EVs is crucial for the layout of retired batteries. Therefore, we first briefly analyze whether EVs represent an environmentally friendly substitute for traditional fuel vehicles through a comprehensive environmental assessment and economic evaluation and consideration of policy support (section 2). The structure and failure mechanism of LIBs and new-generation rechargeable batteries are introduced in section 3, conducive not only to design a battery structure that is easy to disassemble and recycle but also to help recyclers safely, simply, and efficiently recycle batteries. Considering the large volumes of batteries soon to be retired, we comprehensively assess the existing secondary use technology and recycling technologies for retired power batteries from environmental, economic, and safety perspectives (sections 4 and 5). ___________ (5) ___________. The sustainability of LIBs and future battery technologies is analyzed through a discussion of the life cycle assessment, strategic resource, and economic analysis (section 6). Next, we briefly examine the laws and regulations of secondary battery use and recycling in the USA, Germany, Japan, and China (section 7). Moreover, the possibilities and difficulties of recycling rechargeable batteries are considered (section 8), from which we discuss whether rechargeable batteries are environmentally friendly, safe, and sustainable and can form a closed loop from production to recycling.

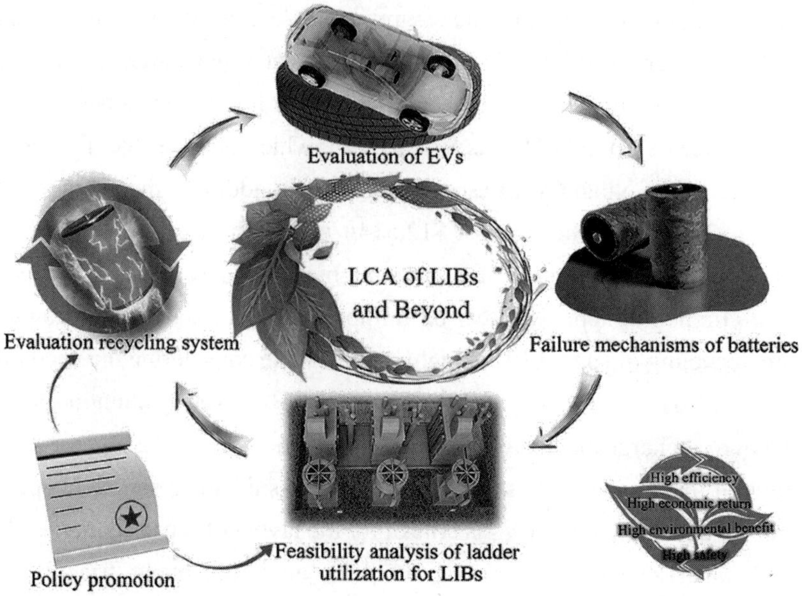

Figure 3 Closed loop from production to recycling of LIBs and beyond

Evaluation of EVs

5 Intensification of urbanization has been driven by the rapid growth of the social economy. ___________ (6) ___________. Global oil demand was 99.2 million barrels per day (mb/d) in 2018 and is set to increase by 5.5 mb/d from 2018 to 2023 at an average of 1.2 mb/d a year.[14] Following unprecedented expansion in 2018, oil production in the USA has increased by a record 2.2 mb/d and will account for 70% of the increase in global production capacity until 2024, ranking first in the world and adding a total of 4 mb/d.[15] Fundamentally, oil demand depends on the strength of the global economy. It is predicted that China and India will account for 44% of the 7.1 mb/d growth in global demand by 2024._Global oil reserves as of the end of 2018 are estimated to be 1,729.7 billion barrels and approximately, 48.3% of this oil is located in the Middle East.[2] The crude oil price of Brent Spot of 2018 reached 74 dollars per barrel ($/b), and it is predicted that the crude oil price of Brent Spot will exceed 100 $/b by 2035.[16] Global CO_2 emissions from fuel combustion in 2017 were 32.79 Gt CO_2, of which oil combustion contributed 12.77 Gt CO_2.[17] Asia is the largest source of emissions, particularly China, which has reached 9.10 Gt CO_2, followed by India at 2.08 Gt CO_2.[17] Transport was the main source of emissions in 2017, accounting for 49% of the global total. Thus, energy security, high oil prices, and high CO_2 emissions are a growing problem throughout the world.

6 ___________ (7) ___________. EVs are considered to be one of the most promising alternatives to traditional internal combustion engine vehicles (ICEVs). Such vehicles are becoming capable of good energy efficiencies and reduced carbon dioxide (CO_2) emissions, and are sometimes termed "zero-emission vehicles". Three types of EVs have appeared on

the current market, including PHEVs, hydrogen fuel-cell electric vehicles (HFCVs), and BEVs. It is unclear if EVs will become a popular means of transport and replace ICEVs. Uptake will depend on consumer purchase behavior and implementation of policies and infrastructure in individual countries. The economic benefits of EVs influence the purchasing behavior of consumers, and the environmental benefits of EVs influence the degree of government implementation. The state of EV promotion also affects consumers' purchasing behavior. Therefore, we analyzed whether EVs as a replacement for ICEVs represent a more environmentally friendly alternative and whether they save energy.

Economic Competitiveness of EVs

7 Economic competitiveness analyses of EVs, i.e., benefit-cost analyses, can be categorized into two types. The first considers the life cycle cost of ownership from a consumer standpoint, including initial costs and likely future running costs associated with activity over time, i.e., manufacturing, depreciation, operating and maintenance, insurance, and registration costs. The second category refers to the life-cycle social costs (LCSC), which are based on social or government positions and take into account the social cost of CO_2 and air pollutant emissions.

8 The economic competitiveness of EVs in China has been analyzed in detail by Zhao et al.[18] including the life-cycle private costs (LCPC) from the consumer perspective and LCSC from a social perspective (Figure 4a). These economic analyses are based on the EV market and policies before 2015. Two different sized BEVs were selected from the Chinese EV market. They are the compact vehicle shanghai-GM Sail Springo and the multipurpose vehicle (MPV) BYD E6. They found that even with central government subsidies, the LCPC for BEVs is approximately 1.4 times as high as those of comparable ICEVs. The LCSC of EVs are also higher than ICEVs in the current Chinese vehicle market. Therefore, from a social perspective, promotion of electric vehicles in China might not achieve the goal of reducing air pollution and greenhouse gas emissions in a cost-effective manner. This is because the benefits of reducing CO_2 emissions from EVs do not offset the LCPC differences and the government subsidies on EVs. ___________ (8) ___________. Tatari et al.[19] investigated the life cycle cost (LCC) of five vehicle types in the USA, without considering battery replacement costs. These results showed that for a variety of reasons, such as changes in the future price of electricity, the future electricity mix in the region, and future gasoline price, the life cycle costs of the studied vehicle types will change. The ICEV is the most cost-effective vehicle type with an average LCC of $87,028, followed by BEVs, with an average LCC of $89,244. As shown in Figure 4b, the life cycle cost of an ICEV varies over a greater range and the uncertainty is also higher than that of BEVs. ___________ (9) ___________. Similar results have appeared in recent research by Breetz et al.[20] In almost all cities in the USA, the high purchase price of BEVs and their rapid depreciation far offset their potential fuel savings.

9 On the basis of the above analysis, there remains room for improving the competitiveness of EVs compared with ICEVs in most regions from a consumer standpoint. __________ (10) __________. For example, O'Mahony et al.[21] found that high levels of use and appropriate government incentives would make large EVs more economically competitive than ICEVs. Rises in gasoline prices would also affect the fueling costs for ICEVs to a considerable extent, whereas a rise in electricity prices would only slightly affect fueling costs of EVs (Figure 4c). According to analysis by the IEA, the global electricity demand for all EVs in 2018 is estimated to be 58 terawatt-hours (TWh), of which China will account for 81% (Figure 5a).[6] The estimated demand for EVs in 2018 increased by 17%, slightly less than in the two previous years, which is equivalent to 0.5% of total global electricity consumption.[6] EV electricity demand in China and Norway, which respectively have the largest fleet and the largest market share of EVs (Figure 5b), account for less than 1% of those countries' total electricity demand.[6] Therefore, current EVs have a limited effect on electricity demand but will encourage greater electrical use. As the deployment of EVs increases, the effects of additional electricity demand on power grids should be considered.

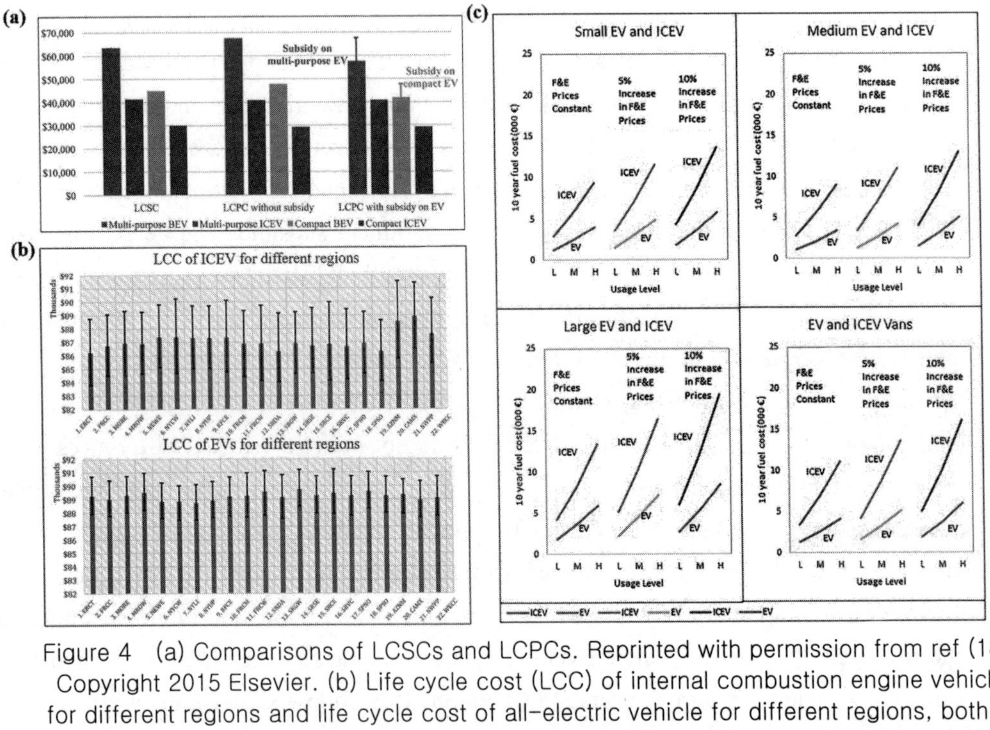

Figure 4 (a) Comparisons of LCSCs and LCPCs. Reprinted with permission from ref (18). Copyright 2015 Elsevier. (b) Life cycle cost (LCC) of internal combustion engine vehicle for different regions and life cycle cost of all-electric vehicle for different regions, both in thousand dollars. Reprinted with permission from ref (19). Copyright 2015 Elsevier. (c) 10-year fuel costs for all EVs and ICEVs (€). Reprinted with permission from ref (21). Copyright 2018 Elsevier

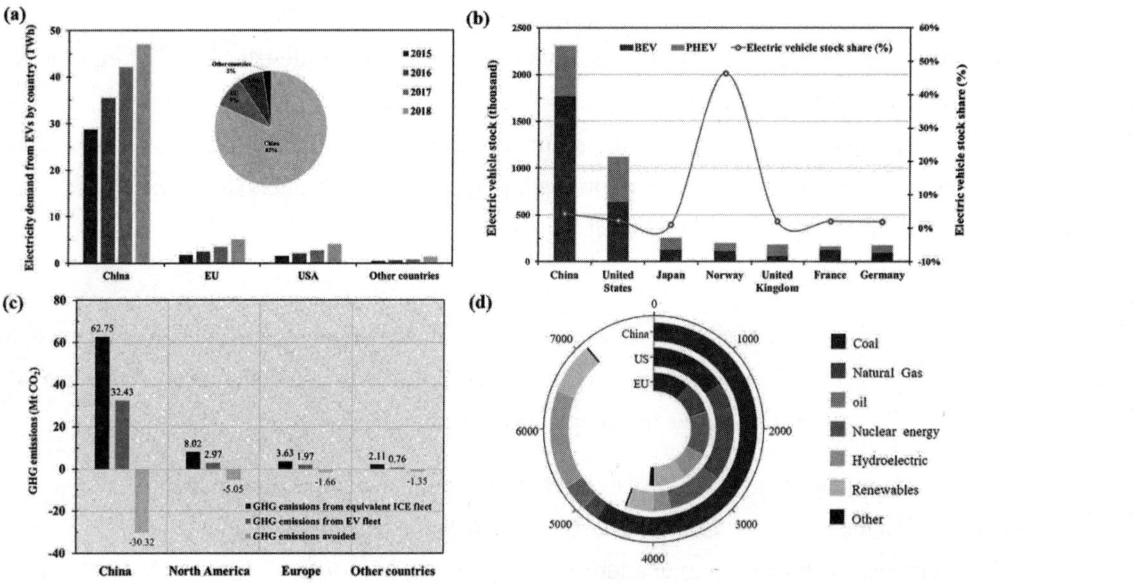

Figure 5 (a) EV electric demand from EVs by country, 2015–2018; (b) EVs stock and share by country, 2018. (c) GHG emissions avoided by EVs compared to equivalent ICE fleet by mode and region, 2018. (a), (b), and (c) Adapted with permission from ref (6). Copyright 2019 International Energy Agency. (d) Percent share of electricity generation by fuel (TWh) in China, EU, and US, 2018. Data derived from ref (2)

...

(The end of the passage)

A) Vehicle demand has also risen dramatically, leading to a sharp rise in global oil consumption, oil prices, and air pollution.

B) Such huge LIB demand will result in considerable consumption of resources for their manufacture.

C) However, owing to the great uncertainty of the social costs of pollutants and GHG emissions, EVs might become socially more economically competitive.

D) Guided by conventional recycling processes for spent LIBs, the recycling technologies are proposed for future battery technologies.

E) Many factors including government incentives, vehicle usage levels, battery replacement, electricity mix, and fuel costs (oil and electricity prices) might affect or change the economic competitiveness of EVs.

F) Therefore, considerable efforts have been made to develop advanced electrochemical energy storage technologies and electric vehicles (EVs).

G) Simultaneously, retiring of the current generation of EVs will create a need to recycle spent LIBs.

H) This uncertainty arises because of difficulty in predicting future gasoline prices versus electricity prices, and the future electricity mix.

I) Over the past decade, researchers have been committed to developing high efficiency, low cost, and pollution-free spent LIB recycling processes.

J) Increasing concerns about energy security, urban air pollution, and high oil prices driven up by the transport sector have prompted global policy-makers and researchers to seek alternative fuels for transportation.

Task 20: Compose a research article introduction based on one scientific research/experiment you have conducted following the outline below. Present your writing orally in groups and assess each other's performance.

(1) Claiming importance of a research area while emphasizing central issue(s)
(2) Reviewing previous research clearly relevant to the present one
(3) Pointing out research gap unaddressed by previous literature
(4) Announcing research purposes of present study
(5) Elaborating on details of the present study (analytical procedures, paper structure, etc.)

Unit 3
Methods Sections

Objectives

- Understand informative and descriptive communicative purposes of research article methods sections;
- Identify the generic structure and characteristic moves in research article methods sections;
- Acquaint with methods, practices, and devices prevalent in engineering research;
- Understand the precision and accuracy of scientific language especially in the use of technical and semi-technical vocabulary;
- Identify complex and compressed noun phrases expressing propositional contents;
- Identify commonly used reporting verbs and verb-noun collocations in presenting research methods and procedures;
- Identify commonly used sentence patterns in reporting and synthesizing experimental procedures;
- Distinguish different types of adverbial phrases and clauses along with their meanings and functions;
- Raise awareness of phrasal and syntactic complexity in research article methods sections.

Task 1: Discuss in groups which of the following communicative purposes typical research article methods sections need to fulfill.

Communicative Purposes	Yes/No?
(1) Briefly describing research methodology and experimental settings of present study	

continued

Communicative Purposes	Yes/No?
(2) Clearly relating research methodology and data analysis to research objective(s) of present study	
(3) Emphasizing theoretical/methodological principles behind present study	
(4) Stating key theoretical/methodological assumption/simplification /hypothesis made in present study	
(5) Reviewing in detail previous research utilizing the same methodology as present study	
(6) Claiming feasibility and effectiveness of present research methodology and procedures	
(7) Highlighting creativity and uniqueness of experimental design and procedures of present study	
(8) Systematically describing research objects (e.g. material samples, processing techniques, mathematical models) of present study	
(9) Justifying representativeness of research objects (e.g. material samples, processing techniques, mathematical models) and specifying the characteristics required	
(10) Elaborating on the system and mechanism of research tools or equipment used in present study	
(11) Clearly describing each experimental procedure of present study	
(12) Presenting data collection methods and processes used in present study	
(13) Reporting data analysis methods and steps adopted in present study	
(14) Pointing out types of data obtained from main experimental procedures	
(15) Discussing methodological or procedural limitations of present study	
(16) Pointing out main results or findings built upon data analysis	
(17) Defining key variables and parameters measured, calculated, or inferred in present study	
(18) Presenting symbol(s) and formula(s) representing theoretical/ methodological principles behind present study	

continued

Communicative Purposes	Yes/No?
(19) Stating in what units key variables and parameters of present study are measured and recorded	
(20) Illustrating the overall research design using visual aids (e.g. tables, figures, images)	
(21) Describing main obstacles and difficulties encountered in present study and discussing how the problems were confronted	
(22) Discussing the selection of comparison groups and matching criteria and/or describing the sampling procedures	
(23) Outlining the statistical methods/techniques used for data analysis	
(24) For qualitative approaches (e.g. in-depth interviews, questionnaires), specifying in sufficient detail how data were analyzed	
(25) Describing the measures taken to ensure that the present study was carried out in accordance with relevant ethical guidelines especially for studies involving human biological materials or animals	
(26) Itemizing and justifying the personnel costs, operating expenses, major equipment, and other specified expenditure	

Task 2: Observe the underlined noun phrases (head nouns in bold) with a varying degree of complexity in the research article methods section exemplified below, and discuss with each other different types of pre- and post-modifiers of head nouns (e.g. nouns, adjectives, prepositional phrases, participial clauses, etc.), so as to better understand the precision and specificity of scientific language. Finally, identify the head nouns of noun phrases and their modifiers in the given excerpts of research methods.

Characterization of iron (Fe) nanoparticles (NPs)[1]

The crystalline **structures** of the resulting NPs were studied by an X-ray **diffractometer** (XRD, Rigaku MiniFlex600) using a Cu Kα source (λ = 1.5418 Å). The mean crystalline **size** was estimated from the obtained XRD pattern using the Scherrer formula. The mean **size**, size **distribution**, and **morphology** of the NPs were examined with a transmission electron **microscope** (TEM, Hitachi H-7650) operated at an acceleration voltage of 100 kV. TEM **samples** were prepared by dropping the **suspension** of NPs onto a carbon-coated TEM **grid** and dried in air. A scanning **TEM** (STEM) equipped with a high-angle annular dark-field (HAADF) **detector** and energy-dispersive X-ray **spectroscopy** (EDS) was also used to further analyze the nanoscale **structure** of the NPs. HAADF-STEM **imaging** and EDS elemental **mapping** were performed on a JEOL JEM-ARM200F **microscope** operated at 200 kV with a spherical aberration **corrector** and a nominal **resolution** of 0.8 Å.

The magnetic **properties** of the NPs were evaluated with a hybrid superconducting quantum interference **device**–vibrating sample magnetometer (SQUID-VSM, Quantum Design). The **compositions** of the NPs were further analyzed using X-ray photoelectron **spectroscopy** (XPS) on a high-performance XPS system (Shimadzu Kratos, Axis-Ultra DLD) using Al Kα radiation.

The colloidal **stability** of the Fe NPs in an aqueous medium over a range of pH was investigated by monitoring the hydrodynamic size and zeta potentials of the NPs using the dynamic light scattering (DLS) **technique** with a Zetasizer Nano (Malvern Panalytical, UK). The **effect** of ionic strength in the medium on the stability of the NPs was also assessed by using **NaCl** at various concentrations.

Excerpt (1)[2]

Photocatalytic Activity Experiments[2]

The photocatalytic activities of the samples were evaluated in terms of the photocatalytic removal of NO, which was performed at ambient temperature in a continuous flow reactor. The volume of the rectangular reactor, which was made of stainless steel and covered with quartz glass window, was 4.5 L (30 cm × 15 cm × 10 cm). One sample dish with a diameter of 12.0 cm containing 0.10 g of catalyst powders was placed in the middle of the reactor, and 30 W visible LEDs (General Electric) with a light intensity on the catalyst surface of 50 mW cm^{-2} were vertically placed outside the reactor as a light source. The NO gas was supplied by a compressed gas cylinder at a concentration of 120 ppm NO (N_2 balance), then mixed with the air generated by air gas generator (BCHP-A10), and diluted to about 600 ppb by using a gas calibrator (Sevenstar D08F) in combination with two mass flow controllers supply. The gas flow rate through the reactor was controlled at 1000 mL min^{-1} by a mass flow controller. After the adsorption-desorption equilibrium was achieved, the lamp was turned on. The concentration of NO was continuously measured using a NOx analyzer model T200 (Teledyne API).

Excerpt (2)[3]

Sample Preparation and STM Measurement[3]

We used high quality WTe_2 crystals (HQ Graphene Company, The Netherlands). A piece of WTe_2 sample of ~5 mm^2 area and ~0.1 mm thick was stuck on a Cu plate and then mounted on the sample holder. A carbon tape was used for *in situ* exfoliating WTe_2 sample to get a clean surface in an ultrahigh vacuum (UHV) chamber. The base pressure of the UHV chamber was kept under 2×10^{-10} Torr. The STM images, spectroscopic images, and tunneling spectra were acquired at 5 K. The modulation voltages used for the lock-in detection were set 10 mV and 4 mV of 616

Hz for measuring spectroscopic images and tunneling spectra, respectively.

Calculation of Band Structure

All the DFT calculations were done by using Vienna *ab initio* Simulation Package (VASP) (1, 2) with the projected augmented wave (PAW) method. (3) Exchange and correlation functional are described at the level of generalized gradient approximation (GGA) parametrized by Perdew–Burke–Ernzerhof. (4) The cut off energy of the plane wave expansion used was 400 eV. The lattice constants and atom positions in the unit cell are taken from the literature. (5) In the self-consistent calculations, the Brillouin zone was sampled with (24 × 12 × 8) k-points.

Excerpt (3)[4]

Experimental Procedure[4]

Commercial pure α-alumina nano-powder (Baikalox-BMA15), with an average particle size of 100-150 nm, specific surface area of 16 g/m^2 and impurity levels of 2Ca, 10Si, 12Na, 6Fe and 20K ppm, was used for this study. Spark plasma sintering was performed using a hybrid SiC-Carbon tool setting described elsewhere [①], although an inner die diameter of 8 mm was used here. Sintering was conducted at temperature of 1000 and 1,050 °C under applied pressures from 500 to 800 MPa. A heating rate of 50 °C/min was employed up to 800 °C, followed by a heating rate of 12.5 °C/min up to the designated temperature. Samples at 1,050 °C were sintered with no dwell period (except one sample sintered with a 1 h dwell under 500 MPa). Samples sintered at 1000 °C were provided with a 15 min dwell time, so as to reach full densification. Pressure was applied at room temperature and increased linearly to reach a maximum at 800 °C. The discs obtained were 8 mm in diameter and about 1.2 mm thick. Density was determined using the Archimedes method. Cross-sections of specimens were ground and polished, followed by thermal etching. Microstructure was examined with a high-resolution scanning electron microscope (HRSEM; FEI, Verios-460L) at the center of the cross-section. Starting powder particle size distribution was determined by laser diffraction (Malvern Mastersizer 3000). Grain size and grain size distribution were determined by Thixomet® software for image analysis [②]. At least five images were analyzed for each specimen. Samples for TEM characterization were prepared with a focused ion beam (FIB; FEI, Helios-460F1) and examined using a high-resolution transmission electron microscope (TEM; JEOL, JEM 2100F).

Excerpt (4)[5]

Experimental Details[5]

The ten polycrystalline $Yb_{1-x}Sc_xMn_6Sn_6$ samples (with x = 0.00, 0.10, 0.15, 0.30, 0.35, 0.50, 0.65, 0.75, 0.90 and 1.00) were prepared from high-purity (at least 99.9%) commercially available elements. A 10% excess of Yb was used to compensate for the Yb losses due to the high vapor pressure of Yb at elevated temperature. Pellets of the elements were compacted using a steel die, and enclosed into silica tubes sealed under purified argon (300 mmHg), and then placed in a tube furnace. A preliminary homogenization treatment was conducted for one week at 500 °C. The samples were then ground, compacted and enclosed again, then annealed for ten days at 740 °C before to be quenched in water.

The precise chemical composition of the alloys was checked from microprobe measurements (JEOL JXA 8530-F) and was found close to the nominal one. Room temperature X-ray powder diffraction patterns were recorded using a Philips X'pert Pro Diffractometer ($\lambda_1 = 1.54056$ Å and $\lambda_2 = 1.54439$ Å). Diffraction patterns were analyzed with the Rietveld method using the Fullprof software [1]. Besides the peak profile parameters, the refinements comprised scale factor, zero shift, cell parameters, crystallographic position parameters (z_{Mn}, $z_{Sn(2e)}$) and the Yb/Sc occupancy ratio at the R site.

The DC magnetic measurements were performed using two different apparatus: a Physical Property Measurement System (PPMS) from Quantum Design (from 5 K to 350 K, in fields up to 9 T) and a Vibrating Sample Magnetometer (VSM) from MicroSense (from 300 K to 500 K, in fields up to 2 T). Thermomagnetization curves were recorded upon cooling under constant applied magnetic field (0.05 T). The magnetic entropy change ($-\Delta S_M$ given in units of $mJ.cm^{-3}.K^{-1}$) was calculated from magnetization isotherms, using the method proposed by Pecharsky and Gschneidner Jr. [2], with field steps of 0.2 T and temperature increments of 5 K.

Task 3: Read the two exemplary methods descriptions first, paying attention to the main clauses highlighted in bold and adverbial phrases/clauses underlined. Then quickly identify the main clauses and adverbial phrases and clauses in the following methods excerpts.

Example (1)

The test specimens were cut into samples of dimensions 25 × 8 mm, **ground** by 1000 grid emery paper and **polished** using diamond solution. **The boriding heat treatment was carried out** in a solid medium containing an Ekabor-II powder mixture placed in an electrical resistance furnace operated at the temperature of 1173 K under atmospheric pressure. **Test specimens were sealed** in a stainless steel container together with the Ekabor-II powder mixture. **The holding time** for the samples **was 4 hours**. Following the completion of the boriding process, **the test specimens were removed** from the sealed container **and allowed** to cool in still air.

Example (2)

Every experiment was repeated at least twice with triplicates or sextuplicates for each condition. **Data are represented** as mean ± SD and **were** statistically **analyzed** with IBM SPSS software (version 22) or GraphPad Prism (version 6.05) using analysis of variance (one-way or two-way ANOVA) with $p < 0.05$ considered significant. ***In vivo* data were analyzed** in GraphPad Prism using the Kruskal-Wallis test with *post hoc* Dunn's multiple comparisons test with $p < 0.05$ considered significant.

(1) Materials were prepared following a procedure described previously based on a modified Hummers method and allowing the production of endotoxin-free materials.

(2) Bright-field microscopy using a PrimoVert inverted optical microscope (Carl Zeiss, UK) using an objective with 200× magnification was performed to determine the lateral dimension distribution of the l-GO (sheets in the microsized range) and to verify the size reduction of the s-GO (below Abbe's diffraction limit).

(3) TEM was performed using a FEI Tecnai 12 BioTwin microscope (FEI, The Netherlands) at an acceleration voltage of 100 kV. Images were taken with a Gatan Orius SC1000 CCD camera (GATAN, UK).

(4) UV-vis absorbance spectra were obtained for GO samples at 7.5 to 20 μg/mL using a Varian Cary 50 Bio UV-vis spectrophotometer (Varian Inc., Agilent Technologies, UK).

(5) Continuous-wave EPR measurements were carried out at room temperature, using an EMX-Micro X-band spectrometer (Bruker, UK) operating at a frequency of ~9.86 GHz, center field at 3,520 G, and attenuator at 30 dB.

(6) First, CNTs were deposited using the chemical self-assembly method reported previously. Briefly, trenches with different dimensions were patterned on an HfO_2 blanket film, followed by 7 nm SiO_2 evaporation and a lift-off process. Electron-beam lithography was used to define all of the patterns mentioned here, either in PMMA or HSQ. Before self-assembly of the NMPI monolayer, the substrate was cleaned in oxygen plasma (0.3 torr, 5 min).

(7) The completed device was annealed at 150 °C in HMDS ambient for 40 min to improve device performance. The highly doped silicon substrate was used as a back-gate. All the electrical measurements were carried out in a Cascade summit semi-automated probe station in air.

(8) The stoichiometric mixture of Bi, Se, Sb, and Co was melted in an evacuated quartz tube at 1195 °C. Later the melt was cooled to 550 °C at the rate of 2 °C/hr and retained at 550 °C for 24 h, followed by cooling to 30 °C in 3 h. The structural properties of the grown BSS and BCSS crystals were studied by X-ray diffraction (XRD) on powdered samples at Cu K_α edge from a lab x-ray source.

(9) Ingots with a nominal composition of $Cu_{47}Zr_{47}Al_6$ and $(Cu_{47}Zr_{47}Al_6)_{99}Sn_1$ (at. %) were prepared by arc melting of high purity elements mixture (purity ⩾ 99.9%) in a Ti-gathered high purity argon atmosphere. The ingots were inverted and remelted four times to generate a homogeneous alloy and followed by suction casting into the copper mold to form cylindrical rod samples with 2 mm in diameter and 30 mm in length. The amorphous nature of these rods was confirmed by X-ray diffraction (XRD, Philips Xpert pro) and high resolution transmission electron microscopy (HRTEM, FEI Tecnai G2 F30 equipped with energy dispersive X-ray analysis detector). Room-temperature compression tests were performed on samples with an aspect ratio (height: diameter) of 2:1 at an initial strain rate of 10^{-4} s^{-1}. Lateral surfaces of deformed specimens were investigated by high-resolution scanning electron microscopy (HRSEM).

(10) $Bi_{1-x}Sb_x$ single crystals were grown using a modified Bridgman method. Flat crystals up to 1 cm in length were obtained by cleaving a crystal boule. The $Bi_{0.97}Sb_{0.03}$ crystal was mechanically exfoliated onto a SiO_2/Si^{2+} substrate. The flakes were about 300 nm thick, on average. The Josephson junctions were fabricated using standard electron-beam lithography followed by a RF Ar^+ ion etch, and in situ sputter deposition of 100-nm-thick Nb electrodes and a few nanometres of Pd as a capping layer.
ARPES measurements were carried out at the I05 beamline of the Diamond Light Source. Crystals of $Bi_{1-x}Sb_x$ ($x = 0.03$ and 0.04) were cleaved in ultrahigh vacuum at a temperature of 30 K, and the ARPES data were recorded at a temperature of 10 K in the low 10^{-10} mbar pressure range, with an overall energy resolution of 15 meV and k resolution of 0.015 $Å^{-1}$. The Fermi surface map shown in Fig. 2 was recorded using circularly polarized radiation of energy 60 eV from an $x = 0.04$ crystal.

Task 4: Identify the main idea of each paragraph in the Methods and Experimental section in Text I and discuss in groups what communicative purposes listed in Task 1 are achieved, after figuring out the main purpose of the study in the Introduction section.

Text I

Live Imaging of Label-Free Graphene Oxide Reveals Critical Factors Causing Oxidative-Stress-Mediated Cellular Responses[6]

INTRODUCTION

1 The interest in graphene and its translation into commercial products have been expanding during the past few years. Graphene-based materials (GBMs) are a large family of different materials [1-2] varying tremendously in their lateral dimensions, number of carbon layers (*i.e.*, thickness), and surface properties. Today, GBMs are already contained in various commercial products (sportswear, tennis rackets, bicycle frames, tires, or innovative paints), while conductive inks, paints, or filaments for 3D printing are expected to become components of flexible displays and wearable electronics in the near future. (3) Longer-term applications are also anticipated in the healthcare sector, either as drug delivery systems, as biosensors for health monitoring and e-health, or in the form of biomedical implants. (2) However, in order to be fully adopted by both industry and society, new enabling technologies need to demonstrate not only their benefits but also their long-term safety and sustainability. [2], [4]

2 In consideration of the previously described pulmonary health and safety concerns for carbon nanomaterials (*e.g.*, carbon nanotubes) [5-6] and the fact that the lungs will be one of the primarily exposed organs following inhalation of nanomaterials (in particular airborne GBMs), there is a great need to understand the critical parameters impacting interactions between GBMs and the cells of the pulmonary system.

3 For the vast majority of GBMs, interactions with mammalian cells and their overall safety profile remain largely unknown. By far the most studied type of GBMs in biological research has been graphene oxide (GO), the heavily oxidized form of graphene in the form of sheets. This popularity can be explained by its ease of production and its hydrophilic surface character that allows its colloidal dispersion in the aqueous and physiological milieu.

4 Hazard assessment studies have indicated that structure–function relationships should be based on specific physicochemical properties of the studied nanomaterial. Both intrinsic (*e.g.*, physicochemical properties) and acquired (*e.g.*, aggregation state in local milieu or secondary coating with components from the cell culture medium, blood, or different organs) features need to be taken into consideration before obtaining a clear understanding of the parameters that could lead to adverse effects. [7] In addition, the oxidative stress paradigm has been shown to successfully explain and predict the outcomes of the cellular response to many nanomaterials. [8-10] According to this paradigm, cellular response to nanomaterials is governed by induction of oxidative stress, which can either be counterbalanced by the activity of antioxidative enzymes (tier 1 response), lead to subsequent activation of pro-inflammatory pathways (tier 2 response), or result in cell death in the case of excessive levels of reactive oxygen species (ROS) production (tier 3 response).

5 Previous *in vitro* studies have compared graphene oxide sheets of large (2 μm) and small (350 nm) lateral dimensions to conclude that larger materials were more inflammogenic than smaller ones, with an almost flat inflammatory response for the small GO even at the highest concentration (6 μg/mL) tested. [11] A different study by Ma et al. reported that large GO (750–1300 nm) induced more cytokine production than two smaller GO types (50–350 or 350–750 nm), but concluded that all GOs were inflammogenic. [12] GO sheets were also found to be pro-fibrotic in cells and animals, with larger GO (1676 nm) sheets being more potent than their smaller counterparts (179 nm). [5] Currently, only a limited number of studies have reported that small GOs may cause more damage than larger GO (above 1 μm). Orecchioni et al. found smaller GOs (<1 μm) to be more inflammogenic than large GOs (1–10 μm), even though large GOs was still inducing elevated levels of a subset of cytokines. [13] In relation to surface coating, different studies have demonstrated that bovine serum albumin [14-16] or the level of serum in the cell culture medium [17-18] could greatly alleviate the toxic response observed without coating, suggesting that the shielding effect of adsorbed proteins was able to mitigate adverse responses to materials. In most cases, however, the two parameters of lateral dimensions and serum protein coating were considered separately, which precluded

any analysis of their interrelation and did not allow determination of the importance that each parameter holds in the context of cellular response induced by GOs. Moreover, the effect of protein coating on the effect of micrometer-sized GO sheets (>3 μm) is not well documented, as most studies have only considered GO nanosheets (<500 nm).

6 In the present study, we aimed to comparatively determine the importance of these two parameters (*i.e.*, lateral dimensions and serum protein surface coating) in relation to induction of oxidative stress, inflammation, and cytotoxicity in a human lung epithelial cell line (BEAS-2B). On the basis of previous reports describing the individual role of each criterion, we postulated that while surface adsorption of proteins may alleviate the adverse effects induced by smaller GO sheets, by shielding their most reactive parts, it may not be sufficient for larger materials. Therefore, we also hypothesized that lateral dimensions may have a more dominant role than protein shielding on the biological impact of the GO. [19] In addition, based on the results obtained with various nanomaterials, we wanted to test whether cellular responses to GOs can be explained and predicted using the oxidative stress paradigm. [10]

7 To address these questions, we produced endotoxin-free materials with two clearly distinct lateral dimensions—small, nanometer-sized GO (50-200 nm; s-GO) and large, micrometer-sized GO (5-15 μm; l-GO)—using a modified Hummers method. [20] We demonstrated that these two materials differed only by their lateral dimensions, allowing us to determine the role of one physicochemical parameter at a time. The role of serum protein coating in the initial phase of the interaction between nanomaterials and cells was addressed by controlling the presence of fetal bovine serum (FBS) (10%) during the first 4 h of exposure. Exploiting the intrinsic fluorescence of the two GOs without introducing additional surface modifications and/or without attaching fluorescent dyes, we were able to follow in real time the interactions of GOs with cells at a single-cell level over a 24 h period using confocal live-cell imaging. Furthermore, correlation between the lateral dimensionality and serum coating of the material with cytotoxicity, induction of oxidative stress, and inflammation was attempted.

2. Results and Discussion

…

3. Conclusions

…

4. Methods and Experimental

4.1 Production and Characterization of the l-GO and s-GO

Production

8 Materials were prepared following a procedure described previously (20) based on a modified Hummers method and allowing the production of endotoxin-free materials. (22) The starting graphite used in the reaction was graphite powder (product code #282863, <20 μm, synthetic, Sigma-Aldrich, Merck Sigma, UK). Following this chemical exfoliation procedure, materials corresponding to the "large" GO sample (l-GO) were produced with final concentrations

ranging between 1 and 2 mg/mL, achieving a yield of *ca.* 15–20%. (22) Large GO materials were then sonicated in a bath sonicator (VWR, 80 W) for 5 min and centrifuged at 13 000 rpm for 5 min at room temperature to prepare the "small" GO sample (s-GO).

Optical Microscopy

9 Bright-field microscopy using a PrimoVert inverted optical microscope (Carl Zeiss, UK) using an objective with 200× magnification was performed to determine the lateral dimension distribution of the l-GO (sheets in the microsized range) and to verify the size reduction of the s-GO (below Abbe's diffraction limit). Hundreds of particles were manually measured by determining the Feret diameter using ImageJ (NIH, USA).

Transmission Electron Microscopy

10 TEM was performed using a FEI Tecnai 12 BioTwin microscope (FEI, The Netherlands) at an acceleration voltage of 100 kV. Images were taken with a Gatan Orius SC1000 CCD camera (GATAN, UK). A 20 μL sample (100 μg/mL) was placed on a Formvar/carbon-coated copper grid (CF400-Cu) (Electron Microscopy Services, UK). Filter paper was used to remove excess liquid.

Atomic Force Microscopy

11 AFM images were acquired using a Multimode 8 AFM (Bruker, UK) in tapping mode in air with a J-type scanner, Nanoscope V8 controller, and an OTESPA silicon probe (Bruker, UK). Samples were prepared by depositing 20 μL of a 100 μg/mL GO suspension on a freshly cleaved mica surface (Agar Scientific, Essex, UK) coated with poly-L-lysine 0.01% (Sigma-Aldrich, Merck Sigma, UK) and allowed to adsorb for 2 min. Excess unbound materials were removed by gently washing with Milli-Q water and then allowed to dry in air. Lateral dimension and thickness distributions of GO sheets were carried out using Nanoscope Analysis software (version 1.40, Bruker, UK).

UV/Visible Spectroscopy

12 UV–vis absorbance spectra were obtained for GO samples at 7.5 to 20 μg/mL using a Varian Cary 50 Bio UV–vis spectrophotometer (Varian Inc., Agilent Technologies, UK). Dual beam mode and baseline correction were used throughout the measurements to scan the peak wavelength and maximum absorbance between 200 and 800 nm.

Raman Spectroscopy

13 Samples were prepared for analysis by drop casting ~20 μL of GO (100 μg/mL) dispersion onto a glass slide. Samples were left to dry for at least 2 h at 37 °C. Spectra were collected using a DXR micro-Raman spectrometer (Thermo Scientific, UK) using a 50× objective lens and using a $\lambda = 633$ nm laser with an exposure time of 25 s at an intensity of 0.4 mW. Spectra were averaged over five independent locations and considered between 250 and 3500 cm^{-1}, enabling visualization of the D and G scatter bands. The average I_D/I_G ratio for each sample was then calculated.

Zeta-Potential Measurements

14 Electrophoretic mobility (μ) was measured using a ZetaSizer Nano ZS (Malvern Instruments, UK) after dilution of samples with water (100 μg/mL) in disposable cuvettes (Malvern Instruments, UK). Default instrument settings and automatic analysis were used for all measurements, performed at room temperature with a backscattering angle of 173°. The equipment software converted automatically the μ to zeta-potential (ζ) values by Henry's equation. All values for samples prepared are triplicate measurements, and values were mean ± SD.

Thermogravimetric Analysis

15 The weight loss of GO samples was performed by TGA using a Pyris 6 (PerkinElmer Ltd., UK). Lyophylized GO (1–2 mg) was weighed into a ceramic crucible and analyzed from 25 to 995 °C at 10 °C/min, under a nitrogen flow of 20 mL/min.

X-ray Photoelectron Spectroscopy

16 The composition of GO surfaces was studied by XPS at the NEXUS facility (the UK's National EPSRC XPS Users' Service, hosted by nanoLAB in Newcastle-upon-Tyne). XPS was recorded using a Thermo Theta Probe XPS spectrometer with a monochromatic Al Kα source of 1486.68 eV. The survey XPS spectra were acquired with a pass energy (PE) of 200 eV, 1 eV step size, and 50 ms dwell time and averaged over five scans. The etching was 90 s. High-resolution C 1s XPS spectra were acquired with a PE of 40 eV, 0.1 eV step size, and 100 ms dwell time and averaged over 20 scans. Spectra from insulating samples have been charge-corrected by shifting all peaks to the adventitious carbon C 1s spectral component binding energy set to 284.6 eV. CasaXPS software (Casa Software Ltd., UK) was used to process the spectra acquired at NEXUS. For the deconvolution of the different components, the CasaXPS software was used and the different regions were assigned according to NIST's XPS and laSurface databases, after subtracting the background using a Shirley algorithm:

$$\pi - \pi^* : 292.0 - 290.0 \text{ eV}$$

$$\mathrm{O-C=O} : 290.0 - 288.6 \text{ eV}$$

$$\mathrm{C=O} : 287.8 - 286.8 \text{ eV}$$

$$\mathrm{C-O} : 286.6 - 285.5 \text{ eV}$$

$$\mathrm{C-C} \text{ and } \mathrm{C=C} : 284.6 - 284.5 \text{ eV}$$

All peaks were deconvoluted using a Gaussian:Lorentzian (70:30) function, apart from the peak for C–C and C═C, which was fitted by an asymmetric Lorentzian function. Each deconvoluted peak, except that of the $\pi - \pi^*$ contribution, was constrained to the same full width at half-maximum value.

Endotoxin Content of the Samples

17 Materials were tested for their endotoxin content based on the method described previously. (21) In brief, cells were exposed to l-GO and s-GO for 24 h in order to allow the production of TNF-α in the presence or absence of polymyxin B. When added into the interaction of GO with cells, polymyxin B would prevent the production of TNF-α if this production is mediated by endotoxin and not the result of sterile inflammation due to the GO sheets.

Electron Spin Resonance Spectroscopy

18 Continuous-wave EPR measurements were carried out at room temperature, using an EMX-Micro X-band spectrometer (Bruker, UK) operating at a frequency of ~9.86 GHz, center field at 3520 G, and attenuator at 30 dB. Approximately 1.3 mg of lyophilized powder of each GO material dispersed in the cell culture medium in the presence or absence of FBS (50 μg/mL) and then freeze-dried was placed in the bottom of an EPR quartz tube. The field was calibrated using 2,2-diphenyl-1-picrylhydrazyl (DPPH) as a standard sample with $g = 2.0036$. The carbon radical concentration was estimated by calculating the area of the high-resolution EPR spectra by double integration and calculating the number of spins relative to the DPPH standard. Obtained spin concentrations were normalized by the respective masses of GO samples and standard, yielding radical concentrations in mmol/g.

Agglomeration of GO in Cell Culture Medium

19 GO samples were incubated (100 μg/mL) for 24 h at room temperature in RPMI 1640 cell culture medium (Sigma-Aldrich, Merck Sigma, UK) supplemented with 20 mM glutamine (Sigma-Aldrich, Merck Sigma, UK), 10% FBS (Gibco, Thermo Fisher Scientific, UK), 1000 units penicillin, and 1 mg/mL streptomycin (Sigma-Aldrich, Merck Sigma, UK). GO agglomerates were then obtained by centrifugation at room temperature, for 30 min at 13 000 rpm. The supernatants containing the cell culture medium were discarded, and the GO pellets were gently resuspended in Milli-Q water. This washing step was repeated once in order to remove excess unattached biomolecules and electrolytes, enabling the characterization of GO agglomerates by AFM, dynamic light scattering, and zeta potential measurements, as described above.

4.2 Cell Culture

20 Human epithelial bronchial immortalized cells (BEAS-2B, CRL-9609, ATCC, LGC standards, UK) were maintained in RPMI 1640 cell culture medium (Sigma-Aldrich, Merck Sigma, UK) supplemented with 20 mM glutamine (Sigma-Aldrich, Merck Sigma, UK), 10% FBS (Gibco, Thermo Fisher Scientific, UK), 1000 units penicillin, and 1 mg/mL streptomycin (Sigma-Aldrich, Merck Sigma, UK) at 37 °C in a humidified 5% CO_2 incubator. Cells were passaged twice a week using a 0.05% Trypsin–EDTA solution (Sigma-Aldrich, Merck Sigma, UK) when reaching 80% confluence. Activity of trypsin was stopped using 10% FBS. All experiments were done using cells with a passage number between 25 and 35.

4.3 Cell Culture Treatment

21 Depending on the experiment, cells were seeded in 96- (modified LDH and DCF-DA assay), 12- (cell count, modified LDH assay, HE oxidation), or six- (PI/annexin V staining, PCR experiments) well plates (Corning, Costar, Sigma-Aldrich, Merck Sigma, UK) and treated with GO when reaching 60–80% confluence. Cell treatments were performed after dispersing the GO in RPMI 1640 cell culture medium in either the absence of FBS (w/o FBS) or the presence of 10% FBS (w/FBS). Material suspensions were thoroughly vortexed shortly

before the treatment. For cells treated with GO dispersed in FBS-free medium, the cell culture medium was supplemented with 10% FBS 4 h after treatment. Incubations with GOs were maintained for 2, 4, 24, or 72 h. All experiments were repeated at least twice.

4.4 Cell Count Experiment

22 For cell counting, cells were seeded and treated in triplicates in 12-well plates. After 24 h of treatment using the indicated concentrations of GOs (w/FBS or w/o FBS), supernatants were removed and cells were collected using 0.05% Trypsin–EDTA. After 5 min of incubation, the action of Trypsin–EDTA was blocked using 10% FBS. Cells harvested from each well were transferred to a separate 1.5 mL microtube, and 10 μL of Trypan Blue (0.4% solution, Sigma-Aldrich, Merck Sigma, UK) was mixed with 10 μL of cells. Live cells (unstained for Trypan Blue) were counted using a Neubauer counting chamber. For each condition, cells were treated in triplicates and cell count was repeated twice.

4.5 PI/Annexin V-Alexa Fluor488 Conjugate Assay

23 For the PI/annexin V staining experiment, cells were seeded and treated in six-well plates after reaching 60-80% confluence. After 24 h of treatment, supernatants were removed and cells were gently washed once with phosphate-buffered saline (PBS) containing Ca^{2+}/Mg^{2+} (Sigma-Aldrich, Merck Sigma, UK). Annexin V staining was performed according to the instructions of the manufacturer (Molecular Probes, Thermo Fisher Scientific, UK). In brief, cells were trypsinized for 5 min, centrifuged at 1500 rpm for 5 min, then resuspended in 50 μL of annexin binding buffer (Molecular Probes, Thermo Fisher Scientific, UK), and stained with 1 μL of annexin V–Alexa Fluor488 conjugate for 20 min at 25 °C. Propidium iodide (1 mg/mL, Sigma-Aldrich, Merck Sigma, UK) was added shortly before the analysis to the final concentration of 1.5 μg/mL. At least 2500 cells were analyzed using the Amnis ImageStream platform (Amnis ImageStream MKII, Merck, UK) and Inspire system software (Amnis, Merck, UK). Camera magnification was 60×; the 488 nm excitation laser was set to 60 mW; the 785 nm excitation laser was set to 0.02 mW. Images were acquired with a normal depth of field, providing a cross-sectional image of the cell with a 2.5 μm depth of focus. The results were analyzed by IDEAS software (Amnis).

4.6 Confocal Microscopy

Plasma Membrane Staining

24 Cells were seeded in a Cellview cell culture dish (627870, Greiner Bio-One Ltd., UK) and treated when reaching 60%-80% confluence with 50 μg/mL s-GO (w/FBS or w/o FBS) or l-GO (w/FBS or w/o FBS). For w/o FBS conditions, 10% FBS was added to each well after 4 h. CellMask green plasma membrane stain (C37608, Thermo Scientific, UK) was added to the cell culture medium containing GO sheets before treatment (dilution 1:2500) and added to the cells at the same time as the GO. After 2 h of treatment, time lapses were set to start the live-cell imaging. Five positions were chosen for each condition, and the experiment was repeated three times. Cells were examined under a Zeiss 780 multiphoton confocal laser

scanning microscope using a 40× objective with a time lapse mode. Excitation wavelengths for the CellMask green plasma membrane stain and GOs were 488 and 594 nm, respectively. Emission maximum for the CellMask green plasma membrane stain was 520 nm, while emission wavelength for the GOs was 620-690 nm. Time lapse videos and images were processed using Zeiss microscope software ZEN.

DCF-DA Staining

25 Cells were seeded in a Cellview cell culture dish (627870, Greiner Bio-One Ltd., UK), washed once with PBS (with Ca^{2+}/Mg^{2+}, Sigma-Aldrich, Merck Sigma, UK), and preloaded with 20 μM DCF-DA dye for 45 min at 37 °C, in a humidified 5% CO_2 incubator. After preloading, cells were washed in PBS (with Ca^{2+}/Mg^{2+}, Sigma-Aldrich, Merck Sigma, UK) and treated with 50 μg/mL of s-GO and l-GO in the presence or absence of FBS for 4 h or with 1 mM H_2O_2 for 2 h (as a positive control). After 4 h of treatment, cells were washed in PBS (with Ca^{2+}/Mg^{2+}, Sigma-Aldrich, Merck Sigma, UK) and imaged using a Zeiss 780 confocal laser scanning microscope using a 40× objective. Excitation wavelengths for the DCF-DA dye and GOs were 488 and 594 nm, respectively. Emission maximum for the DCF-DA dye was 520 nm, while emission wavelength for the GOs was 620–690 nm. Images were processed using Zeiss microscope software ZEN.

Lysotracker Blue Staining

26 Cells were seeded in a Cellview cell culture dish (627870, Greiner Bio-One Ltd., UK) and treated when reaching 60%–80% confluence with 100 μg/mL s-GO (w/FBS or w/o FBS) or l-GO (w/FBS or w/o FBS) for 24 h. After treatment, cells were stained with 75 nM LysoTracker Blue DND-22 (L7525; Thermo Fisher Scientific, UK) for 30 min, in a humidified 5% CO_2 incubator. Imaging of the cells was performed using a Zeiss 780 confocal laser scanning microscope and a 40× objective. Excitation/emission wavelengths of 373/422 nm (DAPI filter set) were used. Images were processed using Zeiss microscope software ZEN.

Lipid Peroxidation Assessed by Bodipy 581/591 C11 Staining

27 Cells were seeded in a Cellview cell culture dish (627870, Greiner Bio-One Ltd., UK) and treated when reaching 60–80% confluence with 50 μg/mL of s-GO or l-GO in the presence or absence of FBS for 2 h. Cumene hydroperoxide treatment (100 μM, for 2 h) was used as a positive control for lipid peroxidation. After treatment, cells were stained using 10 μM Bodipy 581/591 C11 (C10445, Image-iT lipid peroxidation kit, Thermo Fisher Scientific, UK) for 20 min at 37 °C, in a humidified 5% CO_2 incubator. Imaging of the cells was performed using a Zeiss 780 confocal laser scanning microscope and a 40× objective. Excitation/emission wavelengths of 581/591 nm (Texas Red filter set) and 488/510 nm (traditional FITC filter) were used to detect oxidized (green) and nonoxidized (magenta) signal from the probe. The ratio of the emission fluorescence intensities at 590 to 510 nm gives a read-out for cellular lipid peroxidation. Images were processed using Zeiss microscope software ZEN.

Actin Filament Staining

28 Cells were seeded in Cellview cell culture dish (627870, Greiner Bio-One Ltd., UK) and treated when reaching 60-80% confluence with 50 μg/mL of s-GO or l-GO in the presence or absence of FBS for the first 4 h of treatment. After treatment, cells were fixed using 3.7% PFA for 10 min at room temperature, permeabilized using 0.1% Triton X-100 in PBS for 5 min, and stained with 4.5 nM Alexa Fluor 488 Phalloidin dye (A12379, Thermo Fisher Scientific, UK). After staining cells were washed three times in PBS and observed using a Zeiss 780 confocal laser scanning microscope and a 40× objective. Excitation wavelengths for the Alexa Fluor 488 Phalloidin dye and GOs were 488 and 594 nm, respectively. Emission maximum for the Alexa Fluor 488 Phalloidin dye was 520 nm, while emission wavelength for the GOs was 620–690 nm. Time lapse videos and images were processed using Zeiss microscope software ZEN.

4.7 Modified LDH Assay

29 For the LDH assay, cells were seeded and treated for 24 h in triplicates or sextuplicates in 12- or 96-well plates, respectively. The LDH assay was modified to avoid any interference coming from the interaction between GO and the assay reagents. (33) Briefly, LDH content was assessed in intact cells that survived the treatment, instead of detecting the amount of LDH released in the media upon treatment. Media was aspirated and cells were lysed with 100 μL of lysis buffer for 45 min at 37 °C to obtain cell lysates, which were then centrifuged at 4000 rpm for 20 min in order to pellet down GOs. A 50 μL amount of cell lysate supernatant was transferred to a new 96-well plate and mixed with 50 μL of LDH substrate mix (CytoTox 96 nonradioactive cytotoxicity assay, Promega, UK) and left to react for 7-15 min at room temperature, after which 50 μL of stop solution was added.

$$\text{Cell Survival\%} = (\alpha_{490\text{mm}} \text{ of treated cells}/\alpha_{490\text{mm}} \text{ of untreated cells}) \times 100$$

30 The absorbance was read at 490 nm using a spectrophotometer plate reader (FLUOstar Omega, BMG Labtech, UK). The amount of LDH detected represented the number of cells that survived the treatment. The percentage cell survival was calculated using the equation above.

4.8 DCF-DA Assay

31 For the DCF-DA assay, cells were seeded and treated in sextuplicates in 96-well plates for 4 h. After treatment, cells were gently washed with 100 μL per well of prewarmed PBS (with Ca^{2+}/Mg^{2+}, Sigma-Aldrich, Merck Sigma, UK) in order to remove the materials and subsequently incubated with 20 μM DCF-DA (Sigma-Aldrich, Merck Sigma, UK) diluted in PBS (with Ca^{2+}/Mg^{2+}) for 45 min at 37 °C, in a humidified 5% CO_2 incubator. After incubation with the dye, fluorescence intensity was read using a spectrofluorimeter microplate reader (FLUOstar Omega, BMG Labtech, UK) with a 488 nm excitation laser and collecting emission at 520 nm.

4.9 HE Oxidation

32 For the HE oxidation experiment, cells were seeded and treated in triplicates in 12-well plates

using the indicated concentrations of GO sheets for 4 h. After treatment, supernatants were aspirated and cells gently washed once with 1 mL per well of prewarmed PBS (with Ca^{2+}/Mg^{2+}, Sigma-Aldrich, Merck Sigma, UK). Cells were detached using 0.05% Trypsin-EDTA solution (Sigma-Aldrich, Merck Sigma, UK) for 5 min, then centrifuged for 5 min at 1500 rpm; supernatants were then aspirated, and pellets containing cells were resuspended in 1 μM hydroethidine (Sigma-Aldrich, Merck Sigma, UK) for 20 min. Ten thousand cells were analyzed on a BD FACSVerse flow cytometer using 488 nm excitation and 620 nm band-pass filters for HE detection. GOs alone were run in order to set up the gates including the cell population for the analysis and eliminate the signal coming from free materials or cell debris. Cells treated with the materials, but unstained with HE, were also run in order to ensure that the detected signal was not due to the inherent fluorescence of GOs. Percentages of unstained cells and cells stained with HE were calculated.

4.10 RT-qPCR Analysis

33 BEAS-2B cells were seeded in six-well plates and treated in triplicates with the indicated concentrations of either l-GO or s-GO for 24 h. After treatment, supernatants were removed and total RNA was extracted with an *ad hoc* kit (Aurum Total RNA mini kit, Bio-Rad, UK) according to the manufacturer's instructions. The concentration of total RNA was determined by measuring the optical density on a Biophotometer Plus spectrophotometer (Eppendorf AG, Germany); the purity was checked measuring both absorbance ratios 260 nm/280 and 260 nm/230 nm, with expected values between 1.8 and 2.0. First-strand cDNA was then prepared from 1 μg of RNA in a total volume of 20 μL using the iScript cDNA synthesis kit (Bio-Rad, UK). Real-time PCR was performed using the CFX96 real-time PCR detection system (BioRad, UK). The cDNA reactions contained 2 μL of Fast SYBR Green Master Mix (BioRad, UK), each primer at 200 nM, and 2 μL of cDNA from reverse transcription PCR in a 25 μL reaction. After an initial denaturation step at 95 °C for 10 min, amplification was carried out with 40 cycles of denaturation at 95 °C for 10 s and annealing and elongation at 60 °C for 30 s. Amplification was followed by a melting curve analysis to confirm PCR product specificity. No signals were detected in no-template controls. All samples were run in triplicate, and the mean value of each triplicate was used for further calculations. Relative gene expressions were calculated using the ΔΔCT method. The quantity of GAPDH (housekeeping) transcript in each sample was used to normalize the amount of each transcript, and then the normalized values were further normalized to the expression value in untreated cell samples to calculate a fold change value.

4.11 ELISA Assay

34 For the ELISA assay, cells were seeded in 96-well plates and treated in triplicates at 70% confluency, with the indicated concentrations of either s-GO or l-GO for 24 h. After the treatment, supernatants were collected and centrifuged at 10000*g* for 10 min at 4 °C to pellet down the GO. Supernatants were collected after centrifugation, and concentration of

cytokines was determined using human IL-6 and IL-8 kits (BD Biosciences, UK) according to the manufacturer's recommendations.

4.12 *In Vivo* Experiments

Experimental Animals

35 Six- to 8-week-old female C57BL/6 mice (Envigo, UK) were used in the present study, in accordance with the ARRIVE guidelines and after ethical approval from the UK Home Office, under Project License no. 70/7763. Animals were kept in IVC cages in groups of 5 with free access to food and water, on a normal 12 h light and dark cycle. All experiments were conducted using 3 animals per group, except at longer time points, where 5 animals were assigned to each group.

Intranasal Instillation

36 For the intranasal instillation of GO, mice were kept under light anesthesia (2.5% isoflurane with oxygen flow of 2 L/min). Instillation of 50 μL was performed by pipetting approximately half of the volume in each nostril. Mice were held in a supine position, tilted to about 60°, in order to allow for the efficient entry of the whole volume. Mice were observed until full recovery, which occurred within 5 min after instillation. Both l-GO and s-GO materials were diluted to 1 mg/mL in an aqueous solution of 5% (m/v) dextrose in ultrapure water, totaling an instilled dose of 50 μg. The same volume of aqueous solution of 5% dextrose was administered as a vehicle control. The 5% dextrose solution was sterile filtered prior to dispersion of GO.

Dissection of Lungs

37 Mice were sacrificed at 1, 7, and 28 days after exposure to a single dose of GO by terminal anesthesia, using intraperitoneal injection of 0.2 mL of pentobarbitone. The thoracic cavity was then carefully excised and lungs were dissected. After gently rinsing in Hank's balanced salt solution (Gibco, Thermo Fisher Scientific, UK) to remove excess blood, the lungs were split in two different containers both undergoing an overnight fixation step in 4% (m/v) paraformaldehyde at 4 °C. The left lung used for histopathological analysis was then transferred to 70% (v/v) ethanol before paraffin embedding. The right lung used for Raman mapping was transferred to an aqueous solution of 30% (m/v) sucrose before snap-freezing in optical cutting temperature (OCT) compound.

Lung Histopathology

38 Paraffin-embedded lung sections with a thickness of 5 μm were obtained for hematoxylin and eosin (H&E) staining. Images were collected using a 20× objective under a Pannoramic 250 Flash slide scanner (3D Histech, Hungary), in bright-field mode. Images were processed and analyzed using Pannoramic Viewer (version 1.15.4, 3D Histech, Hungary), with manual segmentation of cell infiltration performed using ImageJ software (version 1.51, National Institutes of Health, Bethesda, MD, USA). Cell infiltration was calculated as the relative area of lung parenchyma where visible alveolar thickening and formation of granulomas occurred. Granuloma size accounted for the area of the manually segmented features. At least 3 mice

were analyzed in each condition and time point.

Raman Mapping of Lungs

39 Snap-frozen lungs embedded in OCT compound were cryo-sectioned and gently rinsed with PBS 1×, Milli-Q water, and methanol, in order to remove any excess OCT compound. Sections were then dried at 37 °C before imaging under a DXRxi Raman microscope (Thermo Scientific, UK), with a 50× objective. Raman maps were obtained using a 633 nm laser operating at 0.4 mW, through a 50 μm pinhole aperture with an exposure time of 0.125 s. Correlation maps were calculated using the OMNICxi software (Thermo Scientific, UK), after comparing to a reference spectrum of GO sample (Figure S21).

4.13 Statistical Analysis

40 Every experiment was repeated at least twice with triplicates or sextuplicates for each condition. Data are represented as mean ± SD and were statistically analyzed with IBM SPSS software (version 22) or GraphPad Prism (version 6.05) using analysis of variance (one-way or two-way ANOVA) with $p < 0.05$ considered significant. *In vivo* data were analyzed in GraphPad Prism using the Kruskal–Wallis test with *post hoc* Dunn's multiple comparisons test with $p < 0.05$ considered significant.

Table 2. Sequence of the Primers for the Genes Analyzed Using RT-qPCR

gene	accession no. (GenBank)	primer	sequence (5′ → 3′)
IL-6	NM_000600.4	forward	AGTGAGGAACAAGCCAGAGC
		reverse	GTCAGGGGTGGTTATTGCAT
CXCL8 (IL-8)	NM_000584.3	forward	CGGAAGGAACCATCTCACTG
		reverse	AGCACTCCTTGGCAAAACTG
TNF	NM_000594.3	forward	TGGGATCATTGCCCTGTGAG
		reverse	GGTGTCTGAAGGAGGGGGTA
IL1β	NM_000576.2	forward	AGCTGATGGCCCTAAACAGA
		reverse	CCTGAAGCCCTTGCTGTAGT
IL1α	NM_000575.4	forward	ACTGCCCAAGATGAAGACCA
		reverse	CCGTGAGTTTCCCAGAAGAA
HMOX1 (HO1)	NM_002133.2	forward	AGCTCTTTGAGGAGTTGCAGGA
		reverse	AGCTGAGTGTAAGGACCCATCG
CSF2 (GM-CSF)	NM_000758.3	forward	GCTGCTGAGATGAATGAAAC
		reverse	AGTCAAAGGGGATGACAAG
GAPDH	NM_002046.5	forward	CCACATGGCCTCCAAGGAGTAAGAC
		reverse	AGGAGGGGAGATTCAGTGTGGTGG

5. Supporting Information

...

References

...

Task 5: Transform the following high-frequency verb-noun collocations used in research article methods sections into active/passive voice. Determine what verb-noun collocations are typically used for each of the following communicative purposes, namely 1) describing the experimental setting, 2) presenting preparation or preprocessing of material samples, 3) describing experimental procedures, 4) describing data collection processes, and 5) describing data analysis methods and steps.

No.	(Phrasal) Verbs	Noun collocates	Examples
(1)	use	method, software, microscope, apparatus, spectrometer, machine, analyzer instrument, radiation, solution, technique, model, equation, microscopy, package, procedure, reader, filter, camera, test, formula, meter, detector, algorithm, spectroscopy, approach	· using Cu Kα radiation · using a scanning electron microscope · using the Vienna ab initio simulation package · the Ewald summation method was used · RF magnetron sputtering technique was used · a high-speed camera is used
(2)	perform, carry out, conduct	measurement, test, analysis, experiment, calculation, imaging, simulation, testing, characterization, procedure, treatment, process, microscopy, scanning, reaction, observation, synthesis, diffraction, calibration optimization, deposition, comparison, preparation	· X-ray photoelectron spectroscopy (XPS) was performed · electrochemical measurements were performed · transmission electron microscopy (TEM) was conducted · performing Vickers micro-hardness measurements · performing high angle annular dark field scanning transmission electron microscopy · conducting Harman's single-factor post-hoc test

continued

No.	(Phrasal) Verbs	Noun collocates	Examples
(3)	wash	cell, sample, product, particle, medium, mixture, membrane, powder, solution, plate, substrate, film	· the combined organic phase was washed · the hydroxide-exchanged cellulose membrane was then washed · the resulting peptide-grafted plates were washed
(4)	record	spectrum, image, pattern, curve, datum, signal, temperature, measurement, change, weight, process, intensity, characteristic, response, force, spectroscopy, time, emission, current, voltage, behavior, resistance, micrograph	· UV-vis absorption spectra were recorded · powder-ray diffraction patterns were recorded · the transmission electron microscope (TEM) images were recorded · record the induced transient voltage signal · recording the radial deformation behavior · record the dynamic receding contact angle
(5)	incubate	cell, sample, mixture, solution, section, plate, membrane, particle, antibody, spheroid	· incubating the drug-loaded hydrogel solution · transferrin-coated polystyrene nanoparticles were incubated · the treated skin mixtures were then incubated · substrates not functionalized with S1P were incubated
(6)	scan	microscopy, microscope, sample, beam, pattern, image, surface, area, specimen	· the specimens for the experiment were scanned · the collected patterns were scanned · scan the harvested specimens · scanning the surface specimen

continued

No.	(Phrasal) Verbs	Noun collocates	Examples
(7)	prepare	sample, solution, film, electrode, specimen, composite, electrolyte, particle, cell, layer, gel	· prepare the working electrode · preparing thin (80 nm) cross-sectional samples · preparing a heparin-cell suspension mixture · a glycerol-water solution was prepared · wool keratin solution extracted from fibers was prepared · perovskite samples for XRD measurement were prepared
(8)	stir	mixture, solution, suspension, reaction, dispersion, powder	· the reaction mixture was stirred · this reaction dispersion was magnetically stirred · stirring the solution
(9)	obtain	image, spectra, solution, sample, value, curve, film, powder, pattern, datum, product, measurement, property, distribution	· obtain a homogeneous solution · obtain the final product · obtain the stress-strain curves · obtain a larger measurement thermoelectric potential · SEM images were obtained · the images and signals were obtained · the mechanically exfoliated single layer graphene films are obtained
(10)	culture	cell, structure	· mouse bone marrow mesenchymal stem cells (MSCs) were cultured · culturing human osteoblast cell
(11)	add	solution, water, amount, powder, mixture, medium, acid	· adding a chemiluminescent substrate · adding then and precursor solutions · samples containing 50 μg were added · the nanobipyramid solution (3.65 mL) was added

continued

No.	(Phrasal) Verbs	Noun collocates	Examples
(12)	take	image, measurement, sample, value, spectra, photograph, micrograph	· scanning electron microscopy images were taken · all bright field and fluorescent images were taken · X-ray photoelectron spectra were taken · took a digital photograph · took four consecutive high-resolution images
(13)	follow	procedure, protocol, method, instruction, process, step, approach, guideline, formula, standard	· following a previously reported procedure · following the similar synthetic approach · follow the life-cycle assessment method · the classical three-step method with continuous heating mode was followed · thermal imidization procedure was followed
(14)	acquire	image, spectra, datum signal, measurement	· photoluminescence spectra were acquired · acquiring SEM and micrographs
(15)	measure	spectra, property, absorbance, concentration, size, thickness, intensity, density, conductivity, content, temperature, resistance, diameter, angle, value, area	· measuring their fluorescence emission spectra · measuring the reflection scattering parameter · Raman spectra were measured · the electrical conductivity was measured
(16)	collect	spectra, datum, sample image, product	· the Raman spectra were collected · XRD patterns were collected · collecting 20 pre-bleach and 200 post-bleach images

continued

No.	(Phrasal) Verbs	Noun collocates	Examples
(17)	deposit	layer, film, electrode, bead, solution, droplet, sample	· deposit the dielectric layer · depositing the first bead · a 15 nm thick layer was deposited · a 10 nm adhesion layer and 100 nm contact layer were deposited
(18)	observe	morphology, microstructure, cell, surface, sample, change, structure	· the morphology of the samples was observed · the microstructures of the samples were observed · observe the microscopic structures · observing the impact facture surface
(19)	treat	cell, sample, surface, substrate, solution	· treating the natural fiber surface · the obtained ultrathin layer was then treated · the surfaces in contact with the air were treated
(20)	determine	concentration, content, composition, amount, size, value, property, distribution	· the specific surface area was determined · the Young's modulus was determined · determine the tensile properties · determine the particle size distribution
(21)	replace	medium, solution, atom	· replacing the electrolyte solution · the culture medium was replaced
(22)	form	solution, film, layer, structure	· a clear colorless solution was formed · no interfacial carbide layer is formed · form a homogeneous solution · forming wet carbon films

continued

No.	(Phrasal) Verbs	Noun collocates	Examples
(23)	characterize	morphology, structure, property, composition, phase	· characterize the mechanical properties · characterize the crystalline phases · the morphology of the samples was characterized · thin film microstructure and crystallinity were characterized
(24)	inject	solution, cell, mixture	· the zinc precursor solution was continuously injected · all samples were intravenously injected
(25)	remove	solvent, medium, layer, residue, water, solution	· the medium containing the nanoparticles was removed · the organic solvent was removed · remove the residual solvent · remove the native oxide layer
(26)	assemble	electrode, battery, layer	· assemble the Li-ion battery · the working electrodes and counter electrodes were assembled
(27)	place	sample, mixture, film, substrate, solution, plate	· the prepared ANF films were placed · the polymer aqueous solution (0.5 mL) was placed · placing the replica/adhesive film
(28)	heat, cool	mixture, sample, solution, furnace	· cool to room temperature · heat the silicon/PVA/PMMA substrate
(29)	dissolve	powder, polymer, sample, crystal	· dissolving the sufficiently dried powder · the resin samples before curing were dissolved

continued

No.	(Phrasal) Verbs	Noun collocates	Examples
(30)	fix	cell, sample, tissue	· monolayer and spheroid samples were fixed · tissues harvested for H&E were immediately fixed
(31)	calculate	value, volume, rate, concentration, energy, area, density, ratio, modulus, percentage, angle	· the specific surface areas were calculated · specific capacity values were calculated · the angle between the two lines was then calculated
(32)	cut	sample, specimen, section, film	· rectangular samples with nominal surface areas of 0.2 cm2 were cut · micrometer thick sections were cut
(33)	repeat	process, procedure, experiment, test	· the coating/drying process was consecutively repeated · the procedures of washing and filtering were repeated · repeating the dip-coating procedure
(34)	capture	image, process, spectra	· capturing the-visible absorption spectra · the top-view and cross-sectional images of the samples were captured · the dynamic process of the contact-separation was captured
(35)	transfer	solution, mixture, sample	· the reaction mixture was transferred · then the aqueous solution was gently transferred
(36)	apply	voltage, condition, field, pressure, load, force	· periodic boundary conditions were applied · a distributed tensile load was applied · applying an external magnetic field · applying uniaxial tensile load

continued

No.	(Phrasal) Verbs	Noun collocates	Examples
(37)	clean	substrate, sample, surface	· cleaning the glass substrates · clean the particle surface
(38)	increase	temperature, pressure, voltage	· increase the electrical conductivity · increasing the external radial pressure · the temperature of the system was gradually increased
(39)	test	specimen, sample, property, performance	· at least five specimens were tested · three individual aerogel samples were tested · testing the bacterial resistance property
(40)	grow	cell, film, crystal, layer, graphene	· grow GaN buffer layer · the smooth films were epitaxially grown · the perovskite oxide films were grown · the graphene used in this work was grown
(41)	centrifuge, dilute	solution, sample, mixture	· the mixture solution was centrifuged · the samples were carefully centrifuged · the solution of purified nanorods was then diluted
(42)	analyze	datum, image, sample composition	· the diffraction patterns were analyzed · the surface morphology and chemical composition of the specimens were analyzed · analyze the chemical composition
(43)	disperse	powder, particle, catalyst	· dispersing 50 mg catalyst powder · the obtained powders were dispersed

continued

No.	(Phrasal) Verbs	Noun collocates	Examples
(44)	solve	equation, problem, structure	· the scalar relativistic equations without spin-orbit interactions were solved · the crystal structure was solved
(45)	purify	product, mixture	· the nanoparticle mixture was purified · purify the crude product
(46)	control, set, maintain, change, keep, raise	temperature, rate, pressure, condition	· setting subsonic boundary conditions · controlling the sulfurization temperature · control the water evaporation and stirring rate
(47)	develop	model, method, algorithm, structure	· develop a thermodynamically consistent model · an automated image analysis method was developed
(48)	evaluate	effect, performance property, efficiency	· the photocatalytic performance was evaluated · the therapeutic effects were evaluated
(49)	purchase	chemical, ethanol	· all the chemicals and reagents were purchased · purchase energy-efficient products
(50)	fabricate	device, electrode, film, sample, composite	· fabricating densely packed thin films · fabricate a prototype photoelectric conversion device · three nanocarbon samples were fabricated · the gradient-structured porous film was fabricated
(51)	immerse	sample, substrate	· the metal-coated samples were also immersed · the quartz substrate was immersed

continued

No.	(Phrasal) Verbs	Noun collocates	Examples
(52)	adjust	pH, concentration, value, ratio, distance	· the pH of the solution was adjusted · concentrations of copper and sulphate were adjusted · adjusting the flow rate ratio
(53)	filter	solution, mixture, signal	· filtering the mixture · the recorded noise signal was filtered
(54)	evaporate	solvent, layer	· a 50-nm-thick Au layer was thermally evaporated
(55)	synthesize	particles, polymer	· Cobalt polysulfide nanoparticles were synthesized
(56)	process	image, datum, signal	· the camera images taken with a wire-grid polarizer were processed
(57)	coat, rinse	sample, surface	· the Au-coated samples were rinsed · the cryofractured surface was coated
(58)	fit	datum, curve, spectra	· the scattering curve recorded at 680 °C was also fitted · the resulting spectra were fitted
(59)	extract	phase, sample	· the organic phase was extracted
(60)	calibrate	intensity, energy	· the light intensity was calibrated
(61)	report	value, result	· report the average value
(62)	polish	surface, sample	· the sample surface with scratches were polished
(63)	minimize	effect, error, damage	· minimizing a least-squared error
(64)	initiate	reaction, synthesis	· initiate the photocatalytic polymerization · initiating graphene synthesis
(65)	represent	value, number, energy	· values of focal adhesion size frequency were represented
(66)	consider	value, difference, *P*	· *P* values of less than 0.05 were considered
(67)	define	area, pattern, electrode	· electrode patterns are first defined
(68)	indicate	temperature, value	· the values of intraband and interband QYs are indicated

Task 6: Read the Methods and Experimental section of Text I carefully and identify the typical verb-noun collocations used in the description of material preparation, experimental procedures, and data collection and analysis.

Task 7: Read Text II carefully and identify the differences between the methods sections of a research article and a lab manual in the use of technical vocabulary, complex noun phrases, sentence patterns, and adverbial phrases/clauses.

Text II

Lab Manual: Rockwell Hardness Test[7]

Objective

To measure the Rockwell hardness of ferrous and non-ferrous metals such as hard alloy, carbon steel, alloy steel, cast iron, brass, aluminium, etc.

Theory and Scope

Hardness may be defined as resistance to penetration or resistant to abrasion. The test involves determination of the depth of dent on the specimen caused by the penetration of certain indenter under certain standard load.

In Rockwell hardness test, an indenter is forced into the surface of a test specimen in two operations. In the first operation, a small initial or datum (minor) load of 10 kg is applied to the penetrator to take care of the roughness of the surface of the specimen; whereas in the second operation a major or standard load (60, 100, 150 kg) is added. The permanent increase in the depth of penetration from the depth reached under the datum load due to the standard load is measured after removing it. The reading on dial which is inversely proportional to the depth of penetration represents the hardness of the material; so that the greater the penetration, the lower the hardness number and vice versa. The indenter used is either a steel ball or diamond cone having an angle of 120 degrees made of black diamond. The indentor is selected depending on the nature and condition of the material. This indentation test which is used on smaller specimens and harder materials is conducted as per IS 1586:2000.

Rockwell hardness testing machine is more extensively used because of its simple testing procedure, direct reading of hardness number. This machine can be used to test materials like hard steel, mild steel, aluminum, cast iron, brass, etc. The indentator or penetrator is either a steel ball or diamond cone with slight rounded point. A steel ball is used with a load of 100 kg to test softer materials like brass and hardness number is found on B scale. A diamond cone is used with a load 150 kg to test harder materials like hard steel and hardness number is found on C scale.

Apparatus

Rockwell hardness testing machine; Ball and Brale (diamond) Indentors; 0.0 Emery paper.

Description of Apparatus

Rockwell hardness testing machine

Rockwell hardness testing machine impacts a standard load on a steel ball or Brale (diamond) indentor. The depth of indentation is recorded on a dial gauge in terms of Rockwell hardness numbers. The dial gauge of the machine is provided with red and black scales with a long pointer. Red scale is used for hardness readings obtained with Ball indentor and black scale is used for Diamond indentor. Hardened steel is tested on C scale with diamond indentor and 150 kg major load. Softer materials are tested on the B scale with 1.587,5 mm diameter steel ball and 100 kg major load. Application of major load is preceded by the minor load of 10 kg. Typical Rockwell hardness testing machines are shown in Figure 1.

Indentors

Brale indentor also called black diamond indenter, a conical-shaped diamond penetrator with 120 degrees apex angle and 0.2 mm radius tip is used for hard steel and cast iron. A hardened steel ball 1.587,5 mm diameter is used for non-ferrous metals.

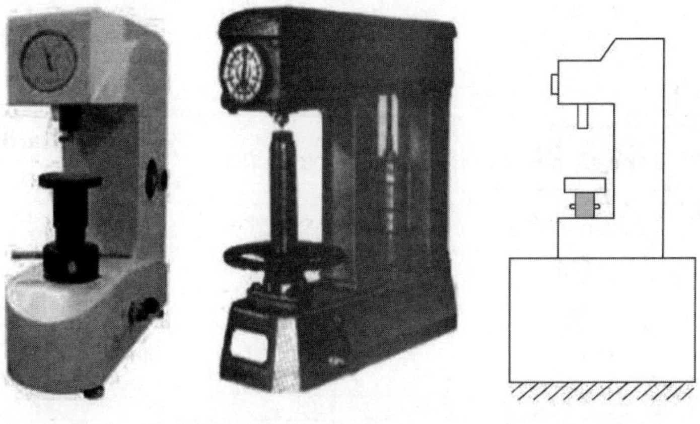

Figure 1 Typical Rockwell hardness testing machines

Procedure

Step 1: Select standard load of 100 kg or 150 kg to be applied to the specimen depending on the nature of the material to be tested.

Step 2: Select the penetrator to be used corresponding to the load selected in Step 1.

Step 3: Remove any oxide scale and foreign materials from the surface of the specimen by rubbing with emery paper.

Step 4: Place the specimen on the hardened anvil of the machine and turn the hand wheel to move the anvil upwards till the surface of specimen just touches the indentor.

Step 5: Apply the datum or minor load of 10 kg to the specimen by rotating the hand wheel slowly until the smaller needle (pointer) on the dial reaches the red mark (dot); in some machines at this point the pilot lamp goes off.

Step 6: Actuate the lever or handle to apply major load of 100 kg for scale B and 150 kg for scale C to the specimen by rotating the hand wheel in about 5 to 8 seconds.

Step 7: On completion of penetration, remove the major load by pulling backward the loading handle to the original position; the initial load 10 kg will still be on the specimen.

Step 8: Read the position of the pointer on the C or B scale dial which gives Rockwell Hardness Number (RHN) of the specimen; record the hardness number read from the dial of the machine.

Step 9: Release the minor load of 10 kg by rotating the hand wheel anticlockwise and lowering the screw.

Step 10: Repeat the Steps 4 to 9 for five times on the same specimen selecting different points for indentation.

Step 11: Take average of five values of indentation of each specimen to obtain the hardness number of the material sample. Plot the bar charts separately for B and C Scales.

Step 12: Compare the results obtained from other hardness tests and draw conclusions.

Observations and Calculations

For different materials guidelines for selecting indentor, major load and hardness scale are listed in Table 1.

Material of the specimen is …

Table 1 Observation table

Sr. No.	**Material**	**Type of Indentor**	**Applied load, kg**	**Scale (S)**	**Rockwell Hardness (RH) No.**		
					Trial No.	***RHN***	***Average RHS***
1.	Hard Steel	Diamond-Cone (120°)	150	RC	1	…	
					2	…	
					3	…	
					4	…	
					5	…	
2.	Mild Steel	Steel ball 1.5875 mm radius	100	RB	1	…	
					2	…	
					3	…	
					4	…	
					5	…	
3.	Brass	Steel ball 1.5875 mm radius	100	RB	1	…	
					2	…	
					3	…	
					4	…	
					5	…	

Result

Rockwell hardness number for material is …

Precautions

1. Surface of the specimen should be well polished, free from oxide scale and any foreign materials.

2. Thickness of the specimen should not be less than eight times the depth of indentation to avoid the deformation to be extended to the opposite surface of specimen.

3. Indentation should not be made nearer to the edge of a specimen to avoid unnecessary concentration of stresses. In such a case, distance from the edge to the center of indentation should be greater than 2.5 times diameter of indentation.

4. Rapid application of load should be avoided. Sudden application of load on the ball may increase the effective indentation force. Also rapidly applied load will restrict plastic flow of a material, which produces effect on size of indentation.

5. Give at least 10 seconds after the lever comes to rest position before recording the reading.

Informative Comments

The Rockwell hardness test utilises the depth of penetration of a test indenter under a standard load as a measure of hardness. The indentor is selected depending on the nature and condition of the material. The standard load of 100 kg is applied in about 5 to 6 seconds whereas the 50 kg load is applied in about 6 to 8 seconds.

Rockwell hardness test is also commonly used to determine the hardness of ceramic materials, plastics such as nylon, polycarbonate, polystyrene, and acetyl. The dial gauge indicates the hardness based on the difference in depth of indentation produced by the datum and standard loads.

For general reference, the combinations of major load and indentor for some materials are given in Table 2.

Table 2 Selection of combination of major load and indentor for some materials

Scale	Major load, kg	Indentor	Application materials
A	60	Cone	Cemented carbide, thin steel, hardened steel
B	100	Steel ball	Brass, copper alloys, soft steels, aluminum alloys, malleable iron
C	150	Cone	Mild steel, cast iron, malleable iron, deep case hardened steel

Task 8: Determine whether each of the following descriptions of methods is taken from a research proposal, a lab report/research article, a lab manual, or a textbook.

(1) During mechanical testing, relevant data points including load, displacement, and time will be recorded for each sample. Additionally, high-resolution imaging techniques such as scanning electron microscopy (SEM) will be employed to capture the microstructural

characteristics of the samples. This information will enable a comprehensive understanding of the sample's mechanical behavior and provide valuable insights into the relationship between microstructure and mechanical properties.

(2) The collected data will be subjected to statistical analysis to determine the significance of any observed differences between the samples prepared using different techniques. The mean values, standard deviations, and confidence intervals will be calculated, and appropriate statistical tests, such as t-tests or analysis of variance (ANOVA), will be employed to evaluate the significance of the observed variations.

(3) The microstructural analysis will involve examining the SEM images to identify key microstructural features such as grain size, grain boundaries, and the presence of any defects or inclusions. Quantitative analysis techniques, such as image processing and grain size distribution analysis, will be employed to extract numerical data from the images. These data will be correlated with the mechanical properties to establish potential relationships between microstructure and material performance.

(4) Hardness testing will provide insights into the material's resistance to localized deformation. The Vickers hardness test will be performed using a microhardness tester. A diamond indenter will be pressed into the sample surface, and the resulting indentation will be measured. The hardness value will be determined based on the indentation dimensions.

(5) The first step in the determination of phase compositions (in terms of the concentrations of the components) is to locate the temperature–composition point on the phase diagram. Different methods are used for single- and two-phase regions. If only one phase is present, the procedure is trivial: the composition of this phase is simply the same as the overall composition of the alloy. For example, consider the 60 wt% Ni–40 wt% Cu alloy at (point A, Figure a). At this composition and temperature, only the phase is present, having a composition of 60 wt% Ni–40 wt% Cu. For an alloy having composition and temperature located in a two-phase region, the situation is more complicated. In all two-phase regions (and in two-phase regions only), one may imagine a series of horizontal lines, one at every temperature; each of these is known as a tie line, or sometimes as an isotherm. These tie lines extend across the two-phase region and terminate at the phase boundary lines on either side. To compute the equilibrium concentrations of the two phases, the following procedure is used: 1) a tie line is constructed across the two-phase region at the temperature of the alloy; 2) the intersections of the tie line and the phase boundaries on either side are noted; and 3) perpendiculars are dropped from these intersections to the horizontal composition axis, from which the composition of each of the respective phases is read.

(6) Forming operations are those in which the shape of a metal piece is changed by plastic deformation; for example, forging, rolling, extrusion, and drawing are common forming techniques. Of course, the deformation must be induced by an external force or stress, the magnitude of which must exceed the yield strength of the material. Most metallic materials

are especially amenable to these procedures, being at least moderately ductile and capable of some permanent deformation without cracking or fracturing. When deformation is achieved at a temperature above that at which recrystallization occurs, the process is termed hot working; otherwise, it is cold working. With most of the forming techniques, both hot- and cold-working procedures are possible. For hot-working operations, large deformations are possible, which may be successively repeated because the metal remains soft and ductile. Also, deformation energy requirements are less than for cold working. However, most metals experience some surface oxidation, which results in material loss and a poor final surface finish. Cold working produces an increase in strength with the attendant decrease in ductility, since the metal strain hardens; advantages over hot working include a higher quality surface finish, better mechanical properties and a greater variety of them, and closer dimensional control of the finished piece. On occasion, the total deformation is accomplished in a series of steps in which the piece is successively cold worked a small amount and then process annealed; however, this is an expensive and inconvenient procedure.

(7) The term annealing refers to a heat treatment in which a material is exposed to an elevated temperature for an extended time period and then slowly cooled. Ordinarily, annealing is carried out to 1) relieve stresses; 2) increase softness, ductility, and toughness; and/or 3) produce a specific microstructure. A variety of annealing heat treatments are possible; they are characterized by the changes that are induced, which many times are microstructural and are responsible for the alteration of the mechanical properties. Any annealing process consists of three stages: 1) heating to the desired temperature, 2) holding or "soaking" at that temperature, and 3) cooling, usually to room temperature. Time is an important parameter in these procedures. During heating and cooling, there exist temperature gradients between the outside and interior portions of the piece; their magnitudes depend on the size and geometry of the piece. If the rate of temperature change is too great, temperature gradients and internal stresses may be induced that may lead to warping or even cracking. Also, the actual annealing time must be long enough to allow for any necessary transformation reactions. Annealing temperature is also an important consideration; annealing may be accelerated by increasing the temperature, since diffusional processes are normally involved.

(8) Conventional heat treatment procedures for producing martensitic steels ordinarily involve continuous and rapid cooling of an austenitized specimen in some type of quenching medium, such as water, oil, or air. The optimum properties of a steel that has been quenched and then tempered can be realized only if, during the quenching heat treatment, the specimen has been converted to a high content of martensite; the formation of any pearlite and/or bainite will result in other than the best combination of mechanical characteristics. During the quenching treatment, it is impossible to cool the specimen at a uniform rate throughout—the surface will always cool more rapidly than interior regions. Therefore, the austenite will transform over a range of temperatures, yielding a possible variation of microstructure and properties

with position within a specimen. The successful heat treating of steels to produce a predominantly martensitic microstructure throughout the cross section depends mainly on three factors: 1) the composition of the alloy, 2) the type and character of the quenching medium, and 3) the size and shape of the specimen. The influence of each of these factors is now addressed.

(9) The experimental procedure for observing HOLZ lines is quite straightforward but, since the lines themselves can be rather elusive, as we suggested previously, you should practice with a specimen such as Si or stainless steel in which the lines are almost always visible. The best way to search for the lines is:

- Select the largest C2 aperture and go to the smallest L (~ 3-500 mm) at which you can see the full angular view of the BFP.
- Examine the Kossel/LACBED pattern (make 2α large) which should reveal Kikuchi bands intersecting at many poles, spanning a good fraction of the stereographic triangle as shown in Figure A.
- If you tilt to such a pole, as shown in Figure B, you should see the ring of HOLZ intensity.
- Tilt to a suitable zone axis for optimizing HOLZ effects. Remember, the best orientation for seeing the HOLZ lines in the 000 disk is not a low-index, high-symmetry pole such as <100> or <111> , but a higher-index, lower-symmetry one such as <114>.
- To see the deficient lines, increase *L* to look in detail at the 000 region of the pattern, and if necessary, put in a smaller C2 aperture, center it and look for the fine dark lines crossing the bright disk as in Figure C. Usually, you'll just need this deficient-line distribution.

(10) Lime reactivity test. 1) Adjust the temperature of about two litres of distilled water to 20° C (25 ± 0.5° C ASTM). Add 600 ± 1 ml (400 ml ASTM) of this water to the reactivity container or flask. 2) Set the mechanical stirrer with the stirring blade or stirring rod in the flask revolving at 300 ± 10 rpm (400 ± 50 rpm ASTM). Insert the thermometer; the temperature of the water in the flask must remain at 20 ± 0.5° C (25 ± 0.5°C ASTM). 3) Weigh 150 ± 0.5 g (100 g ASTM) of the prepared sample for the individual test. With the stirrer running, immediately introduce the weighed quantity of sample into the water by means of the feeding vessel. Start timer. This moment marks the start of the lime slaking period. 4) Record the slaking temperatures at 30 second intervals. Continue readings until less than 0.5 °C temperature change is noted in each of the three consecutive readings. The time at which the first of three consecutive readings was taken marks the end of the lime slaking period. If the period of slaking exceeds 10 minutes record the temperature at one or two intervals after this period till the maximum temperature T_{max} is reached. 5) Draw the wet slaking curve by plotting the measured slaking temperature in °C as a function of time in minutes. Determine

the time t necessary to reach the required slaking temperature T. 6) Calculate RDIN values in each case by dividing 2400 (40 °C temperature rise × 60 sec/min) by the time in seconds in which the temperature rise occurred. 7) Conduct at least three repetitions of each test run to obtain reasonably repeatable values.

(11) The ASTM C127 and C128 procedures were followed to determine the density, specific gravity, and absorption capacity of the fine and coarse aggregate. To determine the gradation of the candidate Michigan aggregate, dried fine and coarse samples were analyzed using the ASTM C136 method. To determine the unit weight of the candidate Michigan coarse aggregate sample, the ASTM C29 procedure was followed.

(12) Test procedures for single shear and double shear tests are similar except that specimen holding fixtures are different. 1) Measure the diameter d of the specimen using sliding vernier calipers taking the average of three readings taken at different points along the length of the specimen. 2) Calculate the maximum load expected to be applied on the specimen using the equation given. 3) Set the universal testing machine for the selected load range. 4) Select an appropriate set of shear shackles to assemble the double shear device. Mount the top part of the shear attachment on the middle adjustable cross-head and the bottom part of the shear attachment shall be mounted on the lower compression crosshead by the mechanism provided for the purpose so that the holes provided for inserting the specimen in the shear attachment are in line. 5) Insert the specimen into the shear device to pass through the holes centrally so that it projects equally on either side. The inner diameter of the hole in the shear test device is slightly greater than that of the specimen. Thus, the arrangement is made so that the specimen will be sheared off at two cross sections due to the applied load, and the test performed is known as double shear test. 6) Move down the intermediate cross slide till it makes contact with the top of the center plate, through which the load is applied to the specimen. 7) Set the load pointer to zero. 8) Start the machine and apply the load gradually till the specimen ruptures by shearing at two cross sections completely. Ensure that the load remains within the range without any need for alteration. At this point record, the load applied on the specimen as indicated on the dial gauge. 9) Remove the load to take out the shear attachment and broken pieces of the specimen. 10) Examine the nature of the failure of sheared surface, sketch and note down the salient features. 11) Calculate the shear strength of the given material specimen. 12) Repeat the experiment with other specimens.

(13) Metal fatigue test procedure. 1) Measure and record the dimensions of the specimen. 2) Document the testing parameters to be used namely, the waveform, maximum and minimum loads, frequency, etc. 3) Clamp one end of the specimen to a motor of the testing machine and fit the other end to the bearing. 4) Hang a predetermined weight at the bearing end for stressing the specimen to the required stress level. 5) Measure the distance from the load end to the minimum diameter of the specimen l mm. 6) Start the motor to rotate the specimen at a constant speed and continue till the specimen fails. 7) Record the number of cycles at which

the specimen failed from the revolution counter. 8) Test the other specimen of the sample with changed weights as outlined in the preceding steps. 9) Construct the S-N curves for the material using fatigue data provided by the test specimens. 10) Examine the fracture surfaces of the broken fatigue specimen and sketch the result with details. 11) Analyse and discuss the obtained results.

(14) Experimental details including full synthetic procedures, FT-IR, elemental analysis and powder X-ray diffraction measurements, thermogravimetric analysis, gas sorption properties, CV measurements, UV-vis diffuse reflectance spectra (DRS) and photoluminescence spectra, scanning electron microscopy (SEM), transmission electron microscopy (TEM), hydrogen evolution experiments, X-ray photoelectron spectroscopy, water wetting properties, theoretical calculations and NMR spectra are available in the Supporting Information.

(15) We performed CVD growth using MoO_3 and sulfur solid precursors without requiring a growth promoter. The growth was performed using a custom-made setup at 850 °C with a nitrogen flow of 10 sccm. The optimal quantities of MoO_3 and sulfur precursors are about 6 mg and 100 mg. In all the experiments, the films were grown in the presence of excess sulfur. The 2D MoS_2 array was fabricated using an e-beam lithography step followed by patterning in an CF_4/O_2 plasma. The 2D arrays were stimulated using a green laser for producing an optical response.

(16) The experiments were carried out with a custom-built low-temperature ($T = 5$ K) STM under ultrahigh vacuum conditions (base pressure $<10^{-10}$ mbar). Single-crystal gold substrate was cleaned by repeated cycles of sputtering with Ar ions and annealing at 723 K. Asymmetric starphene precursor molecules were sublimated from a Knudsen cell at a temperature of 500 K onto a Au(111) surface kept at room temperature. After sublimation, the sample was cooled to cryogenic temperature and transferred to the STM without breaking the vacuum.
For manipulation of single Au atoms, a tunneling resistance between 0.2 and 1 MΩ was used. Scanning tunneling spectroscopy measurements were performed using a lock-in technique with a bias modulation amplitude of 40 mV at a frequency of 707 Hz.

(17) Two types of $La_{0.7}Sr_{0.3}MnO_3/BiFeO_3$ heterostructures were grown on Nb-doped TiO_2-terminated $SrTiO_3$ (001) substrates. The first sample had a layer stacking sequence at the interface of $La_{0.7}Sr_{0.3}O$–MnO_2–BiO—FeO_2. The layer termination of the second sample was MnO_2–$La_{0.7}Sr_{0.3}O$–FeO_2–BiO.(5, 7, 19) Both types of heterostructures were then transferred to an ultrahigh vacuum STM chamber (7×10^{-11} mbar) and cleaved in situ at 80 K to obtain the cross-sectional slice of the Nb-$SrTiO_3$/$La_{0.7}Sr_{0.3}MnO_3$/$BiFeO_3$ heterostructrues. The scanning tunneling spectra (STS) were acquired at 80 K by using the current imaging tunneling spectroscopy (CITS) mode, where a series of tunnel current images was obtained at different sample bias voltage Vs. In this work, Vs was varied from +3.5 V to -3.5 V for STS measurements.

(18) The PMN-29PT single crystals (CTG Advanced Materials, LLC, USA) used in this study

have the rhombohedral $3m$ crystal symmetry at room temperature. The platelet samples have the size of $2.5 \times 2.5 \times 0.113$ mm^3, and coated with gold electrodes on the two larger surfaces. The $P-E$ loops were measured using a modified Sawyer-Tower circuit at room temperature with E_0 ranging from 1.43 kV/cm to 7.5 kV/cm at a fixed frequency (f = 1 Hz, 10 Hz, and 50 Hz), and with different f from 0.1 Hz to 100 Hz at a fixed E_0 ($1E_c$, $2E_c$ and $3E_c$), where the coercive field (E_c) is defined as 2.5 kV/cm according to the 0.1 Hz $P-E$ loop measured at room temperature. The hysteresis area, remnant polarization and coercive field are obtained directly from these $P-E$ loops. The loop area is obtained by the integration of the closed curve using a self-written Matlab code.

(19) The hyperlens fabrication started from the glass substrate with half-cylindrical grooves. A 100 nm Cr was first deposited to a flat glass substrate, and then the FIB was used to mill an array of 50 nm wide slits on the Cr film. A cylindrical grooves array was obtained using the Cr film as a mask, as well as isotropic wet etching in a buffered oxide etch (in a 6:1 ratio). In the next step, Cr film was removed using CR-7 Cr etchant, and then a 50 nm of Cr was deposited by electron beam evaporation to be used as the photomask. Different patterns shown in Figure 4a were made on the Cr film using the FIB. Subsequently, the Cr mask was covered by 60 nm thick PMMA A2 (MicroChem) using spin-coating, on top of which the Ag/Ti_3O_5 multilayered structures were deposited using 2-step method. All the film depositions in this work were done using the SEMICORE e-beam evaporation system.

(20) To fabricate devices, SY-PPV (Merck AG, relative molecular mass M_w 980,000) was dissolved in toluene and spin-coated onto a glass/ITO substrate covered with 60 nm poly(3,4-ethylenedioxythiophene):poly(styrenesulfonic acid) (PEDOT:PSS) as the hole-injecting layer. The layer thickness of SY-PPV is in the range of 100 nm to 180 nm. The cathode was subsequently thermally evaporated on top of the polymer (chamber pressure 10^{-7} mbar), leading to a PLED structure of ITO/PEDOT:PSS/SY-PPV/Ba (5 nm)/Al (100 nm), where ITO is indium tin oxide. For electron-only devices, patterned electrodes of Al were deposited on glass by thermal evaporation to form the anode, resulting in a structure of Al/SY-PPV/Ba/Al. For the blended PLEDs solutions were prepared by dissolving the required quantities of the host PVK (M_w 50,000–100,000) and semiconductor MEH-PPV (M_w 603,000) in chlorobenzene.

Task 9: Briefly describe a material test method in your research field based on the following questions while paying attention to the use of technical vocabulary, noun phrases with modifiers, and appropriate verb-noun collocations.

(1) What theory is the test method based on?
(2) What material properties does the test method determine?
(3) What are the applications of the material test method?

(4) On what factors does the precision of the test method depend?
(5) What are the requirements for the test specimens? What is the minimum number of specimens of a kind to be made for the test? How are the standard test specimens usually prepared?
(6) What are the requirements of the testing apparatus? What is the working principle or mechanism of the main apparatus?
(7) What basic simplifying assumptions are usually made in the design of the test setting? Explain how far these assumptions are valid.
(8) What basic parameters (variables) are measured or observed directly? What parameters (variables) reflecting material properties are calculated or assessed indirectly in the test?
(9) Is there a well-established relationship between the measured/observed parameters (variables) and specific material properties?
(10) What formulas are used for computing specific material properties?
(11) What are the crucial steps in the performance of the test method?
(12) Is it preferable to test the specimens from the underside, topside, cross-section, or others?
(13) What is the accuracy to which the basic parameters (variables) are measured, for example, the length of the specimens?
(14) In what units are the basic parameters (variables) expressed or recorded? How are the units of different parameters (variables) related to each other?
(15) Is it necessary to take the average of or apply correction to the obtained numerical values of basic parameters (variables)? What is the permissible variation while applying a correction or taking the average?
(16) What are the acceptance criteria for the results of the test method? Or what is meant by "good" results?
(17) How are the test results expressed? What kind of graph is often drawn with the test results?
(18) What is clearly undesirable in the test procedures or obtained results?
(19) What precautions need to be taken while performing the method?
(20) What might be the sources of error in the test?
(21) What factors may be responsible for any difference between laboratory results and standard material requirements?
(22) What are the advantages and limitations of the test method over the other methods available to test relevant material properties?

Task 10: Discuss in groups the functions of the underlined adverbial phrases/clauses to better understand the grammatical complexity in the written form of scientific language.

Note:

Adverbial phrases/clauses could be used to communicate information of different natures so as to provide additional context and description to the main clause of a sentence. Adverbial phrases/clauses could be used to express: 1) manner (describing how the action mentioned in the main clause is taking place or previously took place), 2) place (describing where the action in the main clause is taking place or previously took place), 3) time/sequence (describing when the action in the main clause is taking place or previously took place), 4) frequency (referring to the frequency of the action in the main clause), 5) condition (stating the conditions related to the verb, adverb, or adjective in the main clause), 6) reason (giving the reason for the action in the main clause), 7) purpose (explaining the reason to take a specific action), and 8) result (pointing out the outcome of the action in the main clause). An adverbial phrase could be in the form of an adverb, multiple adverbs, a prepositional phrase, a to-infinitive clause, an adjective phrase, a participial (-ed/-ing) clause, etc., while an adverbial clause is always a dependent clause containing a subject and a verb. In scientific language, it is commonly observed that multiple adverbial phrases/clauses are used to elaborate and extend the action in a sentence's main clause from different perspectives, reflecting the necessary precision and formality of knowledge construction.

Example sentence:

To visualize nanoparticle structure **(Purpose)**, transmission electronic microscopy (TEM) was performed by depositing the particles onto a 400-mesh carbon-coated copper grid (Electron Microscopy Sciences) and staining with 1 wt % uranyl acetate (Electron Microscopy Sciences) **(Manner)**, followed by observation under a Zeiss Libra 120 PLUS EF-TEM transmission electron microscope **(Time/sequence)**.

(1) To reflect the slight degree of misalignment of nanocubes within clusters, which can be noted from the TEM image (Figure 1), the particle positions within elementary clusters were randomized using the fractal generating algorithm described previously after setting the fractal dimension Df = 3, which produces cluster structures, as illustrated in Figure 2.

(2) To characterize the structures of different mesh electronics within glass needle-like constrictions to understand the design parameters for successful injection, a pulled glass tube with controlled ID central constriction was positioned under a microscope objective for bright-field and confocal fluorescence imaging, and the mesh electronics were partially injected through the constriction.

(3) This location was selected to place the biofiltration system as close as possible to waste reception hall (thereby ensuring constant supply of air contaminated with bioaersols generated from agitation of the waste heap) without exposing the researcher to the hazards and risks associated with tipping and loading operations as well as moving vehicles within the waste hall.

(4) The transient absorption (TA) was performed using a regenerative amplifier (Libra, Coherent, 1 kHz, 90 fs) with tunable wavelength and analyzed using a high-speed charge-coupled device (S11071-1104, Hamamatsu) with a monochromater (Acton 2358, Princeton

Instrument) at 1 kHz enabled by a custom-built board from Entwicklungsbuero Stressing.

(5) The relationship between the deflection and the distributed load was calculated according to the Euler-Bernoulli equation as FORMULA, where ω and q represent the deflection and distributed load, respectively, x describes the orientation direction of an arm, E is the Young's modulus of the material, and I is the second moment of area of an arm's cross section.

(6) The Raman measurements employed a 20× objective to obtain strong optical coupling to the thin films, resulting in an effective measurement area of approximately 75 μm × 75 μm and motivating the use of the Renishaw Streamline imaging capability to acquire about 250 Raman measurements for each composition sample, which were averaged to obtain the representative pattern for the sample.

(7) The mixture was first ball-milled with a speed of 110 rpm for 1 h to ensure the homogeneity of the obtained raw mixture, after which it was sealed in a quartz tube and annealed at various temperatures (300, 350, 400, 500, 550, and 600 °C) for 10 h to investigate the optimum synthesis temperature to achieve phase-pure Li6PS5Cl with high lithium ion conductivity.

(8) The gravimetric energy and the power densities of the KIC device were calculated by numerically integrating the galvanostatic discharge profiles using the equations below FORMULAS, where I is the constant current (A), U is the working voltage (V), m is the total mass of the electrode, t_1 and t_2 are the start/end-of-discharge time (s) of the KIC device, respectively, and t is the discharge time (s).

(9) Afterwards, composite mixtures were dried in a forced air circulation oven for 12 h at 60 °C, melted at 200 °C for 3 min, and pressed to 5 ton for 2 min followed by instantaneously depressurizing to release the gas due to vaporization of residual oil from CNSP, and pressed again to 8 ton for 3 min in a hot hydraulic press to prepare sheets of about 200 mm × 200 mm × 2 mm, followed by quenching in an ice bath.

(10) The PbSe nanocrystals before and after the heating experiments were imaged using high-resolution TEM (HRTEM), high-angle annular dark-field scanning transmission electron microscopy (HAADF-STEM), and energy dispersive X-ray spectroscopy (EDS) with the aberration corrected electron microscopes (TEAM 1 and ThemIS) operated at 300 kV at the National Center for Electron Microscopy of Lawrence Berkeley National Laboratory.

(11) The sandwich was finally program-cooled at a rate of 40 °C/min until 150 °C, during which the melt crystalline morphology of the sample was observed as a function of temperature on an optical microscope (Leica, DM 2500) under crossed polarizers, with its POM images taken at decreasing temperatures (210, 200, 190, 160, and 150 °C), respectively, using a digital camera connected to the microscope and interfaced with a microcomputer.

(12) Firstly, the Ar gas was injected into the chamber to remove residual air, after the temperature reached the target temperature, a flow of methane gas with small amount of oxygen was introduced into the chamber to yield the carbon black for a certain amount of time, then the flow of methane and Ar gas was turned off until the chamber was cooled to room temperature

by air, and finally the carbon black was collected for use.

(13) Then, the Sn layer was treated by H_2 plasma in a plasma-enhanced chemical vapor deposition (PECVD) system, with H_2 flow, RF power density and chamber pressure of 20 standard cubic centimeter per minute (SCCM), 10 mW/cm^2 and 30 Pa at 200 °C, respectively, for 5 min to form discrete Sn droplets that were later used as catalyst to mediate the vertical growth of SiNWs, via a plasma enhanced VLS process with silane (SiH_4) as precursor at 400 °C.

(14) In our computational analysis, we modelled the CID system in a predefined volume of a subcellular size, discretized the system in each spatial direction, resulting in a given number of equally sized voxels that satisfy the modelling assumptions and constraints, and used experimentally verified kinetic rate values for certain reactions as well as plausible values for the kinetic rates of the remaining reactions.

(15) In order to create bilateral frontal femoral condyle bone defects, after careful exposure of the knee joint, a 1 mm pilot hole was drilled and the defect was gradually widened using different drills of increasing size using a surgical motor (Elcomed 100, W&H Dentalwerk Burmoos, Austria) at low rotational drilling speed (800 rpm) and continuous external cooling with saline, until a final cylindrical defect of 2.5 mm in diameter and 5 mm in depth was reached.

(16) To investigate the fine microstructure and the chemical composition of the synthesized NPs, high-angle annular dark-field scanning transmission electron microscopy (HAADF-STEM), selected area electron diffraction (SAED), and energy-dispersive X-ray spectroscopy in STEM mode (STEM-EDX) were performed using JEM-ARM200F cold FEG probe and image aberration corrected microscope, operated at 200 kV and equipped with large angle CENTURIO EDX detector and QUANTUM GIF.

(17) After the introduction of prior knowledge for parameters (initial values (vector θ^0)), lower and upper bounds, and probability density functions (PDFs) in this algorithm, parameter vectors are randomly sampled from a non-stationary proposal posterior probability distribution, which is an arbitrary multivariate Gaussian distribution with a mean value at θ^0 for the first sampling or the last accepted parameter vector during the next parameter sampling.

(18) The composite film was scraped and crushed into brown powder for pyrolysis treatment in a tube furnace under N_2 atmosphere at 350 °C for 3 h with a ramp of 1 °C min^{-1} and then at 500 °C for 1 h with a ramp of 5 °C min^{-1}, which could generate supporting carbon in the pores to support and retain the porous structure as well as protect the Pt NPs during the crystallization of metal oxide framework.

(19) Moreover, by comparing the experimental data with the theoretical fitting curve, at the low membrane density condition (≤ 300 nM), which is employed later in this study, the apparent locomotion rate constant k'' can be viewed as independent of the initial concentration of the probes, thus validating our approach that extracts the inherent encounter rate differences among various surface ligands.

(20) Finally, the aqueous solution of QDs was filtered to remove free degraded gelatin with a molecular weight of less than 100 000 by using a Millipore ultrafiltration unit equipped with an ultrafiltration membrane with a nominal molecular weight limit of 100 000, and this process was repeated three times by dilution/ultrafiltration to obtain purified, gelatinated QDs in 50 mL of deionized water.

(21) To map internal forces being experienced across the different regions of the simulated axoneme, the length of each 24 nm (at rest) repeating unit of each microtubule doublet was measured computationally and the length of these units relative to the rest length was calculated at each time point, which indicated whether it was under tension (lengthened) or compression (shortened).

(22) To achieve a continuous buffer exchange (to increase the labeling efficiency) and free dye removal in the labeling process, on-line microdialysis systems were fabricated using hollow fibre membranes (MWCO 3 kDa, diameter of 500 μm), which were taken from tangential flow filtration modules (mPES MicroKros Modules, Spectrum) and tube connectors (Harvard Apparatus).

(23) Then, by plotting and linear fitting for the relation between the logarithm of the lifetime values (time needed to the end point) l*nt* and the reciprocal of the elevated test temperatures *T*, the activation energy E_a can be calculated as the product of the slope and the universal gas constant *R*, which makes life prediction under the lower service temperatures possible.

(24) The residual stress in PZT thin films was measured by a Raman spectroscopy at room temperature using a Jobin-Yvon spectrometer with an excitation of 442 nm line (80 mW) from HeCd laser (Kimmon Electric, USA), and X-ray diffraction (XRD) using a Bruker D8 Discover diffractometer (Bruker Corporation, Billerica, USA) equipped with a two-dimensional (2D) detector VNTEC-500.

(25) After the dislocation was introduced, periodic boundary conditions were applied along the *x* and *z* directions and the system was relaxed using the conjugate gradient algorithm at constant volume, followed by another relaxation at constant pressure to reach the minimum energy configuration and the dislocation was dissociated into two Shockley partials.

(26) Well calcined powder was repeatedly ground (for 3 h during each process), pelletized (10 mm diameter disk shape during each process), and heated as sintering processes at 1,050 °C (24 h under air environment), 1,150 °C (48 h under air environment), 1,250 °C (72 h under air environment) and, finally, at 1,350 °C (96 h under air environment).

(27) Then, absorption coefficients and reflectance spectra can be simulated by using the dielectric function expressed by the sum of Lorentz-type oscillators with resonant frequencies equal to vibration frequencies of phonon modes, the oscillator strength obtained by the above method, and the damping factor corresponding to a line width of 20 cm^{-1}.

(28) The stained cells were washed three times in PBS and used for confocal microscopy under a TCSSP2 confocal laser scanning microscope (Leica, Wetzlar, Germany) for visualization of

the reduced and oxidized fluorescent dye at excitation/emission wavelengths of 581/591 nm (Texas Red filter set) and 488/510 nm (traditional FITC filter), respectively.

(29) The reported XAS spectra were acquired at the $L_{2,3}$ edges of Mn and $M_{4,5}$ edges of Tb edges under 30 kOe of magnetic field (applied parallel to the X-ray propagation vector) and using the two circular polarization (left, σ+ and right, σ−) at normal incidence (θ = 0°) and rotating the sample plane θ = 45° respect to the X-ray propagation vector.

(30) The optical detection of the transients was carried out with a detection system consisting of a pulsed (pulser MSP 05 - Müller Elektronik Optik) Xenon lamp (XBO 450, Osram), a SpectraPro 500 monochromator (Acton Research Corporation), a R 9220 photomultiplier (Hamamatsu Photonics), and a 500 MHz digitizing oscilloscope (TDS 640, Tektronix).

(31) The electronic band structure was also obtained by first-principles DFT using the Quantum-Espresso package under the local density approximation (LDA) of the exchange-correlation functional and norm-conserving pseudopotentials following the Trouiller-Martins scheme to model the interaction between core and valence electrons of Mo and Se atoms.

(32) The capillary assembly of nanoparticles was in all instances preceded by a conditioning phase, whereby the colloidal meniscus is dragged across a non-patterned area of the pre-heated substrate for up to 20 min to achieve steady evaporative and flow conditions and stable nanoparticle accumulation prior to crossing over the nanopatterned traps.

(33) Furthermore, the mechanical properties of the as-synthesized scaffolds were studied by evaluating the samples' comprehensive strength using Material testing machine (TMS-pro, Food Technology Corporation, Sterling, Virginia), where the measurement distance value of the device was 1 mm, the measuring speed was 20 mm/min, and the measuring force was 0.

(34) After stirring the mixture for 20 min and sonication for 10 min to facilitate quantum dot dispersion, the homogeneous mixture was poured into a mould made of two low-roughness pieces of tempered glass linked by a poly(vinyl chloride) gasket, and irradiated with 365 nm light from a ultraviolet lamp for 5 min to trigger radical polymerization.

(35) To apply continuous stretching to the central zone of micropatterned cell colonies, microcontact printing was performed to print circular adhesive patterns with a diameter of 400 um onto the deformable PDMS membrane on top of pressurization compartments (with a diameter of 200 um) in a custom-designed microfluidic cell stretching device.

(36) The transmission spectra from deep-ultraviolet region to near infrared one were measured at room temperature by a McPherson VUVas2000 spectroscope and a PerkinElmer Lamda-900 UV–vis–near-infrared spectrophotometer with a transparent crystal plate, which was cut from as-grown crystals and polished to optical grade in thickness of 1 mm.

(37) The sweeps are optimized using node level performance engineering techniques, buffering techniques and performance measuring tools like the Intel architecture code analyzer (IACA), the lightweight performance-oriented tool suite for x86 multicore environments (LIKWID) and the Roofline model to efficiently exploit current HPC hardware.

(38) The pullout test samples were cut into small rectangular blocks, and the longitudinal cross-sections along the direction of the fibers were polished with sand papers and grinding paste to observe the influence of the surface modifications on the morphology of the interface between the Ni-Ti fibers and the vinyl ester resin matrix.

(39) The polymer was exposed to a 30 kV e-beam (VEGA3, Tescan) with a dosage of 1200 $\mu C/cm^2$ and developed in preheated NMP (100 °C) for 3 min (for a clean silicon substrate) or 1 min (for a silicon/graphene substrate) followed by 30 s in air (allowing the sample to cool down gradually), soaked for 30 s in IPA, and blow-dried by nitrogen flow.

(40) The morphology, microstructure, and element chemical states of the products were characterized by field-emission scanning electron microscope (FESEM, JEOL JSM-6700F), transmission electron microscope (TEM, JEOL JEM-2100F), and X-ray energy-disperse spectrometry (XPS, XSAM-800 spectrometer with an Mg Kα radiation source), respectively.

(41) The microhardness test with a load of 50 g for 15 s at an interval of 50 μm was performed across the weld zone to reveal the microhardness distribution in the weld zone in detail, and another microhardness test with an interval of 1 mm was performed across the entire joint to display the microhardness distribution in the entire joint.

(42) The light-responsive properties of the LCs were studied using an Avantes AvaSpec-2048L fiber-optic spectrometer and an Avantes AvaLight-DH-S-BAL balanced deuterium-halogen light source coupled to a vertical-beam custom cavity equipped with a Linkam LinkPad temperature controller and a Thorlabs M405FP1 photoexcitation light source.

(43) Optical signals were separated through a 565 long-pass dichroic, using a 535/20 emission filter for the green signal and a 605/20 emission filter for the red signal, and amplified through photomultiplier tubes (H7422-P40 Hamamatsu) connected to a signal preamplifier (Model 5113, Signal Recovery AMETEK Advanced Measurement Technology).

(44) In addition, the PCA approach determines principal components (PCs) solely based on the directions that describe the highest variability of the data, regardless of their dependencies on the matrix of process variables (X) and with the assumption that the first few principal components are also the most related components to the predictors.

(45) Before mixing different types of tiles for creating finite arrays with designed size, a tenfold excess (relative to the concentration of staple strands) of a full set of 44 negation strands were added to each type of DNA origami tile and quickly cooled down from 50 °C to 20 °C at 2 sec per 0.1°C.

Task 11: Describe one step or procedure in an experimental research you have conducted with reference to the examples and tips below.

Example (1)

Step 1: verb-noun collocation in main clause	monitor evolution
Step 2: node words in main clause	evolution was monitored
Step 3: pre- and post-modification of node noun(s)	the stress-free evolution of radius of curvature, *r*(*t*), was monitored
Step 4: adverbial phrases/clauses with different functions	Subsequently, the stress-free evolution of radius of curvature, *r*(*t*), was monitored for up to $2.6 \cdot 10^7$ s using a digital camera, taking care to align its optical axis perpendicular to the sample plane.

Example (2)

Step 1: verb-noun collocation in main clause	develop a method
Step 2: node words in main clause	a method was developed
Step 3: pre- and post-modification of node noun(s)	An automated image analysis method was developed
Step 4: adverbial phrases/clauses with different functions	An automated image analysis method was developed for curvature determination, significantly reducing the error bars in the strain vs. time curves.

Your description

Step 1: verb-noun collocation in main clause	
Step 2: node words in main clause	
Step 3: pre- and post-modification of node noun(s)	
Step 4: adverbial phrases/clauses with different functions	

Task 12: Skim through Text III quickly and answer the following questions.

(1) What is the purpose of this research article?

____

____

(2) What does the literature review focus on? Why?

____

____

(3) What research gap is identified based on the literature review?

____

____

(4) What raw materials were used and how were they processed in advance to synthesize Hydride-terminated silicon nanocrystals (H-SiNCs)? Is it necessary to describe the preparation of raw materials in such a detailed way?

____

____

(5) What is the purpose of the reference citation in paragraph 6?

____

____

(6) How were the mixed surfaces of methyl ester and polyethylene glycol moieties of AP-SiNCs obtained?

____

____

(7) What properties of the synthesized highly luminescent AP-SiNCs were measured sequentially? What standard method(s) or tool(s) were employed in the measurement of each specific property?

____

____

(8) What conclusions have been drawn with respect to each specific property of the synthesized highly luminescent AP-SiNCs?

____

____

(9) What sentence structures or patterns are frequently used in the Experimental Section, and for what reasons?

____

____

(10) For what reasons do you think the Experimental Section uses significantly less evaluative expressions (adjectives, adverbs, nouns, etc. intentionally used to express authorial stance,

judgment, and attitude) than the Introduction section (see the underlined words)?

____

____

Text Ⅲ

Mixed Surface Chemistry: An Approach to Highly Luminescent Biocompatible Amphiphilic Silicon Nanocrystals[8]

Abstract: Amphiphilic nanoparticles (AP-NPs) are attractive for many far-reaching applications in diverse sectors. Amphiphilic silicon nanocrystals (AP-SiNCs) are particularly promising for luminescence-based bioimaging, biosensing, and drug delivery because of their size- and surface chemistry-dependent photoluminescence (PL), high PL quantum yield, long-term photostability, and robustness to bioconjugation. Numerous studies demonstrated the synthesis of high-quality SiNCs that are compatible with organic solvents. However, preparing water-soluble SiNCs while maintaining their attractive PL properties is very challenging, and to date, only one report of blue-emitting AP-SiNCs has appeared. This report outlines a straightforward one-step thermal hydrosilylation approach that affords AP-SiNCs soluble in aqueous media in high concentrations (i.e., 14.4 mg/mL silicon core-based), exhibit bright long-lived PL in the red/near-infrared spectral region, are biocompatible, and present bioconjugable surface groups.

INTRODUCTION

1 Amphiphilic nanoparticles (AP-NPs) have attracted substantial attention partly because of their potential use in bioimaging and [1–5] biosensing, [2,4,5] as well as drug and gene delivery agents. [2,4–10] AP-NPs have also been investigated for energy capture and storage, photonics, and electronics, [11,12] as well as photocatalysis; [13] they can also be exploited in the stabilization of oil and water emulsions. [14–16] The wide range of potential applications of AP-NPs is facilitated by their exquisitely tailorable size, shape, optoelectronic properties, biocompatibility, and mixed hydrophobic/hydrophilic surface moieties.

2 Silicon is abundant; [17] silicon-based nanomaterials are biocompatible and biodegradable, [17–19] and porous silicon (p-Si)-based drug delivery products from pSivida Corp., MA, USA, have been approved by the FDA. [20] Recent progress in synthetic methods offers tangible quantities of silicon nanocrystals (SiNCs) exhibiting size- and surface-tunable photoluminescence (PL) color [blue to near-infrared (NIR)] and high PL quantum yield (PL QY; 60%–70%). [17,21,22] In addition, SiNCs have been explored as proof-of-concept magnetic resonance imaging contrast agents. [23]

3 It is well established that tailoring SiNC surface chemistry provides a convenient approach toward introducing reactivity, rendering particles resistant to environmental degradation, as well as tailoring optical response. A variety of approaches have been used to render SiNCs hydrophobic and compatible with nonpolar media—some include hydrosilylation, [21-22], [24-29] reactions with organolithium or Grignard reagents, [30-32] and dehydrocoupling. [33-34] Despite these impressive advances in surface chemistry, rendering SiNCs water-soluble while retaining their favorable optical properties (e.g., NIR emission, long-lived excited state, etc.) remains an important challenge. Doubtless, this is partly because of the susceptibility of the SiNC surface to water-mediated oxidation. (35) Still, a variety of approaches for making water-compatible SiNCs have been explored; they include direct functionalization with ligands bearing carboxylic acid, [36-37] amine, [38-40] or hydroxyl [40-41] terminal groups, as well as amino acids, [42] sugars [42-43], and hydrophilic polymers. [44-45] Other approaches have also seen micelle encapsulation of hydrophobic ligand-capped SiNCs using amphiphilic polymers or lipids. [35], [46-47] To date, only one report of blue-emitting amphiphilic SiNCs (AP-SiNCs) exhibiting nanosecond excited states has appeared in which NCs were modified using Pt-mediated hydrosilylation. [48]

4 Combining the optical properties of SiNCs and biocompatibility with an amphiphilic response could lead to far-reaching applications. In the present contribution, we describe the synthesis and properties of highly luminescent AP-SiNCs bearing mixed surfaces of methyl ester and polyethylene glycol moieties.

Experimental Section

Reagents and Materials

5 Hydrogen silsesquioxane (HSQ, trade name Fox-17) was purchased from Dow Corning, and the solvent was removed to yield a white solid that was used directly. Hydrofluoric acid (48–50%) was purchased from Fisher Scientific and used as received. Methyl-10-undecenoate (96%) was purchased from Sigma-Aldrich and used as received. Allyloxy (polyethylene oxide) (35–50 EO, 1,500–2,000 g/mol) was purchased from Gelest Incorporation and heated at 50 °C under vacuum (0.2 Torr) for 48 h prior to use. Toluene was collected from a Pure Solv purification system immediately prior to use. All other solvents were of reagent grade and used as received. Milli-Q water was used throughout. Amicon Ultra-15 Centrifugal Filter Units (30k MWCO) were purchased from Sigma-Aldrich. Hydrophilic nylon filters (0.20 μm) were purchased from Fischer Scientific. HeLa cells (CCL2 line) were obtained from the ATCC company. MTS reagents (for cell colorimetric assays) were obtained from Promega Corporation, Madison, USA. The absorbance at 490 nm was measured using a SpectraMax plate reader (Molecular Devices, USA).

Synthesis of Hydride-Terminated Silicon Nanocrystals

6 Hydride-terminated silicon nanocrystals (H-SiNCs) were prepared using a well-established literature procedure. (49) Briefly, the SiNC/SiO_2 composite obtained from reductive thermal

processing of HSQ at 1,100 °C was etched using ethanolic HF (Caution! HF is dangerous and must be handled with extreme care and in accordance with local regulations). Following extraction from HF into toluene, the H-SiNCs were isolated by centrifugation at 3000 rpm. The orange/yellow solid was redispersed in toluene and activated molecular sieves S4 (4 Å) were added. The suspension was mixed thoroughly using a glass pipet and then transferred to another test tube and isolated by centrifugation at 3000 rpm for 5 min.

Thermal Hydrosilylation of H-SiNCs with Mixed Ligands

7 H-SiNCs were dispersed in 8 mL (ca. 34 mmol) of methyl 10-undecenoate and transferred to a dry argon charged-Schlenk flask equipped with a magnetic stir bar. Subsequently, 4 g (ca. 2.3 mmol) of allyloxy (polyethylene oxide) was added. The mixture was subjected to three freeze–pump–thaw cycles using an Ar-charged double-manifold Schlenk line. After warming to room temperature, the Schlenk flask was transferred to a silicone oil bath preheated to 180 °C and stirred for 24 h under static Ar. The reaction mixture became transparent within 30 min.

8 After cooling to 50 °C, the reaction mixture was transferred to a 50 mL polytetrafluoroethylene centrifuge tube and hexane (40 mL) was added. The mixed ligand-functionalized SiNCs precipitated from the dispersion immediately. The supernatant was decanted, and the precipitate was dispersed in 5 mL of $CHCl_3$ and 20 mL of hexane (antisolvent) was added. After mixing by shaking, the mixture was centrifuged at 5000 rpm for 10 min resulting in two separate layers; the SiNCs resided in the bottom layer as evidenced by visual observation of red PL of AP-SiNCs upon exposure to a standard handheld UV light. The top layer was removed using a glass pipette and discarded. This purification cycle was repeated for a total of 5 times. The AP-SiNCs were precipitated again upon addition of 30 mL of hexane. The supernatant was discarded and 5 mL of ethanol (100%) was added to disperse the precipitate. Then, 15 mL of hexane was added to the ethanol solution with shaking, and the mixture was centrifuged at 5,000 rpm for 10 min. The top layer was removed and discarded using a glass pipette. This purification cycle was repeated a total of 5 times. The SiNCs were precipitated again by adding 30 mL of hexane. Finally, the precipitate was dispersed in toluene or water.

Quantification of the Solubility of AP-SiNCs in Water

9 The aforementioned functionalized AP-SiNCs were dispersed in 10 mL of ethanol, followed by 40 mL of Milli-Q water added. The volume of the mixture was reduced to ~15 mL using a Schlenk line vacuum (0.2 Torr). The resulting mixture was passed through a 0.2 μm nylon syringe filter, transferred to a 30k MWCO ultra centrifugal filter unit, and centrifuged at 5000 rpm for 30 min. The concentrated solution (ca. 1.5 mL) was diluted to 15 mL with Milli-Q water and centrifuged again. The concentrated solution (ca. 1.5 mL) was then transferred to a glass vial, and the volume was adjusted with water to 3.5 mL.

10 To determine the mass concentration of SiNCs in this aqueous solution, 150 μL of this solution was transferred to a preweighed vial, dried using a Schlenk line, and reweighed using

an analytical balance with four decimal place accuracy. This provided an estimation of mass (i.e., SiNCs and surface ligands) concentration. Thermogravimetric analysis (TGA) was performed using the dried sample to estimate the percentage of SiNCs and surface ligands.

Fourier Transform Infrared Spectroscopy

11 Solid allyloxy-PEG-OH was placed on a clean silicon wafer. AP-SiNC samples were prepared as thin films drop-cast from water onto a silicon wafer. Spectra were recorded using a Thermo Nicolet Continuum FT-IR microscope. A spectrum of the identical silicon wafer was used for background subtraction in Fourier transform infrared (FT-IR) data processing.

Thermogravimetric Analysis

12 AP-SiNC samples were prepared as a solid residue and transferred to a platinum pan. Weight loss was monitored in the temperature range of 25–800 °C at a ramp rate of 10 °C/min under an Ar atmosphere using the Mettler Toledo TGA/DSC 1 Star System. The degree of surface functionalization was determined using TGA, assuming that the mass loss resulted from the loss of grafted organic surface groups. [50]

X-ray Photoelectron Spectroscopy

13 AP-SiNC samples were deposited from water dispersions onto a copper foil. After drying in vacuo, the foil was transferred to a Kraton AXIS 165 instrument. The base pressure in the sample chamber was lower than 1×10^{-9} Torr. A monochromatic Al Kα source operating at 210 W with an energy hν = 1486.6 eV was used. Survey spectra were collected with an analyzer pass energy of 160 eV and a step of 0.3 eV. For high-resolution spectra, the pass energy was 20 eV and the step was 0.1 eV with a dwell time of 200 ms. The X-ray photoelectron spectroscopy (XPS) data were processed and fit using CasaXPS (VAMAS) software. All spectra were calibrated by defining the C 1s emission to 284.8 eV. The extrinsic loss structure was removed using a Shirley type background. High-resolution Si 2p XPS spectra were fit by creating a minimum number of Gaussian–Lorentzian curves. The position constraint was increased for each curve by 0.6 or 1 eV to account for spin–orbit splitting or to evaluate different oxidation states. The area constraints for Si $2p_{1/2}$ components were multiplied by 0.5 to account for the ratio of population of Si $2p_{1/2}$ states over Si $2p_{3/2}$ states. The full width at half-maximum (fwhm) was maintained the same for the Si $2p_{1/2}$ and Si $2p_{3/2}$ states by multiplying the fwhm constraints for the former by 1. The fwhm was maintained below 1.2 eV.

Electron Microscopy

14 Bright-field transmission electron microscopy (TEM) images were acquired using a JEOL 2010 TEM with an LaB_6 filament operating at an accelerating voltage of 200 kV. TEM samples were prepared by depositing a droplet of a dilute AP-SiNC toluene suspension on a holey carbon-coated copper grid placed on a filter paper. The excess solvent was drawn into the filter paper, and the grid was dried in a vacuum chamber (0.1 mTorr) for at least 24 h prior to data collection. The NC size was obtained by evaluating at least 300 particles using ImageJ

software (Version 1.45).

Dynamic Light Scattering Analysis

15 Dynamic light scattering (DLS) of the AP-SiNC solutions was measured using Malvern Zetasizer Nano series (Nano-ZS). Aqueous and toluene dispersions of AP-SiNCs were prepared such that their absorbances at 632.8 nm were approximately 0.01.

UV–Visible and PL Spectroscopy

16 Toluene or water dispersions of AP-SiNCs were prepared such that their absorbance at 350 nm was ~1. The UV–visible absorption spectra of the dispersions were recorded using a Hewlett Packard 8453 UV–vis spectrophotometer. The 351 nm line of an Ar laser was used to excite SiNCs suspended in toluene or water. The resulting PL was collected by an optic fiber, passed through a 400 nm long-pass filter to eliminate scattered light from the excitation source and fed into an Ocean Optics USB2000 spectrometer. The spectral response was calibrated to a black-body radiator (Ocean Optics LS1).

PL Lifetime

17 PL lifetimes were acquired upon illuminating a quartz cuvette containing a toluene or water solution of the AP-SiNCs in question using a modulated argon ion laser (351 nm, ~20 mW). The laser was modulated by an acousto-optic modulator operating at 200 Hz. Emission from the SiNCs was channeled into a photomultiplier (Hamamatsu H7422P-50) connected to a photon counting card (PMS-400A). Lifetime decay data were fit to a stretched exponential function in Mathematica (Version 10) given by $I(t) = I_o[\exp(-(t/\tau)\beta)] + C$, where I_o is the initial photon intensity, τ is the effective lifetime, and β is a stretching parameter that can vary between 0 and 1 (smaller values indicate broader lifetime distributions). [51]

PL QY Measurements

18 The AP-SiNCs were suspended in toluene (or water) and diluted such that the solution absorbance was below 0.1 at 425 nm. Absolute PL QYs were measured using a HORIBA K-Sphere Petite integrating sphere (diameter: 3.2 in.), equipped with a xenon lamp (185–850 nm) with a wavelength-selecting monochromator. All measurements were carried out using a 425 nm excitation wavelength. Spectra of the sample suspension and the reference (i.e., toluene or water) were corrected for system response and the reference response was subtracted from the sample (i.e., used as a background). The absolute QY was determined by taking the appropriate ratio of excitation and emission peak areas and applying the following equation.

$$\phi = \frac{\#\ \text{emitted photons}}{\#\ \text{absorbed photons}} = \frac{\int I_{\text{em}(\lambda)}^{\text{sam}} - I_{\text{em}(\lambda)}^{\text{ref}} d\lambda}{\int I_{\text{ex}(\lambda)}^{\text{sam}} - I_{\text{ex}(\lambda)}^{\text{ref}} d\lambda}$$

where, $I_{\text{em}(\lambda)}{}^{\text{sam}}$ = intensity of emitted light from the sample (SiNCs), $I_{\text{em}(\lambda)}{}^{\text{ref}}$ = intensity of background emission, $I_{\text{ex}(\lambda)}{}^{\text{sam}}$ = intensity of excitation light for the sample (SiNCs), $I_{\text{ex}(\lambda)}{}^{\text{ref}}$ = intensity of excitation light for the reference cuvette.

Evaluation of AP-SiNC In Vitro Toxicity

19 AP-SiNC toxicity was evaluated using a manufacturer-developed protocol (Promega Corporation, Madison, WI). Briefly, 25×10^3 HeLa cells were incubated in a clear 96-well plate and cultured to 90% confluency. HeLa cells were then treated with AP-SiNCs of five different concentrations, Cisplatin (10 μM), or untreated (control) and were incubated for 18 h. After incubation, 20 μL of MTS solution was added in each well and incubated for another hour. Absorbance was measured at 490 nm. Three replicate wells were used for each AP-SiNC concentration, Cisplatin, and the control group. Blank absorbance was measured from wells containing cells only (control) and was subtracted from sample absorbance. Normalized viability was calculated based on the following equation

$$\text{Normalized viability}=(\text{sample Abs}/\text{average Abs of the untreated group})\times 100$$

Results and Discussion

…

Conclusions

20 We have successfully prepared AP-SiNCs by straightforward thermal hydrosilylation of H-SiNCs in the presence of mixed ligands. This procedure provides SiNCs with mixed surface functionalities that are soluble in organic and aqueous solvents. The PL maxima of the resulting amphiphilic NCs appears in the red/NIR spectral region and is separated from that of tissue autofluorescence. In addition, the microsecond excited-state lifetimes are expected to enable time-gated imaging. The presented SiNCs exhibit long-term PL stability in water which may facilitate monitoring of disease progression. Finally, the SiNCs did not show detectable toxicity on the HeLa cell line.

Supporting Information

…

Abbreviations

AP-NPs	amphiphilic nanoparticles
PL	photoluminescence
AP-SiNCs	amphiphilic silicon nanocrystals
PL QY	photoluminescence quantum yield
NIR	near-infrared
HSQ	hydrogen silsesquioxane
MWCO	molecular weight cut off
H-SiNCs	hydride-terminated silicon nanocrystals
HF	hydrofluoric acid
TGA	thermogravimetric analysis
FT-IR	Fourier transform infrared

PEG	polyethylene glycol
Ar	argon
XPS	X-ray photoelectron spectroscopy
TEM	transmission electron microscopy
DLS	dynamic light scattering
UV–vis	ultraviolet visible
Abs	absorbance
EPR	enhanced permeability and retention

References

...

Task 13: Select one subsection in the Experimental Section of Text III, and analyze the "input—tool/medium—operation—output" pattern in each experimental step by following the example below. Discuss in groups the discourse patterns and language features used to develop coherence and cohesion. For example, the output or part of the output generated in the previous experimental step forms the input for the next step, linguistically manifested by the use of repeated nouns, transition signals (e.g. first, then), reference words (e.g. aforementioned), synonyms, hyponyms, nominalizations (e.g. "measurement" replacing "measure"), etc.

Example (para. 9):

Sentence (1): The aforementioned functionalized AP-SiNCs were dispersed in 10 mL of ethanol, followed by 40 mL of Milli-Q water added.

Input:	the aforementioned functionalized AP-SiNCs AP-SiNCs
Tool/medium:	ethanol, Milli-Q water
Operation:	disperse
Output:	a mixture (not explicitly stated in this sentence)

Sentence (2): The volume of the mixture was reduced to ~15 mL using a Schlenk line vacuum (0.2 Torr).

Input:	the volume of the mixture
Tool/medium:	using a Schlenk line vacuum
Operation:	reduce
Output:	a ~15 mL mixture (not explicitly stated in this sentence but in the next sentence)

Sentence (3): The resulting mixture was passed through a 0.2 μm nylon syringe filter, transferred to a 30k MWCO ultra centrifugal filter unit, and centrifuged at 5000 rpm for 30 min.

Input:	the resulting mixture
Tool/medium:	a 0.2 μm nylon syringe filter, 30k MWCO ultra centrifugal filter unit
Operation:	pass through, transfer, and centrifuge
Output:	the concentrated mixture/solution (not explicitly stated in this sentence but in the next sentence)

Sentence (4): The concentrated solution (ca. 1.5 mL) was diluted to 15 mL with Milli-Q water and centrifuged again.

Input:	the concentrated solution (ca. 1.5 mL)
Tool/medium:	Milli-Q water
Operation:	dilute, centrifuge
Output:	the concentrated solution (not explicitly stated in this sentence but in the next sentence)

Sentence (5): The concentrated solution (ca. 1.5 mL) was then transferred to a glass vial, and the volume was adjusted with water to 3.5 mL.

Input:	the concentrated solution (ca. 1.5 mL)
Tool/medium:	a glass vial, water
Operation:	transfer, adjust
Output:	the concentrated aqueous solution (not explicitly stated in this sentence but in 1st sentence of next paragraph)

Task 14: Arrange the sentences in each set logically and coherently into a research article methods section.

I

(1) Subsequently, the GO sponge was thermally treated with S powder (the mass ratio of GO:S powder = 1:2) at different temperatures for 2 h in nitrogen to obtain S-RGO sponges.

(2) The structures and surfaces were then measured by energy dispersive spectrum mapping (EDS, JEOL-2100), Fourier transform infrared spectrometer (FTIR, Nicolet, NEXUS 670), Raman spectrometer (Horiba T64000) with a wavelength of 532 nm.

(3) RGO sponges were also synthesized at 600 °C by a similar procedure without S powder.

(4) After freezing, the culture dish was put into a freeze dryer (-50 °C, pressure < 20 Pa) for 3 days to obtain GO sponge.

(5) Brunauer-Emmett-Teller (BET) specific surface area was finally measured by ASAP 2010 surface area system.

(6) For the synthesis of RGO and S-RGO sponge, in a typical process, 100 mg GO was dispersed homogeneously in 50 mL deionized H_2O, and then transferred to culture dishes in a refrigerator.

(7) The morphologies of RGO sponge and S-RGO sponge samples were analysed by field emission scanning electron microscopy (SEM, Zeiss, Ultra 55), transmission electron microscopy (TEM, JEOL-2100).

(8) X-ray photoelectron spectroscopies (XPS) were recorded by an Image Photoelectron Spectrometer (Thermo Fisher Scientific, K-Alpha) with Al Ka X-ray source.

II

(1) Thermal hydrocarbonization of the PSi films was performed by exposing them to a 1:1 (v/v) N_2/acetylene (C_2H_2) flow (1 L min^{-1}) for 15 min at room temperature, followed by a heat treatment for 15 min at 500 °C.

(2) Electrochemical anodization method was used to prepare undecylenic acid modified thermally hydrocarbonized PSi (UnTHCPSi), as described in detail elsewhere.

(3) Briefly, monocrystalline p + -type Si (100) wafers (0.01–0.02 Ω cm resistivity) were electrochemically anodized in a 1:1 (v/v) aqueous hydrofluoric acid (38%)–ethanol electrolyte by applying repeated low and high current density pulsed etching profile.

(4) The obtained THCPSi films were then treated with 10-undecenoic acid for 16 h at 120 °C.

(5) The UnTHCPSi films were finally wet ball milled to produce nanoparticles and separated using centrifugation to obtain the final PSi nanoparticles with desired particle size.

(6) The resulting PSi with high porosity fracture planes and hydrogen-terminated surface was subsequently detached from the substrate as free-standing multilayer films by abruptly increasing the current density to electropolishing region.

III

(1) 3D polydimethylsiloxane (PDMS) scaffolds are free-standing, porous structures made by PDMS (Sylgard 184, Down Corning Co.), an elastomeric material commonly used in biomedical applications.

(2) After cooling, the sugar was dissolved by submersing the sample in distilled water overnight.

(3) The water-driven dissolution of sugar grains produces a network of interconnected micropores inside the PDMS scaffold.

(4) The fabrication procedures involve water dissolution of a sugar scaffold, resulting in a cast PDMS framework consisting of interconnected cavities as previous described.

(5) PDMS was vacuum-forced to percolate inside the resulting sugar framework until all spaces were filled, and cured in an oven at 85 °C for 1 h.

(6) In brief, 500 mg of food-approved sugar was passed through a No. 60 mesh sieve (Sigma-Aldrich) and mixed with 20 μL of deionized water, placed in a silicon mold, and subsequently dried at 65 °C for 30 min.

Ⅳ

(1) The $Ba_{0.5}Sr_{0.5}RuO_3$ bottom electrode layer was grown at a temperature of 750 °C in a dynamic oxygen pressure of 20 mtorr by ablating a $Ba_{0.5}Sr_{0.5}RuO_3$ target (Praxair) at a laser fluence of 1.85 J cm^{-2} and a laser repetition rate of 3 Hz.

(2) This work focuses on 150 nm $0.68Pb(Mg_{1/3}Nb_{2/3})O_3$–$0.32PbTiO_3$ (PMN–PT)/20 nm $Ba_{0.5}Sr_{0.5}RuO_3$ heterostructures grown on (110)-oriented, single-crystalline $NdScO_3$ substrates by pulsed-laser deposition using a KrF excimer laser (248 nm, LPX 300, Coherent), in an on-axis geometry with a 60 mm target-to-substrate spacing.

(3) Following the growth of 150 nm $0.68Pb(Mg_{1/3}Nb_{2/3})O_3$–$0.32PbTiO_3$ (PMN–PT)/20 nm $Ba_{0.5}Sr_{0.5}RuO_3/NdScO_3$ (110) substrates, the electrothermal characterization devices are produced via a multistep, microfabrication process, according to previous literature.

(4) Finally, a 100-nm-thick platinum thin-film resistance heater is sputtered in the shape of a thin strip with four probe pads (two outer and two inner pads) to define the thermal circuit.

(5) Next, a 200-nm-thick blanket layer of SiN_x is deposited on the symmetric ferroelectric capacitor structure using plasma-enhanced chemical vapour deposition ($SiH_4 + NH_3$ based) at 350 °C.

(6) PMN–PT growth was carried out at a deposition temperature of 600 °C in a dynamic oxygen pressure of 200 mtorr using a laser fluence of 1.8 J cm^{-2} and a laser repetition rate of 2 Hz.

(7) Briefly, the relaxor–ferroelectric heterostructure is lithographically patterned and ion-milled to define the bottom electrode and the relaxor–ferroelectric 'active' layer.

(8) Following growth, the heterostructures were cooled to room temperature in a static oxygen pressure of 760 torr at 10 °C min^{-1}.

(9) After this step, the bottom electrode ($Ba_{0.5}Sr_{0.5}RuO_3$) is removed from everywhere except under the active layer and the bottom electrode probe pad.

(10) To define the top electrode, 90 nm $SrRuO_3$ is selectively deposited using an inverse MgO hard-mask process.

Task 15: Read the research article extract below, and work with each other to answer the following questions.

Analysis of austenite-martensite phase boundary and twinned microstructure in shape memory alloys: The role of twinning disconnections[9]

Abstract: An austenite-martensite phase boundary in shape memory alloys (SMA) is associated with a periodic microstructure of martensite twin lamellas. Microscopy studies show that the period, which represents the thickness of the twin lamellas, increases with the distance from the habit plane. This observation is often overlooked when the microstructure and energy of the austenite-martensite interface are evaluated. In this paper we introduce a model that reproduces the variation in the twin lamella period. For this purpose, the overall energy of the phase boundary and the accompanied twinned microstructure is formulated and minimized. In particular, the effect of twinning disconnections, via which twins are tapered or broaden, and the additional energy due to the disconnections, are considered. Fittings of model predictions with measurements based on microscopy images provide evaluations of the twin boundary and twinning disconnection energies. Comparison of the results with expressions based on the theory of dislocations indicates that interactions between disconnections play a dominant role in determining the overall energy of twinning disconnections.

INTRODUCTION

…

Experimental setup and tested material

2.1. Tested material

1 The microstructure associated with the austenite–martensite phase boundary is studied in a $Ni_{50}Mn_{28.5}Ga_{21.5}10M$ single crystal. The crystal is produced by AdaptaMat LTD and is cut along the {100} planes of the cubic austenite phase to a cuboid with dimensions of 20 mm × 3 mm × 2.5 mm. The 10*M* martensite is the stable phase at room temperature, and the martensite to austenite transformation temperature of this alloy is approximately 55 °C. We note that the 10*M* martensite is nearly tetragonal, with a slight monoclinic distortion [31]. The small non-tetragonal distortion gives rise to both type I and type II twins, which are crystallographically different [32]. Based on our optical observations, the twins present within the laminated structure close to the phase boundary were identified as type I. For simplicity, we employ the tetragonal approximation for the martensite phase throughout the analysis presented in this paper.

2 We note that the energy minimization approach is valid under the assumption that the phase boundary and twinned microstructure are very close to equilibrium state. This condition is not necessarily met for all materials, because phase transformation and twinning reorientation in SMA are non-equilibrium processes that are associated with a hysteresis. Considering this issue, Ni-Mn-Ga is especially suitable for this type of analysis due to the very small hysteresis of its phase transformation (about 10 °C [33]) and twinning reorientation (twinning stress as low as 0.1 M Pa [34]).

2.2. Experimental setup and procedure

3 The experimental setup used in this study (see Fig. 1) is the same as that described in details in our recent publication [33], and allows the simultaneous controlled heating and uniaxial loading of the tested sample. The setup consists of a U-shaped thermal bath in which the tested sample is placed, such that its top surface can be observed via the microscope. The thermal bath is glued to the top side of an aluminum plate, while a Kapton® insulated flexible heater foil (OMEGA®) is glued to the opposite side of it. Adjustment of the voltage supplied to the heater allows control over the sample's temperature. In addition, two aluminum rods

constrain both ends of the sample. The left rod is connected to a micrometer displacement stage that allows the adjustment of the initial compression stress along the longitudinal direction of the sample. In this configuration, the sample is compressed throughout the entire phase transformation.

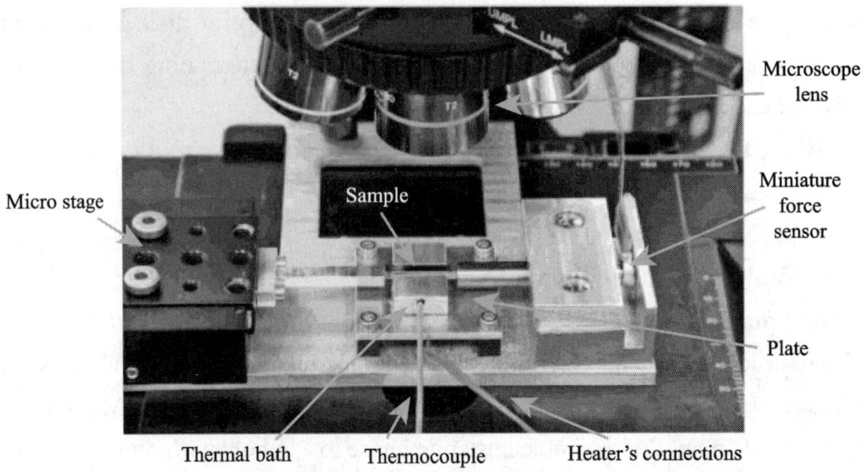

Figure 1 The experimental setup used for controlling the austenite-martensite phase transformation in a SMA single crystal, while simultaneously capturing optical images from the crystal's top surface

4 Initially, the crystal is compressed at the martensite phase along its long 20 mm axis, such that a state of a single variant is created, where the short crystallographic axis of the tetragonal unit cell (*c*-axis) is parallel to the direction of the compression. Next, while the sample is mechanically constrained along its long axis, it is heated above the transformation temperature. The entire setup is placed under an optical microscope (see Figure 1), such that the microstructure evolution throughout the phase transformation can be observed and captured from the top 20 mm × 3 mm surface of the crystal. Images are recorded using an optical camera with a lateral resolution of 1μm.

…

(1) What is the purpose of this research?

____

____

(2) What are the overall results of this research?

____

____

(3) What properties does the selected $Ni_{50}Mn_{28.5}Ga_{21.5}10M$ single crystal show?

____

____

(4) For what reasons was the tetragonal approximation for the martensite phase adopted in this study?

____

____

(5) Why is Ni-Mn-Ga especially suitable for the analysis of austenite-martensite phase boundary and twinned microstructure in SMA?

____

____

(6) What experimental setup was employed in this study? What is the function of each component in the experimental setup?

____

____

(7) How is the process of austenite-martensite phase transformation in a SMA single crystal controlled and monitored?

____

____

(8) Why is the tested material preparation introduced in the present tense here, clearly different from the past tense used in the same sections of Text I and Text III?

____

____

(9) For what reasons is the experimental procedure here described in the present tense, while the same sections of Text I and Text III are presented in the past tense?

____

____

(10) Could you describe the structure and mechanism of one research system or experimental setup you have designed or improved by following the steps below?

- One-sentence overall description of the system (setup): ____________
- Main components of the system (setup) and their functions: __________
- Interaction among main system (setup) components: ______________
- Procedures of system (setup) operation: ____________

Task 16: Fill in the blanks with appropriate verbs in the right form in the following extracts of research article methods sections.

I

The $BiVO_4$ ceramic phase matrix was (1) ________ by solid-state reaction method, using the stoichiometric ratio of the high purity oxide commercially available V_2O_5 (Aldrich, 99%purity and Bi_2O_3 (Vetec, 98%purity). The powders were milled for 6 h with zirconia balls, calcined for

4 h at 500 °C. The composite was (2) ________ from the mixtures of BVO and TiO_2 (Aldrich, 99.8%purity), varying the TiO_2 composition from 0 to 60%wt. The ceramic cylinder was fabricated by the pressing of this mixture, in which the polyvinyl alcohol (PVA 10 %wt.) was (3) ________ as a binder for reducing the brittleness of the ceramic bulk in a metallic mold under the constant uniaxial press, before the being sintered (800 °C/4 h) [1]. The nomenclature of samples is BVO, BV15, BV30, BV45 and BV60 to 0, 15, 30, 45 and 60%wt. of TiO_2 in the composite. The cylinders were (4) ________ to obtain plan uniform faces to minimize air gaps between metal plates used with Hakki-Coleman method and set-up of antenna measurements.

The samples were (5) ________ at room temperature by X-Ray diffraction (XRD) using Cok_α radiation($\lambda = 0{,}178896$ Å) over an angular range of ($20° \leqslant 2\theta \leqslant 80°$) and refined by Rietveld method [2, 3].

The Hakki-Coleman technique was used to (6) ________ the dielectric constant (εr), dielectric loss tangent (tan δ) and quality factor (*Qd*) at the resonant frequency for TE011 mode [4], using the Vector Network analyzer of Agilent model N5230A. The temperature coefficient of resonant frequency (*τf*), by the formula and SFS methodology [5] was also (7) ________, expressed in equation (1).

The reflection coefficient and Smith Chart were measured in Vector Network Analyzer of Agilent N5230A and were (8) ________ with reflection coefficient and Smith chart obtained by numerical simulation. The far field parameter was (9) ________ by numerical simulation. Ansys HFSS software was used to calculate S parameters and three-dimension distribution of the fields inside the passive structure, i.e., the numerical simulations were (10) ________ to approach experimental results.

II

Materials. Metallic powders of Cobalt (Co, 200 mesh), Chromium (Cr), Copper (Cu, 250–300 mesh), Iron (Fe, 100 mesh), and Nickle (Ni, 200 mesh) with a purity of 99.9 wt % were (1) ________ by Sinopharm Chemical Reagent Co. Ltd., China and used as starting materials.

Synthesis procedure. The modified arc-discharge experimental setup is similar to that reported elsewhere [1-3]. Firstly, the well-mixed micro-sized Al, Co, Cr, Cu, Fe and Ni powders were (2) ________ into a piece of 50 mm in diameter and 10 mm in thickness as the targets, Their chemical compositions (listed in Table 1) were (3) ________ to be CoCrCuFeNi, $CoCrCu_{0.5}FeNi$ and $CoCrCuFe_2Ni$, and the total amount of metal powder was 20 g for each target. Then, the compressed target was laid on a graphite crucible serving as the anode, and a graphite needle with a 5 mm diameter (4) ________ as the cathode. Subsequently, the vacuum chamber was (5) ________ to 5.0×10^{-3} Pa, and the mixture gases of Ar and H_2 (the ratio of H_2 to Ar is 1: 4) were introduced into the chamber. During the modified arc-discharge process, the arc discharge was (6) ________ at 100 A for 15 min. At last, the nanoparticles were collected from the inner surface of the chamber after being passivated for 10 h in argon atmosphere. For convenience, the samples were also denoted as CoCrCuFeNi, $CoCrCu_{0.5}FeNi$ and $CoCrCuFe_2Ni$ based on the initial composition.

Characterization. The crystal structures of the as-prepared NPs were (7) ________ using an X-ray diffractometer (Bruker D8 Advance) with a Cu K_α radiation ($\lambda = 1.54$ Å). The morphology, particle size of the samples were (8) ________ using a TEM (JEOL 2010; JEOL Ltd), and subjected to TEM-EDS for elemental analysis. The TEM samples were dispersed in ethanol by sonication, dropped on a carbon-coated copper grid, and then dried at room temperature for observations. Quantitative element analysis was (9) ________ by means of a Varian UltraMass 725 ICP-AES. The magnetic properties were performed (10) ________ a Lakeshore 7410 vibrating sample magnetometer (VSM, Quantum Design Versalab) with a maximum external field $Hm \approx 20\,000$ Oe.

III

The experiments were performed on the LSF-IIIB laser additive manufacturing system established by State Key Laboratory of Solidification Processing with the listed parameters. This system (1) ________ of a 4 kW continuous wave CO_2 laser with a wavelength of 10.6 μm, a five-axis numerical control working table, an inert atmosphere processing chamber (oxygen content ≤ 50 ppm), a powder feeder with a coaxial nozzle and an adjustable automatic feeding device with high precision (type DPSF-2).

As substrate material a C45E4 stainless steel with dimensions of 100 mm × 60 mm × 10 mm (length × width × height) was (2) ________. The surface of the substrate was polished to remove the oxide film by abrasive paper and cleaned with acetone before the LAM process. Spherical powder Inconel 718, which has been manufactured by The Plasma-Rotating Electrode Process (PREP), was (3) ________ as the deposition material. The diameters of the PREP powders range from 45 to 90 μm. Before (4) ________ the LAM experiment, these powders were dried in a vacuum oven at 120 °C ± 10 °C for 2 h.

In order to obtain Laves phases with different morphologies, three types of heat treatments were (5) ________ in this study. Microstructural investigations were (6) ________ by optical microscope (OM) and scanning electron microscopy (SEM) after being etched with an etching solution of 6 ml HCl+2 ml H_2O+1 g CrO_3. The SEM microstructural examinations were performed (7) ________ a Tescan VEGAIILMH microscope. The image processing software Image-Pro-Plus was used for measurements. Ten figures were used for measuring the volume fractions and the morphologies of the Laves phases. And five figures were (8) ________ for measuring the morphologies of γ" phases. Electron probe microanalysis (EPMA) was used to reveal the distributions and contents of the elements. Uniaxial tensile tests were (9) ________ out at room temperature, using a test machine typed INSTRON 3382. The tensile tests were (10) ________ with a constant displacement rate of 1 mm/min. Fracture mechanism analysis was carried out by SEM.

Task 17: Fill in the blanks with appropriate prepositions and adverbs in the following extracts of research article methods sections.

I

The PMN-29PT single crystals (CTG Advanced Materials, LLC, USA) used (1) ________ this study has the rhombohedral 3*m* crystal symmetry (2) ________ room temperature. The platelet samples have the size (3) ________ $2.5 \times 2.5 \times 0.113\ mm^3$, and coated (4) ________ gold electrodes on the two larger surfaces. The $P-E$ loops were measured using a modified Sawyer-Tower circuit at room temperature with E_0 ranging (5) ________ 1.43 kV/cm to 7.5 kV/cm (6) ________ a fixed frequency (f = 1 Hz, 10 Hz, and 50 Hz), and (7) ________ different f from 0.1 Hz to 100 Hz at a fixed E_0 ($1E_c$, $2E_c$ and $3E_c$), where the coercive field (E_c) is defined (8) ________ 2.5 kV/cm according to the 0.1 Hz $P-E$ loop measured at room temperature. The hysteresis area, remnant polarization and coercive field are obtained directly (9) ________ these $P-E$ loops. The loop area is obtained (10) ________ the integration of the closed curve using a self-written Matlab code.

II

A stainless steel tube (1) ________ outer and inner diameters of 1.2 mm and 0.8 mm, respectively, was used as the nozzle (2) ________ which electrolyte was ejected. This nozzle was connected (3) ________ the negative pole of the power supply and used as the cathode. Inconel 718 plates were used (4) ________ workpieces. When the power was on, grooves were fabricated (5) ________ linear movement in the horizontal plane. A 10 wt% $NaNO_3$ aqueous solution controlled (6) ________ 30 ℃ was used as the electrolyte.

The jet shape was observed using a CCD camera. In each experiment, the image of the jet shape was captured (7) ________ the nozzle had scanned at least 5 mm. The surface roughness (Ra) (8) ________ the bottom of the machined groove was measured using a roughness tester (Hommel-Etamic T8000, Jenoptik). A scanning electron microscope (S-3400 N, Hitachi) and a 3D optical profiler (neox, Sensofar) were used to study the ECMed grooves.

III

The stoichiometric mixture of Bi, Se, Sb, and Co was melted (1) ________ an evacuated quartz tube (2) ________ 1195 °C. Later the melt was cooled to 550 °C (3) ________ the rate of 2 °C/hr and retained at 550 °C for 24 h, followed by cooling (4) ________ 30 °C in 3 h. The structural properties of the grown BSS and BCSS crystals were studied (5) ________ X-ray diffraction (XRD) on powdered samples at Cu K_α edge from a lab x-ray source.

Resistivity and Hall measurements were carried out (6) ________ Van der Pauw and six-probe geometry in a commercial, 15 T Cryogen-free system (7) ________ Cryogenic Ltd., UK. 20 μm Au wires were used to make the electrical contacts on a cleaved crystal with a single component silver paste curable (8) ________ room temperature. Using a special switching and scanning unit (9) ________ Keithley Instruments, the magnetoresistance (MR) and Hall measurements were recorded simultaneously. Magnetic properties of the sample were measured (10) ________ a commercial Evercool MPMS3 SQUID VSM system from Quantum design, USA.

Ⅳ

To fabricate flexible high quality ZnO thin film (1) ________ radio frequency (RF) sputtering on top of a polyimide film, the fabrication processing is as follows. (2) ________, the polyimide film was washed by deionized water and treated (3) ________ plasma to get a clean and smooth surface. (4) ________, 2 μm Li-doped ZnO layer was deposited by the RF sputtering of ZnO:Li_2CO_3(99.8 at%:0.2 at%) target (5) ________ power of 120 W. For comparation, un-doped ZnO thin film was fabricated by the same technique but using a pure ZnO target. The working gases were the mixture gas of Ar and O_2 (6) ________ a flux ratio of 80:20. During the sputtering processing, the sputtering chamber pressure was fixed (7) ________ 2 Pa and the temperature was set at 100°.

The device fabrication was also done by the RF sputtering. First, a patterned Au was deposited (8) ________ top of the cleaned polyimide film with a mask. The shape of the Au is a square (9) ________ the side length of 100 μm. Each square Au server as the bottom electrode for the in-plane strain sensor. Two adjacent Au (10) ________ longitudinal direction were connected with an Au line with the width of 10 μm. Then 2 μm Li-doped ZnO or un-doped ZnO layer was deposited by the RF sputtering directly. (11) ________, the top electrode was fabricated by RF sputtering another layer (12) ________ a mask. The shape and size of the top electrode is the same as the bottom one, while in this time two adjacent Au along latitude direction were connected by Au line.

Task 18: Compose a short text to describe in detail one test or experiment you have conducted following the outline below, and present it orally and visually in groups.

(1) Purpose of test (experiment): ________________________

(2) Test (experiment) setup and procedure:

- Overall description: ________________________
- Research objects (samples, specimens, model, etc., and selection and preparation): ________________________
- Test (experiment) procedures including data collection: ____________
- Data analysis methods and steps: ________________________

Unit 4
Results and Discussion Sections

Objectives

- Understand descriptive and argumentative communicative purposes of research article results and discussion sections;
- Identify common structural patterns in research article results and discussion sections;
- Identify verbs, phrases, or sentence patterns mostly used to refer to tables or figures;
- Acquaint with common words and expressions highlighting change, similarity, difference, ranking, percentage, etc. in tables or figures;
- Identify hedges, boosters, and evaluative devices in presenting and discussing research results;
- Acquaint with general-specific text structure and exemplification strategy in reporting and synthesizing results;
- Understand the importance of interpreting and commenting research results with reference to previous studies;
- Identify commonly used linguistic features in claiming values of research results, stating research limitations, and directing future research;
- Understand the change of epistemic modality in the process of reporting, interpreting, and discussing research results;
- Learn to report and interpret research results with the aid of tables or figures;
- Raise awareness of phrasal and syntactic complexity in research article results and discussion sections.

Task 1: Discuss in groups which of the following communicative purposes a typical research article results and discussion section needs to achieve

Communicative Purposes	Yes/No?
(1) Restating research purposes of present study	
(2) Summarizing research methodology and experimental procedures of present study	
(3) Claiming feasibility and effectiveness of present research methodology and procedures	
(4) Restating theoretical/methodological hypothesis made in present study	
(5) Defining key theoretical/methodological terms used in the following data analysis	
(6) Describing main obstacles and difficulties encountered in present study and discussing how the problems were confronted	
(7) Presenting overall experimental or analytical results of present study	
(8) Presenting main research results one by one both numerically and qualitatively	
(9) Referring to tables or figures to report the most salient aspects (e.g. increase, decrease, contrast, percentage, etc.) of quantitative data	
(10) Specifying significant or unexpected results obtained in present study	
(11) Exemplifying significant or unexpected results obtained in present study	
(12) Interpreting the meaning of quantitative results	
(13) Explaining possible errors of quantitative results	
(14) Explaining satisfactory/unsatisfactory research results	
(15) Making generalizations about research results	
(16) Relating research results to the critical issue (gap) identified in the introduction section	
(17) Comparing research results with previous studies, whether citing agreement or disclosing discrepancies, and analyzing possible reasons	

continued

Communicative Purposes	Yes/No?
(18) Relating research results to (inter)national industry standards	
(19) Discussing practical applications of research results with a focus on potential benefits	
(20) Admitting perplexities of data analysis and interpretation	
(21) Stating the values and advantages of research results	
(22) Suggesting the applicability of research results	
(23) Pointing out methodological or procedural limitations of present study	
(24) Concluding research results qualitatively	
(25) Indicating research implications	
(26) Making recommendations or stating a practical need for further study	

Task 2: Figure out the research purpose and methods in the abstract of Text Ⅰ summarize the main idea of each paragraph in the Introduction section and Results and Discussion section, and discuss in groups what communicative purposes listed in Task 1 are fulfilled.

Text Ⅰ

Determination and Modeling of Mechanical Properties for Graphene Nanoplatelet/Epoxy Composites[1]

Abstract: Structural components of modern aircraft, such as the fuselage and control surfaces, are commonly constructed using carbon-filled polymer composites. The addition of graphene nanoplatelets (GNP) to traditional fiber-reinforced composites often increases the tensile modulus. In this work, composites were fabricated with epoxy (EPON 862 with EPIKURE Curing Agent W) and 1–4 wt% (0.6–2.44 vol%) GNP. The GNP used in this study was Asbury Carbon's TC307. To the authors' knowledge, mechanical data for composites with TC307 have not been published before. Composite specimens were tested for macroscopic tensile modulus and modulus as determined by nanoindentation. The macroscopic tensile modulus increased from 2.72 GPa for neat epoxy to 2.93 GPa for 4 wt% (2.44 vol%) TC307 in epoxy. The modulus as determined by nanoindentation showed a similar trend. For all these composites, the tensile strength ranged from 76 to 81 MPa. A multiscale modeling approach, using molecular dynamics data and micromechanical modeling, was used to verify the experimental data, and both experiments and modeling demonstrated that a three-dimensional random dispersion of GNP (~3 to 4 layers) in epoxy was achieved. The constant level of strength with GNP loading is important in applications where GNP is added to the epoxy matrix to increase thermal and electrical conductivity.

INTRODUCTION

1 Structural components of subsonic fixed-wing aircrafts, such as the fuselage and control surfaces, are commonly constructed using carbon-filled polymer composites. Epoxy is often used as the matrix material for these applications. Graphene nanoplatelets (GNP) are short stacks of individual layers of graphite (graphene) that often increase composite tensile modulus and are available at a lower cost ($15–$50/lb) as compared with carbon nanotubes [1-5]. Improved composite mechanical properties are associated with the fillers aspect ratio as well as its surface-to-volume ratio [6]. GNP has a higher surface-to-volume ratio than carbon nanotubes due to lack of access to the inner surface of the nanotube [7]. This study focuses on producing, testing, and modeling GNP/epoxy composites that could be used for aerospace applications.

2 For this study, the epoxy system used was EPON 862 with EPIKURE Curing Agent W. Composites containing 1–4 wt% (0.60–2.44 vol%) Asbury Carbons TC307 GNP (<1 μm average particle diameter) in this epoxy were fabricated and tested for macroscopic tensile properties and nanoindentation. The first goal of this research was to determine what effect GNP would have on the composite mechanical properties. The second goal was to use a multiscale modeling approach to verify the experimental observations as well as to provide insight into the orientation and thickness of the TC307 reinforcement. Per the authors' knowledge, material properties and modeling for Asbury TC307 GNP in this epoxy resin system has never been previously reported in the open literature.

EXPERIMENTAL

…

RESULTS AND DISCUSSION

Field Emission Scanning Electron Microscopy (FESEM) Results

3 Figure 2 shows a FESEM image of a tensile fracture surface for 4 wt% TC307 in epoxy. This figure appears to show a three-dimensional random arrangement of TC307 in epoxy. Figure 3 is also a FESEM image of the same fracture surface using a higher magnification. This figure shows the platelet shape of GNP protruding out of the fracture surface (z-direction).

(Figure 1 and Figure 2 omitted)

Table 1 Experimentally obtained properties for TC307/epoxy composites

Material system	Filler wt% (vol%)	Tensile modulus (GPa)	Ultimate tensile strength (MPa)	Strain at ultimate tensile strength (%)	Nano modulus (GPa)	Scaled nano modulus (GPa)	Hardness (GPa)
Neat epoxy	0 (0.00)	2.72 ± 0.04 n = 6	77.6 ± 0.9 n = 6	7.98 ± 0.35 n = 6	3.61 ± 0.02 n = 16	2.72 ± 0.02 n = 16	0.255 ± 0.003 n = 16
1TC307	1 (0.60)	2.84 ± 0.08 n = 6	81.2 ± 1.6 n =6	5.94 ± 0.82 n = 6	3.70 ± 0.03 n = 33	2.78 ± 0.03 n = 33	0.273 ± 0.005 n = 33
2TC307	2 (1.21)	2.88 ± 0.07 n = 7	80.9 ± 1.9 n = 7	5.91 ± 0.51 n = 7	3.74 ± 0.07 n = 33	2.81 ± 0.07 n = 33	0.273 ± 0.009 n = 33
3TC307	3 (1.82)	2.91 ± 0.05 n = 8	78.1 ± 0.8 n = 8	4.74 ± 0.24 n = 8	3.81 ± 0.06 n = 34	2.87 ± 0.06 n = 34	0.278± 0.006 n = 34
4TC307	4 (2.44)	2.93 ± 0.09 n = 7	75.8 ± 1.8 n = 7	4.55 ± 0.31 n = 7	3.85 ± 0.07 n = 33	2.89 ± 0.07 n = 33	0.281 ± 0.009 n = 33

Tensile Results

4 Figures 4-7 show the average tensile modulus, ultimate tensile strength, and strain at ultimate tensile strength for the TC307/epoxy composites measured according to ASTM D638. Error bars shown are ±1 standard deviation (SD). The average tensile modulus, ultimate tensile strength, and stain at ultimate tensile strength with standard deviation and number of samples tested for each are summarized in Table 1. It is clear from Table 1 and Figure 4 that the average tensile modulus exhibited a substantial increase when increasing GNP from 0 to 1 wt% (0.60 vol%). Further increases in GNP loading resulted in diminishing increases in tensile modulus. In general, this is consistent with the results of Chatterjee et al. who reported an increase from approximately 2.65 GPa for neat epoxy to approximately 3.08 GPa for 2 wt% (1.21 vol%) GNP in epoxy [19]. Although the averages show a monotonic increase with GNP loading for the entire wt% range considered, the substantial uncertainty at each GNP loading level suggests some ambiguity in the trend, possibly with a plateau above 1 wt%. It is important to note that a slight concave trend exists with the average tensile modulus values. Given the uncertainty in the current data, it is difficult to decisively conclude that the trend is concave. From Table 1 and Figure 7, it is clear that adding the TC307 to the epoxy decreases the ultimate tensile strain. It is also clear that there is an increase in the tensile strength between the neat epoxy and the 1 wt% (0.60 vol%) loading level. The strength remained constant between 1 wt% (0.60 vol%) and 2 wt% (1.21 vol%) loadings, and

decreased substantially thereafter. Thus, for this material system the optimal strength falls between 1 and 2 wt% GNP.

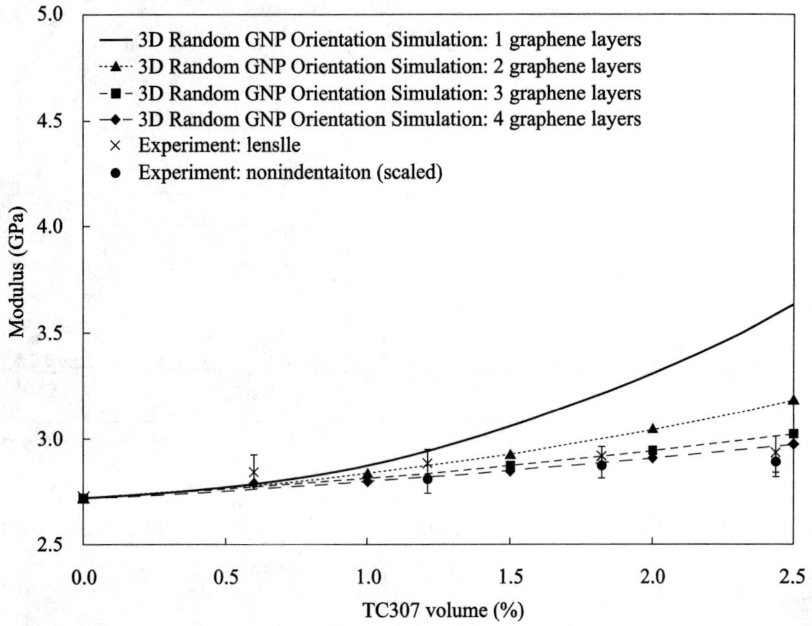

Figure 4 Modulus versus TC307 volume % for 3D random GNP orientation at various GNP thicknesses

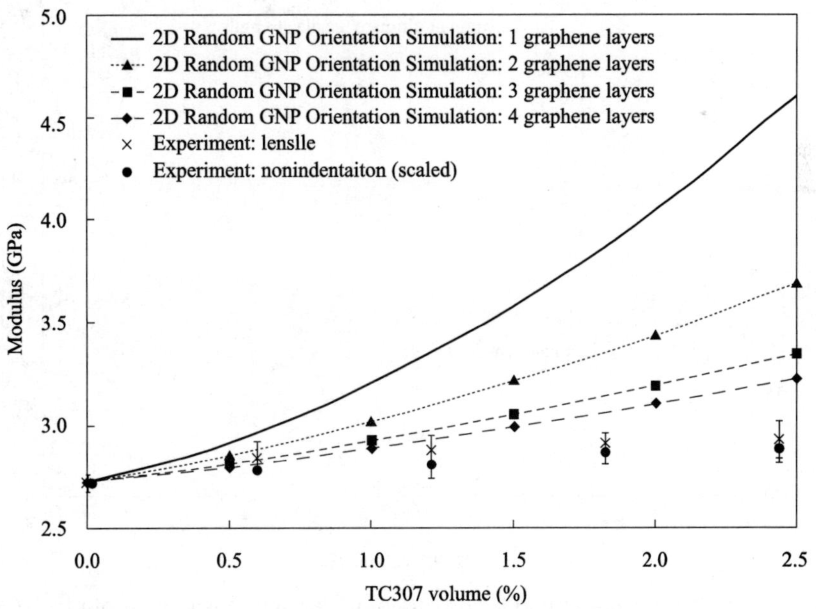

Figure 5 Modulus versus TC307 volume % for 2D GNP orientation (in-plane) at various GNP thicknesses

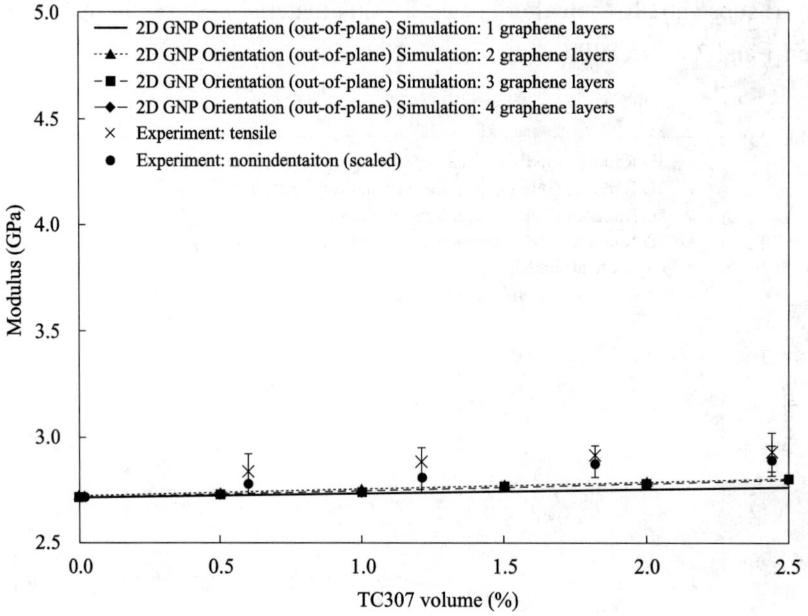

Figure 6 Modulus versus. TC307 volume % for 2D GNP orientation (out-of-plane) at various GNP thicknesses

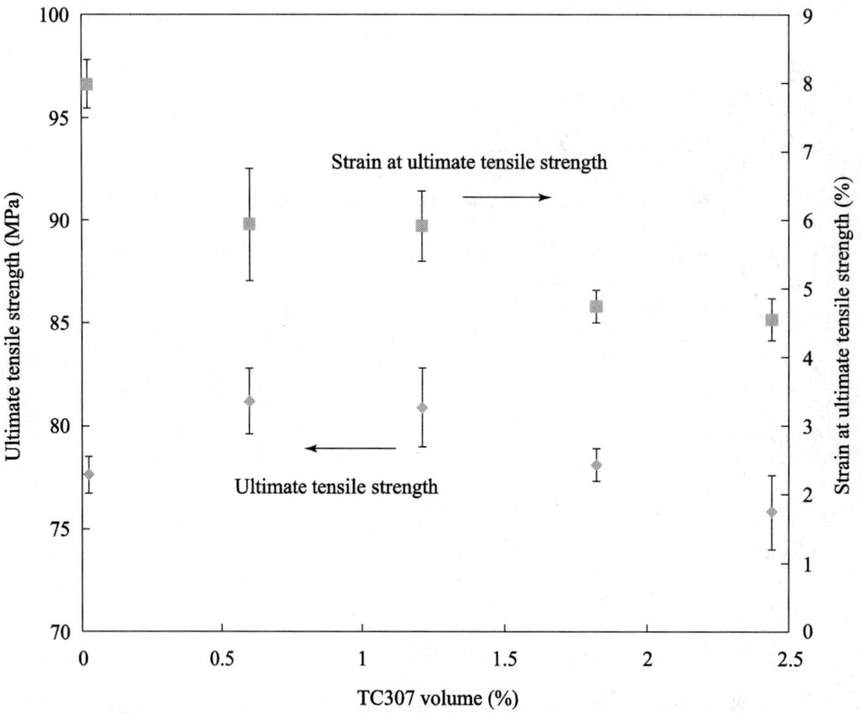

Figure 7 Ultimate tensile strength and strain at ultimate tensile strength for TC307/epoxy composites

Nanoindentation Test Results

5 The modulus (E) and hardness (H) shown in Table 1 were determined by taking the average E and H over a penetration depth of 500–1,500 nm as per the continuous stiffness method. Figure 8 shows a typical E and H curve as a function of indenter depth. Table 1 shows a hardness of approximately 0.26 GPa for neat epoxy and approximately 0.28 GPa for 4 wt% TC307 in epoxy. Figure 9 shows hardness as a function of TC307 volume % (error bars are ± 1 SD). Figure 4 shows the modulus as determined by nanoindentation as a function of volume % of TC307 (error bars are ± 1 SD).

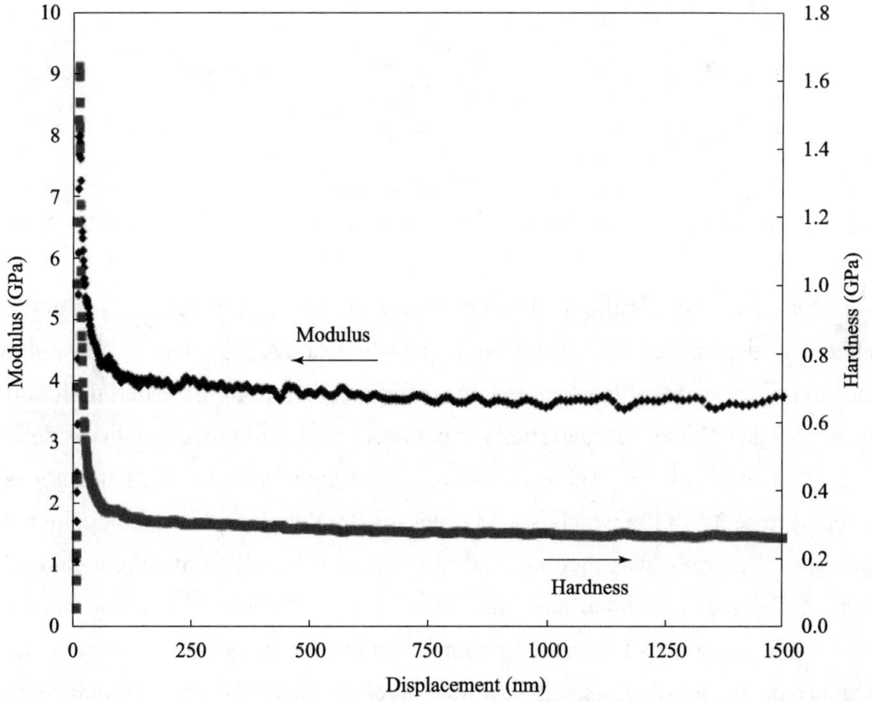

Figure 8 Modulus and hardness as a function of indenter penetration depth

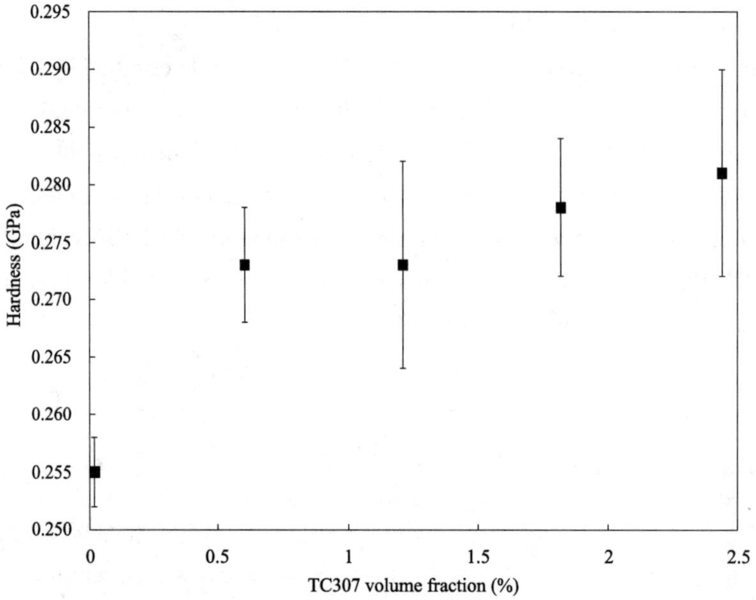

Figure 9 Hardness as determined by nanoindentation for TC307/epoxy composites

6 It has been observed that the modulus as determined by nanoindentation is higher than the modulus as determined by macroscopic tensile test (ASTM D638) for polymer-based composites [20-22]. A possible cause for this difference is pile-up of material around the point of contact. The polymer viscoelasticity is not accounted for in the modulus as determined by Oliver–Pharr method [20-22]. The modulus as determined by nanoindentation for neat epoxy was found to be 3.61 GPa which is 1.33 times greater than that measured by the macroscopic tensile test. This ratio was then applied to the moduli of all formulations and the result is shown in Table 1 as "scaled nano modulus." Figure 4 shows the scaled nanoindentation tensile modulus values. The scaled nanoindentation modulus shows the same trend as the modulus from the tensile tests, with a similar level of uncertainty and a similar concave shape.

Modeling Results

7 Figures 4-6 show the corresponding elastic modulus of the 3D random GNP orientation, the 2D in-plane GNP orientation, and the 2D out-of-plane GNP orientation, respectively, as a function of GNP volume % and GNP thickness (number of layers). The single layer of GNP corresponds to a system with perfect dispersion, while additional layers correspond to decreasing levels of dispersion. It is clear that the elastic modulus increases as the GNP volume % increases for all morphologies, as expected. Increasing levels of dispersion result in increases in the elastic modulus. As expected, the in-plane moduli of the 2D GNP orientation are greater than both the out-of-plane moduli of the 2D GNP orientation as well as the 3D random orientation, for a given level of dispersion.

8 The experimentally obtained ASTM D638 tensile modulus values from Table 1 are also shown in Figures 4-6. The experimental data generally matches the simulated data for the 3D

random GNP orientation (see Figure 4) for low levels of dispersion, particularly at the larger GNP volume %. This is a strong indication that the fabricated composite specimens contained GNPs that were randomly distributed in 3D space (also observed in Figure 2), but had, on average, three to four layers in a single stack. The as-received TC307 GNP contained approximately eight layers so processing the composites reduced this to three to four layers. Thus, it follows that more thorough dispersion could have potentially lead to significant increases in the observed elastic modulus for this composite.

9 As noted above, the experimental trend in the average modulus values for each level of GNP loading exhibits a plateau with the presence of substantial uncertainty, and thus a degree of uncertainty in the convex/concave shape. It is clear that the predicted data for the 3D random GNP orientation shows a convex trend. Beyond the ambiguity associated with the experimental uncertainty, any discrepancy in the trends could also be due to the omission of a physical mechanism in the model. For example, it is possible that the model does not accurately capture the GNP/GNP interactions that could occur at increasing levels of GNP loading. Regardless, the overall matching of the experimental data with the 3D alignment predicted data is excellent considering the complex and multiscale nature of the modeling technique.

CONCLUSIONS

10 In this work, neat epoxy (EPON 862 with EPIKURE Curing Agent W) was fabricated along with 1–4 wt% (0.6–2.44 vol%) TC307 in epoxy composites. Per the authors' knowledge, material properties and modeling for Asbury TC307 GNP in this epoxy resin system has never been previously reported in the open literature. The macroscopic tensile modulus increased from 2.72 GPa for neat epoxy to 2.93 GPa for 4 wt% (2.44 vol%) TC307 in epoxy. The tensile strength appears to reach its highest value between 1 wt% (0.6 vol%) and 2 wt% (1.21 vol%) GNP loading at around 81 MPa, although higher loadings do not substantially change the strength from the neat epoxy level (77.6 MPa for the neat resin vs. 75.8 MPa for 4 wt%). As expected, the strain at ultimate tensile strength decreased from 7.98% for neat epoxy to 4.55% for 4 wt% (2.44 vol%) TC307 in epoxy. Molecular modeling and FESEM micrographs suggest that a random 3D orientation of GNP in epoxy was achieved. The multiscale analysis indicated that the number of GNP layers was reduced from initially approximately 8 layers to 3–4 layers as a result of the composite processing method.

11 It is important that the addition of GNP did not significantly alter the strength of the composite. Because the GNP could be added to the matrix component of a traditional carbon fiber/epoxy composite to increase electrical and/or thermal conductivity, it can be done so without sacrificing the strength of the material. Thus, this material could be used for aerospace structural applications in regions that require heat dissipation and/or electrical shielding.

Main idea of each paragraph:

Para. 1: ____

Para. 2: ______________________________

Para. 3: ______________________________

Para. 4: ______________________________

Para. 5: ______________________________

Para. 6: ______________________________

Para. 7: ______________________________

Para. 8: ______________________________

Para. 9: ______________________________

Para. 10: ______________________________

Para. 11: ______________________________

Task 3: Analyze the structural patterns of reporting results and discussion with the aid of tables and figures in paragraphs 4-9 of Text I by following the example below.

Example

1)Figures 4-7 show the average tensile modulus, ultimate tensile strength, and strain at ultimate tensile strength for the TC307/epoxy composites measured according to ASTM D638. 2)Error bars shown are ±1 standard deviation (SD).3) The average tensile modulus, ultimate tensile strength, and stain at ultimate tensile strength with standard deviation and number of samples tested for each are summarized in Table 1.	(S denotes sentence) S1)-3): Referring to tables and figures to introduce overall results
4)It is clear from Table 1 and Figure 4 that the average tensile modulus exhibited a substantial increase when increasing GNP from 0 to 1 wt% (0.60 vol%).	S4): Highlighting main result
5)Further increases in GNP loading resulted in diminishing increases in tensile modulus.	S5): Highlighting main result
6)In general, this is consistent with the results of Chatterjee et al. who reported an increase from approximately 2.65 GPa for neat epoxy to approximately 3.08 GPa for 2 wt% (1.21 vol%) GNP in epoxy [19].	S6): Discussing results in comparison to previous studies
7)Although the averages show a monotonic increase with GNP loading for the entire wt% range considered, the substantial uncertainty at each GNP loading level suggests some ambiguity in the trend, possibly with a plateau above 1 wt%.	S7): Generalizing main results and providing discussion
8)It is important to note that a slight concave trend exists with the average tensile modulus values. 9)Given the uncertainty in the current data, it is difficult to decisively conclude that the trend is concave.	S8)-9): Generalizing main results and providing discussion

Task 4: Identify the verbs, tense, voices, and sentence patterns often used to refer to tables and figures in research articles based on the following sentences.

(1) Figure 1a shows the typical XRD patterns of as-synthesized nickel chalcogenides/NF.

(2) As shown in Figure 1b and Figure S1(a, b), the as-synthesized Ni_3S_2 presented a granular morphology with a thickness of 1.2 μm coated on the surface of Ni foam intimately.

(3) X-ray spectroscopy (EDS) revealed the Se/S molar ratio in the obtained chalcogenides was 10.6/22.7 (Figure 1c), which coincided with the 3/7 M ratio of Se/S in the reaction.

(4) The average size and thickness of each disk were about 1 μm and 400 nm, respectively (Figure 1f and Figure S1e).

(5) As for the exfoliated NiSe, as shown in Figure 3a, it displayed an approximate hexagonal disk with a size of ca. 1 μm, agreeing with the SEM images (Figure 1f).

(6) Figure 7b depicts their galvanostatic charge-discharge (GCD) curves at a constant current density of 10 mA cm^{-2}.

(7) Then it was expected that the amorphous thin layer wrapping the nanofiber in Figure 5c was the P-doped species.

(8) Figure S8b shows the high resolution Ni 2p spectra of the P-doped samples in comparison with $NiSe/Ni_3S_2/Ni_{12}P_5$/NF-1:1.

(9) The results are in agreement with those for the individual Ni_3S_2 (Figure 2) and NiSe samples (Figure 3).

(10) Figure 7c and d presents the CV and GCD curves of the samples after phosphorization treatment at 5 mV s^{-1} and 10 mA cm^{-2}, respectively.

(11) The electronic band structures as well as DOS of the hexagonal Ni_3S_2, NiSe and the tetragonal $Ni_{12}P_5$ are shown in Figure S10(a, b), Figure S10(c, d) and Figure S10(e, f), respectively.

(12) As plotted in Figure 8e, the composite material showed a superior rate capability of 69.6% when the current density was increased from 5 to 60 mA cm^{-2}.

(13) Figure 10b shows the GCD curves of the HSC at different current densities ranging from 20 to 200 mA cm^{-2}.

(14) The volumetric capacity of the HSC as a function of discharge current density is summarized in Figure 10c.

(15) Thus the relationship between the energy density and power density of the HSC is plotted in Ragone diagram, as depicted in Figure 10d.

(16) The photovoltaic performance of perovskite solar cells with either a CuNW or ITO top electrode, which were fabricated under identical processes except for the transparent electrode, are shown in Figure 6c, d; the performance parameters are summarized in Table 1.

(17) The inset of Figure 1b shows the test devices for probing the resistance of a 5 m wide VO ring.

(18) As a result, with increasingly greater drain electrode width, accordingly more conduction paths are present as shown by the arrows in Figure 1a, and thus IDS increases in proportion to WD.

(19) Scrutinizing Figure 3, the values of the relative resistance changes are higher with increasing humidity, ranging between 0.05% and 0.4% at 0% RH; between 0.2% and 2% at 30% RH; and between 0.5% and 3% at 90% RH.

(20) As depicted in Figure 3c, the photo-generated holes will migrate to the surface along the potential gradient produced by energy band bending and neutralize the negatively charged oxygen molecules adsorbed at the surface.

(21) Statistical analysis (Figure 1) reveals that two types of quadrate networks are preferred at high concentration.

(22) The shape of the droplet was recorded by a digital camera, and the images obtained (Figure 1) were analyzed with ImageJ.

(23) A volume of 200 mL of DME was contained in a three-neck glass bottle (the left side in Figure 1), and 200 mL of the PS solution was contained in the other three neck bottle.

(24) The typical SEM images of the samples after the abovementioned pretreat procedures were shown in Figure 2 in the main text.

(25) Figure 3 represents the corresponding sheet resistance R_s as a function of back gate voltage (V_g) and Hall resistance (R_{xy}) as a function of magnetic field (B) measured at room temperature.

(26) This effect is very clear in the classical calculations (bottom row in Figure 1) and appears less pronounced in the quantum results.

(27) Further examination of Figure 3 shows that the sensors were more selective toward the different VOCs in the arid atmosphere than in the humidified backgrounds.

(28) Our results in Figure 3 demonstrate that particles did not hinder *N* presence at the vascular wall, likely due to a lack of *N* internalization of particles during this time point.

(29) As displayed in Figure 5, the *L/N* ratios extracted from Figure 3 for these eight materials are proportional to their $A_{In/Out}$ indexes, which reveals that 2D materials with large mechanical anisotropy tend to be more exfoliatable.

(30) The HRTEM images in Figure 3 reveal a mechanically driven atomic process of high-angle GB formation through lattice bending, dislocation nucleation and accumulation, large lattice distortion and disordering in a diffuse GB layer about a few nanometer thick.

(31) *In situ* HRTEM images show the formation of a high-angle tilt GB similar to that in Figure 3, but reveal a valuable phenomenon of dynamic recovery in the GB region.

(32) The high-pressure rs phase of CdS (rs-CdS), with a lower I_{111}/I_{200} of about 0.6 (JCPDS 01-071-4151, horizontal line at bottom of Figure 3), has been proposed as an intermediate phase.

(33) In Figure 3, the local heat flux distribution along the three PNRs under 7% tensile strain has also been plotted, which will be discussed in the later section.

(34) Second, the peak at 2.5 V is highlighted by the rose colored bar in Figure 3.
(35) Four representative cases for the PMF calculation are provided in Figure 3.
(36) Figure 3 clearly demonstrates a homogeneous distribution of B in Q-carbon film, while the top layer consists of pure B.
(37) In Table 1 we report formation energies for a chromium-substitutional defect and various substitutional-vacancy complexes for which the position of the Se vacancy with respect to the Cr atom is indicated in Figure 3.
(38) The underlying mechanism resulting in these two different phenomena has been exhibited in Figure 3.
(39) As expected, in the latter case, onion-like structures were observed with PI as the outermost layer, as represented by the dark regions in Figure 3 and as opposed to Figure 2.
(40) According to Fig. 8 and Fig. 9, the increased quantity is very important to parts quality.
(41) This large uncertainty is not surprising; it is the same order of magnitude as for U_{sol} (see Figure 8).
(42) For the other compounds listed in Table 1, uptake was quantified by 1H NMR spectroscopy after isolation and disassembly.
(43) Table 1 indicates that the membrane has a high loading of PAMAM SDPs (~48 wt %).
(44) The composition of the membrane is reported in Table 1.
(45) N_2 adsorption-desorption isotherms and pore size distributions are displayed in Figure 5 and Figure S1, while structural and surface data for all samples are tabulated in Table 1 for comparative analysis.
(46) Data displayed in Table 1 support this trend, with nanoporous Fe exhibiting the thinnest ligaments and the largest discrepancy between measured and theoretical pore volume.
(47) Table 1 summarizes yield values for all solvents, with details described further below.
(48) It is clear from Table 1 and Figure 4 that the average tensile modulus exhibited a substantial increase when increasing GNP from 0 to 1 wt% (0.60 vol%).
(49) An overview of all the steps described so far towards an optimized, fully solution processed device stack is given in Table 1 and the box plot diagram (Figure 1)
(50) The results obtained for the systems studied in this work are presented in Table 5.

Task 5: Discuss in groups what words and expressions are commonly used in the following sentences to describe the tendency of increase, decrease, occurrence, enhancement, etc. as shown in tables and figures.

- Increase:

____

____

- Decrease:

____

____

- Occurrence (e.g. growth, development, appearance, existence, change, shift, etc. excluding increase and decrease):

____

____

- Others:

____

____

(1) It is clear from Table 1 and Figure 4 that the average tensile modulus exhibited a substantial increase when increasing GNP from 0 to 1 wt% (0.60 vol%).
(2) Further increases in GNP loading resulted in diminishing increases in tensile modulus.
(3) From Table 1 and Figure 7, it is clear that adding the TC307 to the epoxy decreases the ultimate tensile strain. It is also clear that there is an increase in the tensile strength between the neat epoxy and the 1 wt% (0.60 vol%) loading level. The strength remained constant between 1 wt% (0.60 vol%) and 2 wt% (1.21 vol%) loadings, and decreased substantially thereafter.
(4) For this material system the optimal strength falls between 1 and 2 wt% GNP.
(5) It is clear that the elastic modulus increases as the GNP volume % increases for all morphologies, as expected.
(6) The curves become steeper and the area enclosed by the hysteresis loops slightly increased with increasing preloading.
(7) The relative shift of preload curves seems to be relatively greater than that in the presence of a magnetic field when a preload was applied.
(8) When the strength of the magnetic field was increased, the influence of preload was found to be rather lesser compared to its influence without the application of a magnetic field.
(9) The natural frequency was shifted to 760 Hz from 280 Hz when a 520 mT magnetic field was applied.
(10) A relatively small change (2.65 times) in the natural frequency resulted in the very high (720 times) change in the stiffness of an MR elastomer when the magnetic field was applied.
(11) The vibration amplitude is reduced by 77% when the applied magnetic field was 520 mT.
(12) The wide scan XPS analysis shows an increase in carbon content from 73.9% for untreated cotton to 82.4% for rGO coated cotton (one pass) and a decrease in oxygen content from 26.1% for untreated cotton to 17.6% rGO coated cotton (one padding pass) (Table 1).
(13) This reduction of the optical band gap (E_{hv}) by ≈0.1 eV from H-DPP to CN-DPP is in accordance with the narrowing of the electrochemical band gap (E_{CV}), as determined by

cyclic voltammetry (CV), due to the larger influence of the substituents on the reduction than on the oxidation potential (Figure 2 and Table 1).

(14) In contrast, the absorption maxima (λ_{max}) are shifted with variation of X over a narrow range from 549 nm for the parent dye H-DPP to 565 nm (Cl-DPP) and 568 nm (Br-DPP) up to 586 nm (CN-DPP), while for the molar extinction coefficients (ε_{max}), only small variations are observable (Table 1).

(15) A rapid initial increase/decrease in fluorescence intensity followed by slower kinetics was repeatedly observed for both HPTS and lucigenin assay which could be due to the higher lipophilicity and stronger acidity of N-H protons (Table 1).

(16) The C/O ratio was also increased from 2.4 (GO) to 6.6 (rGO) after reduction (Table 1).

(17) However, the adsorption isotherms of N_2 showed a small decrease in the BET area (Table 1; Figure 1c, d).

(18) Under the same conditions except a less amount of catalyst (0.025 mol %), the catalytic conversion showed no significant variation, while the selectivity decreased to 94% (Table 1).

(19) In fact, repair of Pep-3 membranes by Pep-2 molecules occurred at a much higher rate than did Pep-3 self-repair under identical conditions (Table 1).

(20) Therefore, a dramatic improvement in ICE from 36.9% for C-800 to 70.9% for PO/C is obtained (Figure 1b and Table 1).

(21) Consequently, the voltage losses in these systems are around or even below 0.6 V (Table 1).

(22) In addition, the γ factor decreases to 1.076 for the second calcination temperature of 950 °C. On the other hand, it is worth noting that the dielectric performance shares a similar trend, as shown in Table 1.

(23) In addition, compared with the C-800 electrode, PO/C-450 and PO/C-800 electrodes exhibit improved ICE of 51.9 and 41.3% and decreased IIC of 484 and 426 mA h g^{-1} (Figure 4c, Table 1).

(24) For DBA-IFD, the introduction of dibutylamino groups results in a rise of the HOMO energies to −5.0 eV, along with a small rise of the LUMO energy to about −3.4 eV (Table 1).

(25) In our attempts to reproduce this result using the phase-transfer ligand exchange method with similarly sized CdSe QDs (4.2 nm), we observed a 4-fold increase in the *g*-factor value at the λ_{CD} from 0.5×10^{-4} to 1.9×10^{-4} (Table 1 and Figure 1).

(26) Comparing spectra of silk fibroin–GO and silk fibroin–rGO bilayers, we find a 48% loss of β-sheet content in silk fibroin–rGO accompanied by a 52% increase in random coils (1635 and 1655 cm^{-1}) and 50 % increase in β-turns (1668 and1684 cm^{-1}) (Table 1).

(27) The results clearly indicate that the increase in GO loading increased the mechanical strength of the samples as shown in Figure 2.

(28) As shown in Figure 2, with the decrease of satellite number, a nearly linear blue shift in peak wavelength appears.

(29) However, TEM-EDS analysis data for cuboctahedron NCs reflected a decrease in the Te atomic % relative to Pb atomic % (Pb/Te = 1.52–3.15), whereas the ratio for the cubes remained at ~1 (Figure 2).

(30) Experimental XRD patterns reported in Figure 2 display a shift of the main diffraction peak from ~15.9 to ~12.4 Å on decreasing RH.

(31) CTR analysis reveals significant changes in the terminal oxygen and water layer structure, as shown in Figure 2 and Figure 3 and detailed in Table S3 in the Supporting Information.

(32) For both macrocyclic molecules, the fluorescence intensity was quenched with increasing concentration of C60 and C70 (see Figure 2 and Figure 3).

(33) Those grown on another pattern with less differentiation than on the flat surface showed decreased length (Figure 2).

(34) Similar to absorption in Figure 2, Au nanorod photoluminescence spectra exhibit a rising intensity at shorter wavelengths.

(35) The heat maps in Figure 3 represent the relative resistance changes of the six sensors after exposure to the different VOC levels (100-1000 ppb) in arid conditions, intermediate humidity, and high humidity levels.

(36) Scrutinizing Figure 3, the values of the relative resistance changes are higher with increasing humidity, ranging between 0.05% and 0.4% at 0% RH; between 0.2% and 2% at 30% RH; and between 0.5% and 3% at 90% RH.

(37) Clear martensite lath can be seen in each figure, and Figure 17 a)-f) shows a downtrend of the martensite, which indicates that the content of martensite decreases with the increase in partitioning temperature.

(38) As expected, the knockdown of ADAM9 resulted in a significant decrease in hydrogel degradation (Supplementary Figure 6 e-g).

(39) Along with the increase of the crystallization time from 48 to 108 h, the intensity of the XRD peaks related to ITE zeolite is gradually enhanced (Figure 1).

(40) However, even for NPCs in the softest hydrogels, decreasing gel degradability resulted in decreased stemness (Figure 6d).

(41) With a decreasing magnetic field of the same polarity as that during TAUM treatment, as in the resistance measurement shown in Figures 2 and 5, the stored charges are gradually released through the electrodes, leading to a decrease in the resistance.

(42) At an even higher strain level of 36%, the twin density increases substantially (Supplementary Figure 14).

(43) Note, however, that IR irradiation (940 nm) does not produce any noticeable increase in conductivity or decrease in WF (Supplementary Figure 5 and Supplementary Table 3).

(44) In all cases, following stretching to 50% strain the maximum increase in resistance is less than 10% (Figure 1d).

(45) However, even up to the 170th cycle (Figure 1d), the surface changes significantly.

(46) Micro-hardness exhibits a decreasing tendency from the PMZ to the HAZ on the 6061 side, as shown in Figure 12.

(47) In comparison to monocultured models, dual cocultures and tri-coculture 3D-MCTS with and without LbL-MPs, presented increased contraction (Figure 6).

(48) The corresponding grain size distributions (Figure 7) show that there was a minor reduction in the number of small grains and that the shift toward larger grain sizes mostly took place in the 200-300 nm range.

(49) As shown in Figure 1 (a), MRR is increased with a higher rotation rate below 200 min-1; however, MRR subsequently shows a dramatic decline at a rotation rate of 300 min-1.

(50) Increasing tenfold the surface density of SA-FasL on microgels had no significant improvement in graft survival (Supplementary Figure 7).

(51) Further improvement in cycling stability can be achieved with SiSPC by increasing the PI binder content to 30%, resulting in a 89.8% capacity retention after 70 cycles.

(52) The significant enhancement of the electrostrain in the composition range $0.25 < y \leqslant 0.32$ has a strong correlation with the onset of the CL phase.

(53) As shown in Figure 1d, the catalyst displays a significant drop in activity with a more negative shift of the overpotential after CN^- interaction, signifying the blockage of Mo sites by CN^- ions.

(54) One hour after the low viscosity LuCI treatment, the reduction in the glucose response was $43.0 \pm 13.6\%$, and three hours after treatment, the reduction decreased to $25.1 \pm 5.4\%$.

(55) With the increase in the amount of zinc solution, the blue shift of PL emission peak becomes obvious, which is attributed to the adequate zinc precursor supplied for ion exchange.

(56) As the illumination intensity increases, more carriers are generated, leading to larger photocurrent and hence larger SNR.

(57) Thus, the reset process can occur more easily at increased temperatures.

(58) Redox cycling shifted the conversion curves toward higher temperatures indicating catalytic activity loss, which is likely because of an initial Ni particle growth over the number of redox cycles.

(59) An increase in μ could arise from a structural change in the oxidant film and an increase in n means an increase in ionic species.

(60) As the temperature increases, the bright area gradually disappears due to the dipole disorder induced by the increased temperature.

(61) The Ni nanoparticles grow with the increase of annealing temperature.

(62) With the increase of grain size, the polarization and strain curves develop pronouncedly, coupled with the increase of all physical parameters.

(63) The conductivity linearly declines with increasing temperature from 5 to 300 K indicating metallic conduction behavior.

(64) At the same time, the rim vanishes and is absent for higher temperatures.

(65) After photoexcitation of MoS_2, holes will transfer to a lower energy state at the VBM of WS_2 while the electrons still stay in MoS_2.

(66) In the case of lower laser powers, a decrease in strength may happen due to a decrease in the wetting length.

(67) In its initial stage of melt penetration, the melt was ready to penetrate down due to a sudden pressure increase at the melt front.

(68) Therefore, it is possible that the liquid metal from the weld pool spreads more toward the horizontal direction.

(69) Given the lower thermal mass of thinner substrates, laser heat input is sufficient to rapidly raise the temperature of the bulk material above the preheat condition.

(70) The day-by-day increment in the CPI clearly shows the increase in cell number in the control sample.

Task 6: Read the following excerpts of research article results and discussion sections, and analyze the structural pattern of reporting and discussing results with the aid of tables and figures by following the outline already given for the first excerpt.

(1)

Compositions and morphologies of nickel chalcogenides/NF[2]

1 X-ray diffractometer (XRD) was used to examine phase purity and crystallographic structures of the obtained samples. Figure 1a shows the typical XRD patterns of as-synthesized nickel chalcogenides/Ni foam (NF). As for Ni_3S_2/NF, apart from three characteristic peaks centered at $2\theta = 45.5°$ (111), 53.0° (200) and 78.3° (220) attributed to those of Ni foam (JCPDS no. 89-7128), the sample showed six obvious characteristic peaks at $2\theta = 21.7°$ (101), 31.1° (110), 37.8° (003), 44.0° (202), 49.7° (113) and 55.1° (212), in good agreement with the hexagonal Ni_3S_2 (JCPDS no. 44-1418). The XRD patterns for NiSe/NF exhibited three characteristic peaks centered at $2\theta = 32.8°$ (101), 44.2° (102) and 49.8° (110), matching well with the hexagonal NiSe (JCPDS no. 65-3425). Furthermore, it was found that the diffraction peaks at 44.0° (202) for Ni_3S_2 and 44.2° (102) for NiSe were very close to the strongest diffraction peak at 45.5° (111) for Ni foam, which probably can account for the successful surficial sulfidation and selenization of Ni substrate.

2 Whereas with simultaneous introduction of Se and S powders into the reaction system, the XRD patterns of the prepared samples all exhibited characteristic peaks of both Ni_3S_2 and NiSe regardless of the Se/S molar ratios ($3/7 \rightarrow 7/3$) (Figure 1a), suggesting the successful syntheses of NiSe/Ni_3S_2 heterostructures. Notably, with the increase of the Se/S molar ratio ($3/7 \rightarrow 7/3$), the intensity of diffraction peak at $2\theta = 32.8°$ (101) for NiSe in the NiSe/Ni_3S_2 composite material was increased. As shown, NiSe/Ni_3S_2/NF-3:7 only exhibited the strongest diffraction peak at 32.8° (101) for NiSe. Whereas for NiSe/Ni_3S_2/NF-1:1, two strongest

diffraction peaks at 32.8° (101) and 49.8° (110) for NiSe were observed. And as for NiSe/Ni_3S_2/NF-3:3, all the diffraction peaks for NiSe can be seen in the XRD patterns. The results suggested the increased selenization extent with the increased Se/S molar ratio.

3 During the solvothermal reaction, diethylenetriamine (dien) was utilized as solvent, the good solubility of Se and S in diethylenetriamine provided good access for the attack of Se and S to Ni substrate. On the other hand, diethylenetriamine is a chelating ligand with good coordination ability to Ni atom, it was supposed that diethylenetriamine acted as a good intermediate media for Ni and Se/S. With the aid of diethylenetriamine, the electrons can be more easily transferred from Ni and Se/S to obtain NiSe/Ni_3S_2 composite material. The possible reaction was as follows:

$$Ni + n(dien) \rightarrow Ni\ (dien)_n$$

$$Ni(dien)_n + Se \rightarrow NiSe + n\ (dien)$$

$$3Ni(dien)_n + 2S \rightarrow Ni_3S_2 + 3n\ (dien)$$

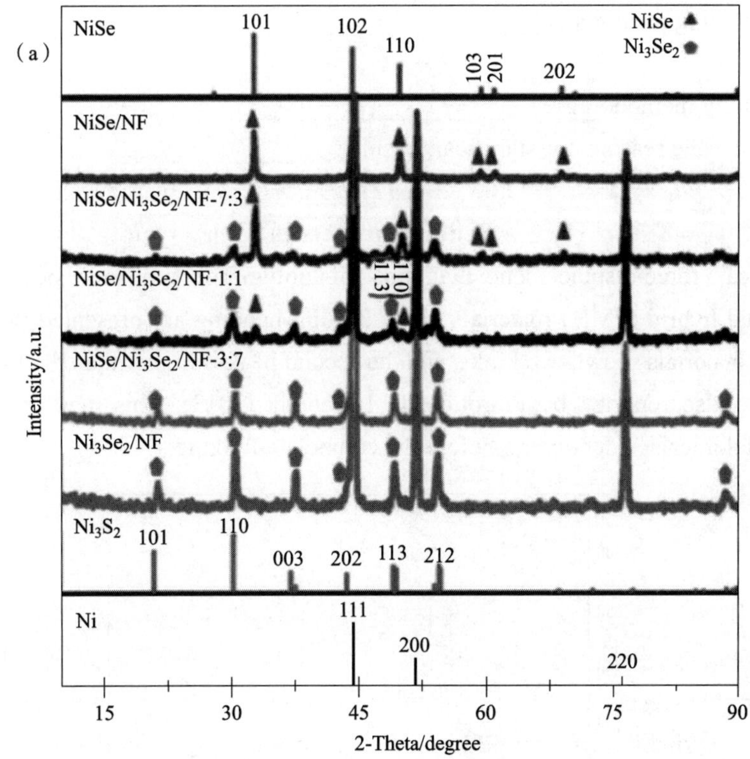

Figure 1 XRD patterns of nickel chalcogenides/NF obtained under different molar ratios of Se/S and the standard profiles for NiSe, Ni_3S_2 and metallic Ni (a)

Outline:

Para. 1:

1) Briefly restating methodology: ____________________

2) Referring to tables or figures to introduce overall results: ________

2.1) Ni_3S_2/NF

- Highlighting main results: ____________________
- Discussing results: ____________________

2.2) NiSe/NF

- Highlighting main results: ____________________
- Discussing results: ____________________

2.3) Similarity of XRD patterns among different samples: ________

- Highlighting main results: ____________________
- Discussing results: ____________________

Para. 2:

1) Briefly restating methodology: ____________________

2) Referring to tables or figures to introduce overall results: ________

- Specifying significant results: ____________________
- Discussing results: ____________________

Para. 3:

1) Briefly restating methodology: ____________________

2) Analyzing possible reasons for significant results: ______________

(2)

Bending properties[3]

The typical force-displacement diagrams of polymer composite, steel and polymer composite-metal hybrid (PMH) materials under 3-point-bending are presented in Figure 1. The curve of PMH materials shows two kinks, and the second peak is higher than the first one, similar phenomenon is also reported by Fiorotto and Lucchetta (2013). This may be caused by the composite-metal interface debonding before the composite fracture.

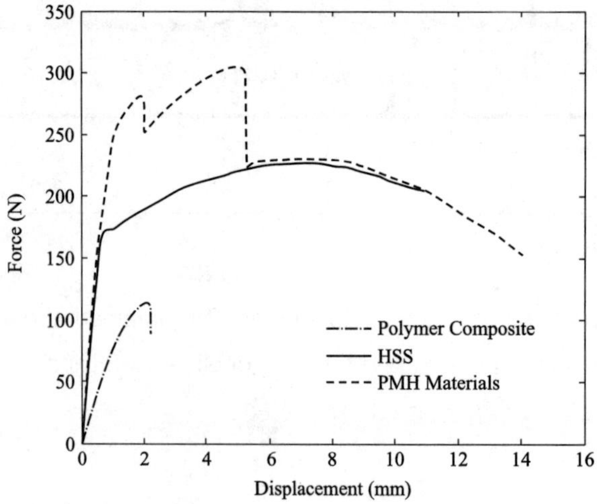

Figure 1 The bending force-flexural displacement curves of the constituent and hybrid materials

The bending modulus and strength of composite, high strength steel (HSS) and PMH specimens are shown in Figure 2. The bending modulus and strength of PMH materials can be calculated using an equation of the same form as Equations (1) and (2), respectively. The calculated value of bending modulus is 148.1 GPa, and the error is 7.73%. The calculated value of bending strength is 407.9 MPa, and the error 13.16%.

Comparing the theoretical calculation with the experimental results, the prediction values are generally higher than the test data for bending. The reason may be that the interface between the composite and metal is an independent phase different from both constituent materials, which is not considered in the theoretical model. The interface is possibly the weakest phase in PMH material systems. Hence, further experiments are needed to characterize the interface and validate the theoretical model considering the interface property.

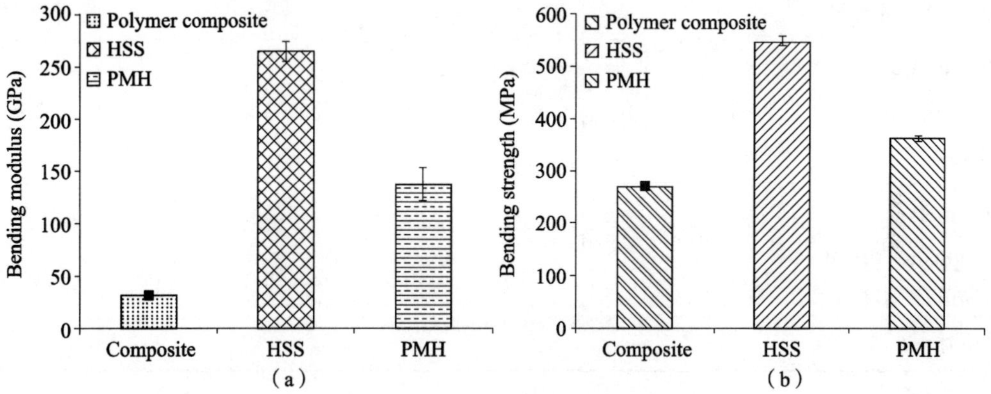

Figure 2 Experimental result histogram of (a) bending modulus, (b) bending strength of composite, HSS and PMH specimens

Task 7: Write a short essay to describe each table or figure below by following the given outline.

Outline:

- Referring to the table or figure to introduce overall results
- Highlighting main results one by one in decreasing order of significance
- Specifying or exemplifying significant results
- Interpreting and discussing results

(1)

Table 1 Comparison of tensile properties between the single and hybrid material specimens

Specimen type	Density ρ (g/cm^3)	Tensile modulus E (GPa)	Yield strength σ_s (MPa)	Tensile strength σ_b (MPa)	Specific modulus E/ρ (GPa/(g/cm^3))	Specific yield strength σ_s/ρ (MPa/(g/cm^3))	Specific tensile strength σ_b/ρ (MPa/(g/cm^3))
35 wt.% short glass-fiber reinforced nylon 66 (PA66-GF35)	1.41	8.5	20.1	139.3	6.03	14.26	98.79
HC340 high strength steel (HSS)	7.85	209.8	362.4	643.5	26.73	46.17	81.97
Polymer composite–metal hybrid (PMH) material	4.63	103.6	171.6	630.9	22.38	37.06	136.26

(2)

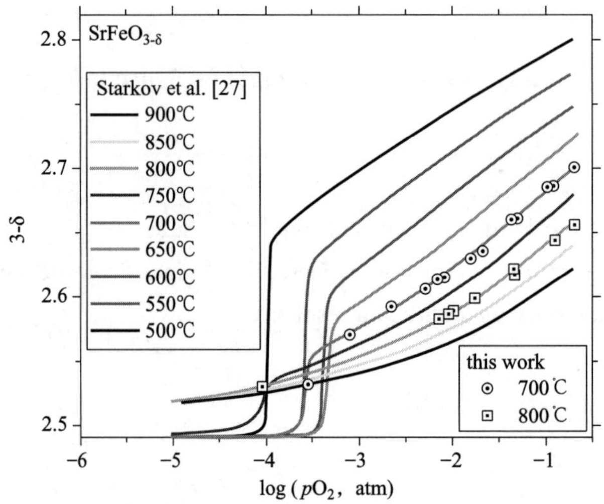

Figure 1 Oxygen content of SF vs pO_2 at different temperatures: points – original data, lines – courtesy of Starkov et al. [1]

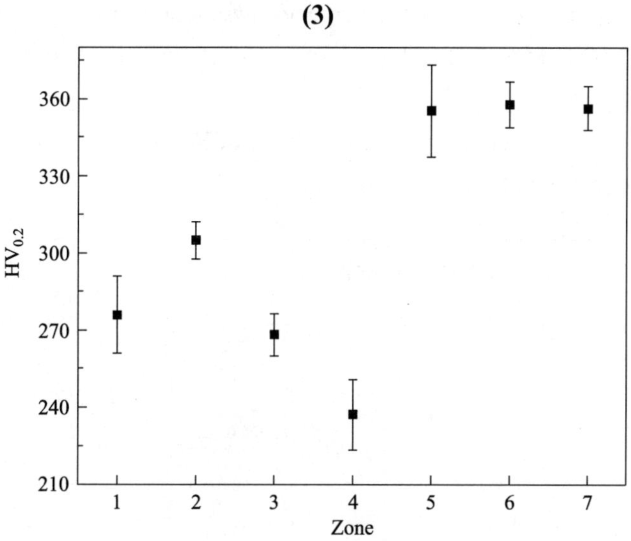

Figure 2 Microhardness distribution in the cross section of the surfacing sample

Task 8: Read Text Ⅱ carefully and answer the following questions briefly.

(1) What is highly essential for the development of efficient MRE-based isolators?

____

(2) What limitations of most MRE-based isolators should be overcome?

____

(3) What aspect in the development of MRE-based isolators or absorbers has not been considered in most previous studies?

____

(4) What is the main purpose of this research?

____

(5) What approach was adopted to study the vibration isolation behavior of an MRE-based isolator?

____

(6) What main results are given in the Conclusion section?

____

(7) What preloading effect for the MR elastomer has been observed and claimed based on Figure 5?

____

(8) What experimental results are obtained directly and indirectly from Figure 6?

____

(9) What features of the proposed MRE-based isolator are presented according to Figure 7?

____

(10) According to Figure 8, why does the magnetic interaction in the developed MR elastomer increase when a preload is applied?

____

(11) What relationshiop between the natural frequency of MRE-based isolator and the stiffness of the MR elastomer is concluded based on Figure 9?

____

(12) What argument is Figure 10 used to support?

____

Text Ⅱ

A New Type of Vibration Isolator Based on Magnetorheological Elastomer[4]

Abstract: In this work, a new type of adaptive vibration isolator based on magnetorheological (MR) elastomer (MRE) is presented. A new method was adopted to develop such an isolator where both a magnetic field and a preload were applied simultaneously. The magnetic attraction force was utilized to change the preload in the single degree of freedom (DOF) system. The system has such a provision that when a magnetic field is applied the preload would be automatically acting to the MR elastomer. In such a combined loading condition, the natural frequency of a single DOF system promptly shifted to a higher frequency and the stiffness of the MR elastomer was significantly increased. The stiffness of the MR elastomer system was found to be increased as high as 730 times of its original stiffness when the magnetic field of 520 mT was applied, which is a significantly higher augmentation than those reported in the literature. The combined effect of the preload and the magnetic field was profound because the magnetic interaction among the magnetic particles was simultaneously boosted by both the magnetic field and the preloading effect. It is often a large difficulty to generate a higher magnetic field in most of the MRE-based isolators. Our study showed that when a suitable preload and a suitable magnetic field are applied together, a highly tunable isolator system can be developed even with the application of a relatively lower magnetic field strength.

INTRODUCTION

1 Base isolation is one of the efficient ways to reduce the unwanted vibration to prevent structural damage. Active and semi-active isolators have been developed by incorporating

the magnetic field-responsive materials such as magnetorheological (MR) fluids and MR elastomers (MRE) into the structures or systems [1-6]. The smart materials, MR fluids and MR elastomers help to mitigate the unwanted vibration by changing their elastic and damping properties under the external magnetic field. Thus, the reduced amplitudes of acceleration or displacement are transmitted to the structure because of the integrated MR materials. However, MR fluids have limitations due to leakage and sedimentation problems. Thus, in recent decades, MR elastomer-based isolation or absorption systems have gained a considerable attention [7-10]. Therefore, understanding the behavior of the MR elastomer under the magnetic field and different loading conditions is very essential to develop efficient MRE-based isolators.

2 MR elastomers are particulate composites and consist of micron-sized magnetic particles embedded within the non-magnetic elastomeric matrix such as natural rubber, silicone rubber, and polyurethane. Based on the distribution of the magnetic particles, MR elastomer could be isotropic or anisotropic. Magnetic particles are randomly distributed in isotropic MREs whereas magnetic particles form a chain because of the applied magnetic field during the crosslinking process in anisotropic MREs. Because of the magnetic particle chains, the anisotropic MREs show a higher MR effect than that of isotropic MREs [1], [11-12]. Nevertheless, in recent years, numerous efforts have been made in order to improve the various properties of MR elastomers such as the development of porous structures, 3D printing and an addition of additives such as magnetic nanoparticles and carbon nanotube [13-17]. Such methods might result in a unique configuration of magnetic particles and increase either zero-field modulus or MR effect or both. Because of the dynamic responsiveness, MREs have been attractive in numerous applications not only as vibration isolators/absorbers but also as electromagnetic shielding, engine mounts for automobile and even soft actuators and sensors [1], [18-19].

3 MRE-based isolators shift the natural frequency or resonance zone of the system upon the application of a magnetic field. Such isolators also have the capability to attenuate the transmitted vibration amplitude by changing the stiffness and damping capability of MR elastomers under the magnetic field. A number of studies dealing with the development of the MRE-based isolation systems are available in the literature [1-4], [8-9], [20-21]. Researchers have developed various kinds of active or semi-active MRE-based isolators/absorbers by coupling the MR elastomer either in a single mode (shear or squeeze) or a shear–squeeze mixed mode [18], [22]. The shear–squeeze mixed mode is found to be a more effective way to change the natural frequency of the MRE-based isolators as compared to a single mode of operation. However, most of the MRE-based devices possess a narrow working frequency range, high power consumption, and a bulky configuration. Thus, there is a need to find an efficient method to enhance the performance of the MRE-based isolator that is free from the inherent problems of current isolators. On the other hand, MRE-based isolators have also

been developed by implementing various control strategies such as fuzzy logic control [23-27], Lyapunov [28] and clipped optimal control [25]. Such control strategies help to increase the response and performance of the isolators.

4 For the single DOF system, there are two ways to shift the natural frequency of the MRE-based isolators; 1) apply higher magnetic fields so that the MR elastomer would change its stiffness until the magnetic saturation of the magnetic particles is reached and 2) increase or decrease the amount of payload. They are illustrated in Figure 1. The first one needs a strong magnetic field that usually makes a bulky system and requires higher power if an electromagnet is used, while the second one needs a dynamic system in order to have a variable payload. In Figure 1, the magnitude transmissibility is defined as a ratio of the output signal (y) and the input signal (x) and it is a function of frequency and the frequency ratio is the ratio of the natural frequency (ω_o) of single DOF system and the excitation frequency (ω).

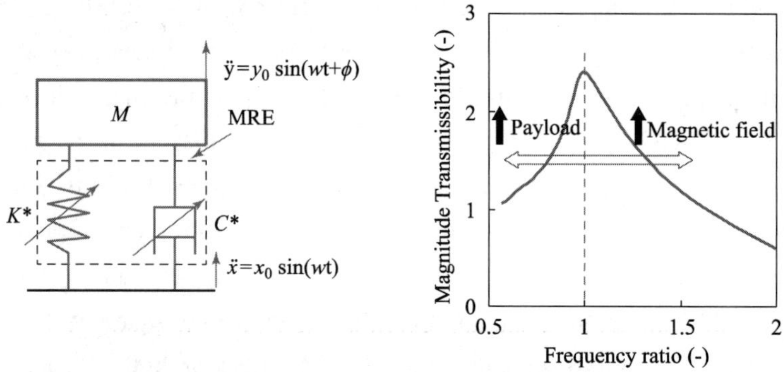

Figure 1 Illustration of a single degree of freedom MRE-based isolation system and for the shifting of the transmissibility curve with increasing magnetic field or payload, where each shifting direction is indicated by the arrow

5 When the magnetic field is applied, the magnetic interaction among the magnetic particles increases the stiffness of the MR elastomer, thus the natural frequency is increased and therefore the resonance zone will shift toward the higher frequency. On the other hand, the increased payload always shifts the natural frequency toward lower frequency range and vice versa. At the same time, the increased payload provides a preloading effect to the MR elastomer and thus MR elastomer would become stiffer and again the resonance zone shifts to the higher frequency range. Therefore, the combined effect might be seemingly neutral or significant. In the previous studies, such a combined effect of the magnetic field and preloading effect have not been considered in developing MRE-based isolators or absorbers.

6 It should be noted that the increasing or decreasing the amount of payload to control the preloading might be a troublesome act in the actual system. Thus, we need an efficient way to change the amount of payload. The magnetic attraction force can be very powerful if the

ferromagnetic materials are brought near to the strong magnets. Thus, the basic idea here is to use the magnetic attraction force between the strong magnet and ferromagnetic material to increase or decrease the amount of payload in the MRE-based isolator. Therefore, the amount of preloading can be controlled easily. At the same time, the MR elastomer has to be exposed to the magnetic field.

7 In this study, the approach is to simultaneously apply a magnetic field and a preloading to study the vibration isolation behavior of an MRE-based isolator. A simple MRE-based isolator has been developed, where various magnetic field strengths can be generated, and at the same time, the amount of payload can be changed to produce different levels of preloading. The magnetic attraction force has been used as a source to change the amount of payload. Therefore, when a different magnetic field is applied, the preload will be changed accordingly. Prior to studying the behavior of the isolator, the preloading effect at the various magnetic field strengths for the MR elastomer is presented under a cyclic compression loading. Thereafter, the combined effect of the preloading and magnetic field in the MRE-based isolator is studied in detail in the frequency range of 200–1000 Hz.

2. Material and method

…

3. Results and discussion

8 First, the preloading effect for the MR elastomer was studied under the cyclic compression loading without and with the application of a magnetic field. Figure 5 depicts the preloading effect on the response of the MR elastomer at different magnetic field strengths and at a constant peak to peak displacement of 0.6 mm and a frequency of 0.1 Hz. A few observations from Figure 5 become evident. First, increasing preload made the force-displacement curves separated. This result signifies that the preloading has a direct effect on the MR elastomer's behavior. Similarly, the curves become steeper and the area enclosed by the hysteresis loops slightly increased with increasing preloading. Thus, it can be claimed that both the stiffness and damping capability of the MR elastomer increased with increasing preload because the slope and area under the force-displacement curve represent stiffness and damping capability, respectively. Second, in the absence of a magnetic field, the relative shift of preload curves seems to be relatively greater than that in the presence of a magnetic field (190, 320 or 520 mT) when a preload was applied. This indicates that the influence of the preload was more prominent when a magnetic field was not applied. This can also be realized from Figure 6 when the maximum forces are compared at various magnetic field strengths and preloads. Finally, the curves without a preload (1 N curves) are continually softer while the curves with a 30 N preload are continually stiffer.

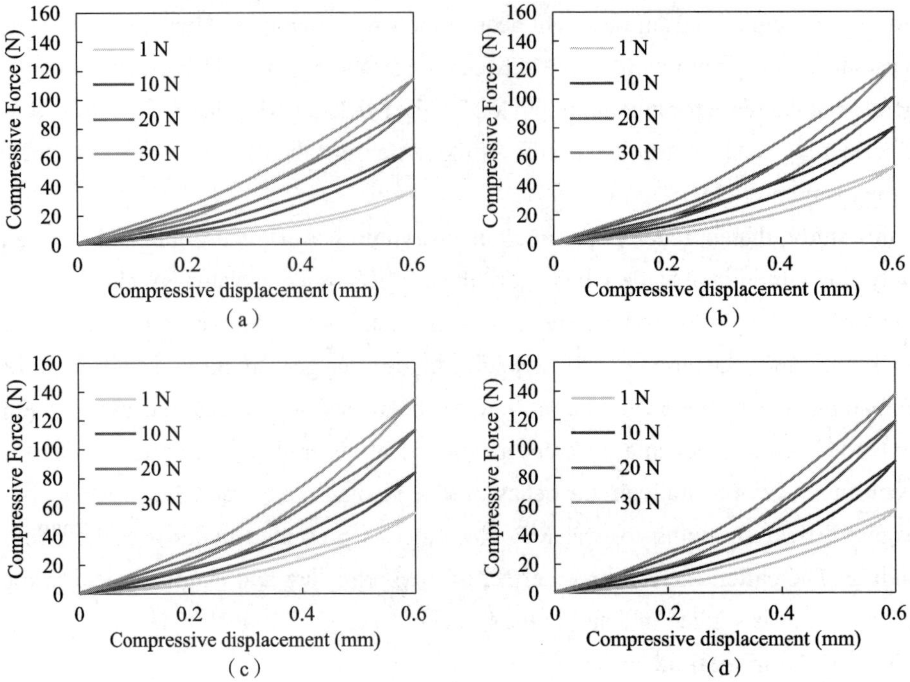

Figure 5 Compressive force as a function of cyclic compressive displacement under different preloads at a 0.6 mm peak-to-peak displacement amplitude of cyclic compression at 0.1 Hz frequency. Various magnetic flux densities applied are: (a) 0 mT, (b) 190 mT, (c) 320 mT and (d) 520 mT

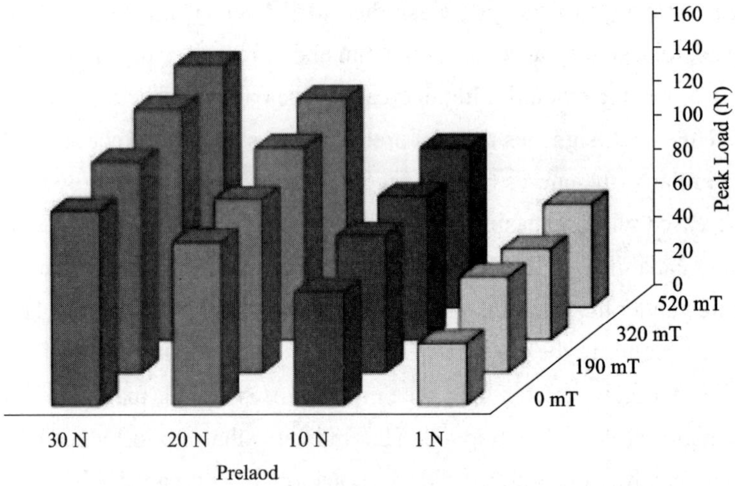

Figure 6 Comparison of the peak force of the MR elastomer to compress by 10% strain at various preloads and different magnetic fields

9 Figure 6 shows the influence of preloads on the amount of force required to compress the MR elastomer by 10% strain under the various magnetic fields. The experimental results

revealed that the preload has a significant effect on the response of the MR elastomer under both conditions, without and with the presence of a magnetic field: the peak load was significantly increased. The maximum increment at 30 N preload and 520 mT magnetic field was observed to be 4 times higher than the load required at no preload and no magnetic field to compress the MR elastomer sample for the same strain level (i.e. 10%). When the strength of the magnetic field was increased, the influence of preload was found to be rather lesser compared to its influence without the application of a magnetic field. This could be explained as the field-induced particle-particle interaction among the magnetic particles would have already made the elastomer stiffer in the presence of the magnetic field. Thus, the higher amount of preload is required to deform the MR elastomer to the same extent as it does in the absence of a magnetic field.

10 Now, the vibration isolation behavior of the MRE-based isolator is presented. Fig. 7 shows the magnitude transmissibility curve for the MRE-based isolator at various magnetic field strengths in the frequency range of 200–1000 Hz. At the peak amplitude of the magnitude transmissibility curve, the phase angle between response signal (y) and excitation signal (x) is $\pi/2$ and the corresponding excitation frequency represents the natural frequency (ω) of the single DOF system.

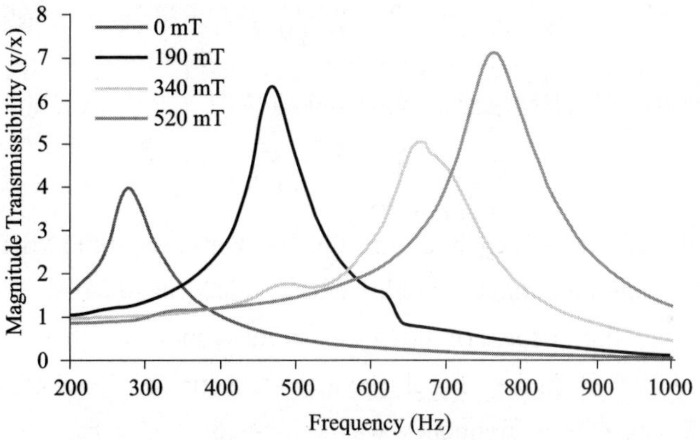

Figure 7 Magnitude transmissibility of the MRE-based isolator as a function of excitation frequency under the various magnetic fields

11 Several features can be noted in Figure 7. The system has one maximum peak on the magnitude transmissibility curve. This signifies that the MRE-based isolator is a single DOF system. The peak of the transmissibility curve is moving toward higher frequency when both magnetic field and payload are increased. Similarly, the peak amplitude of the transmissibility curve increased when the magnetic field and payload were increased. The increase in the peak amplitude signifies that the damping ratio of the isolator is decreased. Moreover, the amplitude of the vibration has been significantly reduced in the resonance zone when the magnetic field is applied.

12 When only the magnetic field is applied, the MRE-based isolator changed the natural frequency due to an increase of the stiffness of the MR elastomer only via particle-particle interaction of the magnetic particles under the magnetic field. In the case when both the magnetic field and preload were applied simultaneously, there is not only the magnetic interaction but also a preloading effect provided by the magnetic attraction force to the MR elastomer, making the MR elastomer even stiffer. Therefore, the natural frequency of the system is boosted by the preloading as well. There is no doubt that the magnetic interaction between the magnetic particles would always increase the natural frequency. However, the magnetic interaction would be even higher at higher preloads as the distance between the magnetic particles decreases when the preload is applied as illustrated in Figure 8. Thus, as a result of a combined effect, the speedy shifting of the transmissibility curve is observed.

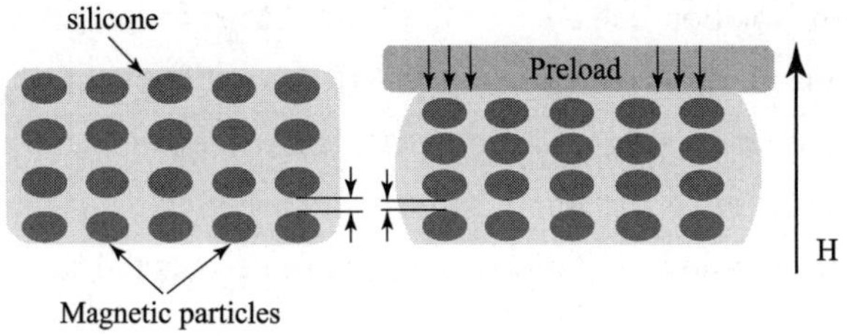

Figure 8 Illustration of the reduction of the particle-particle distance under preloading

13 The stiffness of the MR elastomer is expressed as

$$K = (2\pi f_n)^2 M \tag{1}$$

where f_n is the natural frequency, and M is the absorber mass of the MRE-based isolator.

14 The peak of the magnitude transmissibility curves represents the natural frequency of MRE-based isolator while the stiffness of the MR elastomer can be obtained using Eq. (1) They are presented in Figure 9. Both the natural frequency and stiffness are increased with increasing magnetic field. The natural frequency was shifted to 760 Hz from 280 Hz when 520 mT magnetic field was applied. The relative increase in the natural frequency was found to be 2.65 times higher than that of its original natural frequency. However, the relative increment of the stiffness was significantly higher by 720 times of its zero-field stiffness. Therefore, the relationship of stiffness and magnetic field is expressed in the semi-log graph [Figure 9(b)] in order to have a clear picture of the significant increment of the stiffness. This can be explained with the help of Eq. (2) If the zero-field natural frequency of the MRE based system is high, the smaller change in the natural frequency would make a significant change in the stiffness of the MR elastomer. Thus, a relatively small change (2.65 times) in the natural frequency resulted in the very high (720 times) change in the stiffness of an MR elastomer when the magnetic field was applied.

$$\Delta f_n \approx \frac{1}{2} \frac{\Delta K}{f_n} \tag{2}$$

where f_n is the natural frequency at zero field, Δf_n is the change in the natural frequency of the MR elastomer system and ΔK is the change in the stiffness of the MR elastomer when a magnetic field is applied [30].

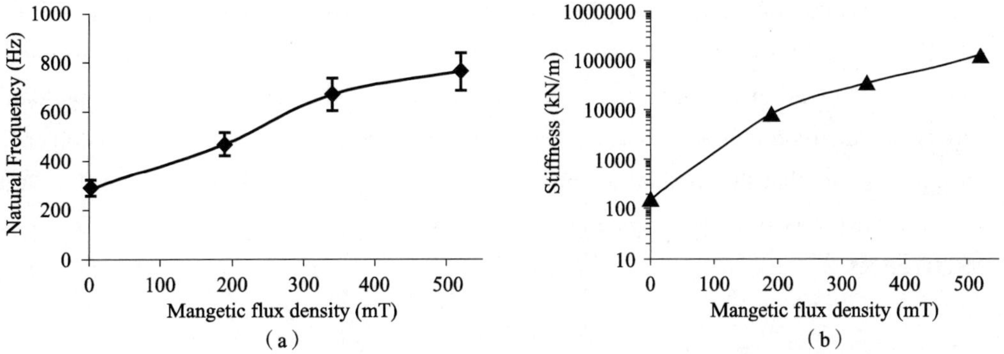

Figure 9 (a) Natural frequency of the MRE-based isolator and (b) stiffness of the MR elastomer under the combined application of both magnetic field and preload

15 Furthermore, a substantial influence of the magnetic field can be seen in the resonance zone. In other words, the acceleration amplitude is significantly attenuated in the resonance zone when the magnetic field is applied. Figure 10 shows the reduction of the acceleration amplitude in the resonance zone when a magnetic field is applied. The vibration amplitude is reduced by 77% when the applied magnetic field was 520 mT. This signifies that the combined loading condition can also significantly attenuate the vibration amplitude at the resonance zone. Such dynamic responsiveness of the new type of isolator demonstrates the ability to be used for the highly adaptive vibration control system.

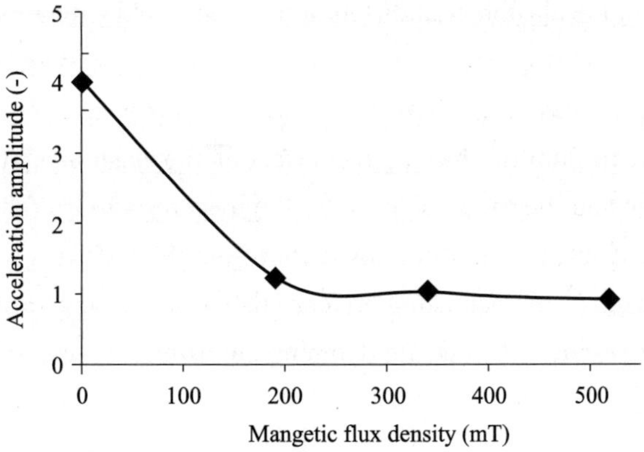

Figure 10 Reduction of the acceleration amplitude at the zero-field resonance frequency (280 Hz) of the MRE isolator when a magnetic field and a preload are applied

4. Conclusions

16 In this work, a simple setup was designed to study the behavior of an MRE-based isolator system where both the magnetic field and preload could be applied simultaneously. It was demonstrated that the MRE-based isolator can quickly shift the natural frequency even at the moderately strong magnetic field. The experimental results proved that this isolator could shift the natural frequency from 280 to 760 Hz when 520 mT of the magnetic field was applied, which correspond to the relative change in the stiffness of the MR elastomer by 73,000% of its original stiffness. The combined effects of the magnetic field and preloading are attributed to providing such a substantial change in the stiffness of the MR elastomer. The result suggests that the method adopted in this work opens a new door to develop highly tunable MRE-based isolation systems.

References

...

Task 9: Read the following excerpt of paragraph 8 from Text II, in which noun phrases in the subject position of main or subordinate clauses are underlined together with node words bolded, predicates also bolded, and adverbials placed in parentheses. Discuss with each other the phrasal complexity, typically reflected in the use of pre- and post-modifiers in noun phrases, and syntactic complexity, typically manifested by adverbials in various forms. Finally, identify the main clauses in the rest paragraphs of results and discussion section in Text II.

Excerpt

First, the preloading **effect** for the MR elastomer **was studied** (under the cyclic compression loading without and with the application of a magnetic field). **Figure 5 depicts** the preloading **effect** on the response of the elastomer at different magnetic field strengths and at a constant peak to peak displacement of 0.6 mm and a frequency of 0.1z. A few **observations** from Figure 5 **become evident**. First, **increasing preload made** the force-displacement **curves separated**. This **result signifies that** the **preloading has** a direct **effect** on the elastomer's behavior. Similarly, the **curves become steeper** and the **area** enclosed by the hysteresis loops (slightly) **increased** (with increasing preloading). Thus, **it can be claimed that** both the **stiffness** and damping **capability** of the elastomer **increased** (with increasing preload) (because the **slope and area** under the force-displacement curve **represent stiffness** and damping **capability**, respectively).

Task 10: Read the results and discussion section of Text II again and discuss in groups what sentences are mainly descriptions of results (e.g. exemplification and specification) and what sentences are more focused on discussions of results (e.g. interpretation and explanation), as shown in the following example.

Example

First, increasing preload made the force-displacement curves separated **(Description)**. This result signifies that the preloading has a direct effect on the MR elastomer's behavior **(Discussion)**. Similarly, the curves become steeper and the area enclosed by the hysteresis loops slightly increased with increasing preloading **(Description)**. Thus, it can be claimed that both the stiffness and damping capability of the MR elastomer increased with increasing preload because the slope and area under the force-displacement curve represent stiffness and damping capability, respectively **(Discussion)**.

Task 11: Transform the following high-frequency verb-noun collocations used in research article results and discussion sections into passive voice. Determine what verb-noun collocations are typically used for 1) restating research purpose, 2) restating research methods, 3) referring to tables and figures, 4) highlighting main results, 5) specifying or exemplifying significant results, and 6) interpreting and discussing results.

No.	(Phrasal) Verbs	Noun collocates	Examples
(1)	show	image, peak, result, curve, spectra, change, value, structure, distribution, pattern, effect, increase, behavior, performance, trend, morphology, profile, difference, dependence, plot, response, property, decrease, stability	· show additional images · show broad spectra · show the almost reversible curves · show any feature current peaks · show a significant change · show a monotonous increase · show a bimodal pore size distribution · show a linear behavior · show a distinct catalytic effect · show a homogeneous and dense diamond structure
(2)	play	role, effect, part	· play critical roles · play a very positive passivation effect · play an important part

continued

No.	(Phrasal) Verbs	Noun collocates	Examples
(3)	have	effect, value, influence, structure, impact, potential, peak, energy, size, diameter, area, property, density, stability, strength, ability, shape, surface, content	· have a beneficial effect · have a remarkable influence · have a bigger raspberry-like structure · have a direct impact · have more energy absorption · have the lowest activation energy
(4)	use	method, model, microscopy, spectroscopy, equation, technique, analysis, approach, electrode, system, cell, value, formula, software, measurement, test, film, solution, process, parameter, theory, structure, procedure, sample	· use a calibration method · use a modified Langmuir adsorption model · use atomic force microscopy · use a nonlinear regressive predictive equation · use X-ray photoelectron spectroscopy · use a reconstruction technique · use elemental scan analysis · use a bottom-up approach · use an active feedback system
(5)	observe	peak, change, increase, difference, trend, behavior, effect, signal, decrease, phenomenon, structure, shift, morphology, fluorescence, reduction, loss, formation, result, value, distribution, enhancement, behaviour	· observe a dramatic increase · observe the sharp decrease · observe a dramatic morphology change · observe clear reduction peaks · observe abundant formation · observe a proliferative effect · observe the adhesion morphology
(6)	perform, carry out, conduct	measurement, analysis, experiment, test, calculation, simulation, study, characterization	· perform repeated intracellular measurements · conduct a fine-tuned analysis · perform multiple cumbersome experiments · perform comprehensive calculations · perform molecular dynamics simulations · carry out a simple statistical test · perform the cyclic photodegradation measurement

continued

No.	(Phrasal) Verbs	Noun collocates	Examples
(7)	investigate	effect, property, performance, mechanism, stability, behavior, structure, influence, morphology, change, composition, role, state, kinetics, ability, characteristic	· investigate the biodistribution and side effect · investigate the charge storage mechanism · investigate the environmental influence · investigate the optical modulating performance · investigate the different catalytic effects · investigate the critical size behavior
(8)	form	structure, layer, network, bond, film, phase, solution, complex, particle, surface, contact, crystal, cluster	· form a bowl-like structure · form a more complete graphite layer · form a partially bundled network · form abundant intermolecular hydrogen bonds · form different ordered structures
(9)	see	section, Figure, Table, information, inset, line, peak	· see experimental section · see green lines in Figure 1 · see supporting information · see Figure 2a inset
(10)	indicate	presence, formation, stability, effect, structure, direction, change, behavior, increase, existence, process, difference, nature, deviation, performance, interaction, position, value, distribution, property, state, reduction, reaction	· indicate an excessive presence · indicate single phase formation · indicate enhanced structural stability · indicate a profound effect · indicate a particularly localized electronic structure · indicate the existence · indicate the different wetting hysteresis behavior
(11)	exhibit	peak, behavior, performance, stability, value, structure, activity, property, capacity, effect, morphology, change, response, increase, characteristic, trend, feature, resistance, density, efficiency, shape, rate, intensity, decrease	· exhibit a completely opposite behavior · exhibit an impressive gravimetric performance · exhibit sharp and explicit diffraction peaks · exhibit high structural and electrochemical stability · exhibit high antibacterial activity · exhibit a broad and asymmetric peak

continued

No.	(Phrasal) Verbs	Noun collocates	Examples
(12)	reach	value, maximum, state, temperature, level, plateau, equilibrium, density, peak, surface, point, limit	· reach a critical value · reach the maximum · reach a relative equilibrium state · reach a maximum saturation temperature · reach a plateau region · reach the saturated level
(13)	increase, decrease	concentration, temperature, content, density, rate, area, number, amount, time, value, thickness, strength, size, resistance, ratio, conductivity, intensity, level, energy, modulus, efficiency, fraction, voltage, volume, power, force	· increase fiber packing density · decrease the already high surface area · increase coordination number · decrease antibody concentration · increase the interfacial shear strength · increase the electrochemical resistance
(14)	confirm	presence, formation, structure, existence, result, effect, stability, nature, distribution, role, hypothesis, composition, observation, behavior, mechanism, performance, synthesis, morphology, activity, property, change, ability	· confirm the abundant presence · confirm the complete formation · confirms the existence · confirm its one-dimensional core-shell structure · confirm the excellent stability · confirm the remarkable exciton quenching suppression effect
(15)	deliver	capacity, density, performance, capacitance	· deliver a high discharge capacity · deliver attractively high volumetric energy density · deliver a better long-term cycling performance · deliver 2.5-fold higher specific capacitance · deliver a remarkable catalytic activity

continued

No.	(Phrasal) Verbs	Noun collocates	Examples
(16)	provide	evidence, site, information, insight, area, pathway, space, value, opportunity, image, force, detail, channel, path	· provide any direct evidence · provide abundant electrochemically active sites · provide both the structural and compositional information · provide a more intuitive insight · provide a highly efficient surface area · provide available internal storage space
(17)	evaluate	performance, effect, property, stability, activity, efficacy, behavior, ability, efficiency, capability, potential, characteristic	· evaluate its dynamic mechanical performance · evaluate the compatibilization effect · evaluate the electro-oxidation stability · evaluate the intrinsic catalytic activity · evaluate the shape memory ability · evaluate the chemophotothermal treatment efficacy
(18)	improve	performance, stability, property, efficiency, conductivity, strength, activity, resistance, quality, adhesion, capacity, efficacy	· improve both charge and discharge performance · improve the electrical conductivity and structure stability · improve minority carrier transport efficiency · improve electrical conductivity and lithium-storage properties · improve the electrochemical activity
(19)	study	effect, property, influence, behavior, structure, mechanism, stability, performance, morphology, interaction, process, change, evolution, dynamics, composition, activity	· study the flexoelectric effect · study its possible influence · study the crystalline or amorphous structure · study the phase transformation behavior · study the heterogeneous nucleation mechanism · study the concentration effects

continued

No.	(Phrasal) Verbs	Noun collocates	Examples
(20)	calculate	value, energy, density, ratio, rate, size, efficiency, parameter, area, number, coefficient, concentration	· calculate the fringe visibility values · calculate the oxygen vacancy creation energy · calculate the exchange current density · calculate the apparent rate constant · calculate the quantitative composition ratio
(21)	present	result, image, curve, value, spectra, data, peak, structure, distribution, performance, pattern, morphology, plot, behavior	· present the computational results · present the total absorptivity spectra · present the low and high magnification images · present the gasification conversion curves · present the averaged volume resistivity values · present a tormented porous structure
(22)	obtain	value, result, image, curve, film, structure, information, spectra, datum, sample, efficiency	· obtain the highest possible values · obtain detailed structural information · obtain more accurate calculation results · obtain superstructured mesoporous titania films · obtain a measurable initial conductivity · obtain a more quantitative insight
(23)	apply	voltage, field, pulse, potential, bias, model, method, pressure, force, strain, stress, load, current	· apply a biasing electric voltage · apply a perpendicular magnetic field · apply a constant potential · apply a constant voltage bias · apply a 3D finite element model · apply two consecutive identical pulses
(24)	follow	trend, procedure, injection, treatment, law, process, method, mechanism, model, protocol	· follow a declining trend · follow a previously reported procedure · follow a subcutaneous injection · follow a 6 h annealing treatment · follow the Lambert law · follow a similar transport mechanism

continued

No.	(Phrasal) Verbs	Noun collocates	Examples
(25)	inhibit	growth, activity, proliferation, formation, aggregation, uptake, expression, interaction	· inhibit bacteria growth · inhibit the crystalline phase formation · inhibit the kinesin motor activity · inhibit platelet aggregation
(26)	gain	insight, understanding, information, access	· gain a comprehensive insight · gain a deeper fundamental understanding · gain additional structural information · gain intracellular access · gain only a slight increase
(27)	enhance	performance, activity, conductivity, stability, property, efficiency, intensity, interaction, effect, rate, strength, ability, response, efficacy, density, capacity, adhesion	· enhance the device photovoltaic performance · enhance more effective photocatalytic activity · enhance mechanical and thermal stability · enhance the catalyst conductivity · enhance the biospecific binding efficiency · enhance the surficial capillary effect
(28)	attribute	peak, increase, loss, difference, behavior, enhancement, formation, effect, performance	· attribute their energy barrier difference · attribute the observed optical effect · attribute the negative behavior
(29)	cause	increase, change, decrease, reduction, shift, effect, damage, degradation, loss, deformation, difference, toxicity, variation, resistance, failure	· cause a continuous increase · cause a negligible decrease · cause a higher volume change · cause detectable ore reduction · cause appreciable permanent damage
(30)	compare	result, performance, spectra, value, property, curve, intensity, effect, activity, image, profile, density, energy, response, efficiency, datum, behavior	· compare the experimental and simulation results · compare its electrical and optical performance · compare the measured absorptance spectra · compare experimental spin-polarized scanning transmission microscopy images

continued

No.	(Phrasal) Verbs	Noun collocates	Examples
(31)	summarize	result, parameter, datum, property, value, performance	· summarize general powder characterization results · summarize the fitted lifetime values · summarize the corresponding performance parameters · summarize the crystallographic data
(32)	demonstrate	effect, stability, performance, capability, ability, property, potential, structure, application, behavior, feasibility, advantage, formation, activity, presence, process, distribution, efficiency	· demonstrate the structural stability · demonstrate the actual application effect · demonstrate the stable long-term cycling performance · demonstrate the antibuoyancy transportation ability · demonstrate the modulation capability
(33)	affect	property, performance, behavior, structure, morphology, activity, formation, efficiency, rate, distribution, process	· affect the conductivity and electrical performance · affect significantly the deformation behavior · affect higher order tertiary structure · affect the segregated morphology · affect the filler network formation
(34)	reveal	presence, structure, peak, distribution, difference, morphology, effect, change, formation, mechanism, nature, increase, existence, stability, evolution, process, behavior, trend, performance, shift	· reveal the uniform presence · reveal a consistent periodic structure · reveal the continuous wave-shaped distribution · reveal a multiplied shear band formation · reveal a better directional effect · reveal no spherical morphology
(35)	find	value, peak, difference, correlation, change, result, increase, effect, signal, ratio, amount, energy, concentration, structure, distribution	· find broad applications · find any significant differences · find a positive correlation · find clear evidence · find a narrow distribution

continued

No.	(Phrasal) Verbs	Noun collocates	Examples
(36)	maintain	structure, integrity, shape, morphology, capacity, property, value, state, stability, concentration, trend, performance, efficiency, activity, density	· maintain a complete porous structure · maintain the internal and overall structural integrity · maintain a cylindrical vessel shape · maintain intact surface morphology · maintain a high reversible volumetric capacity
(37)	promote	formation, growth, reaction, process, transfer, performance, activity, differentiation, release, diffusion, proliferation, separation	· promote 3D spheroid formation · promote faster crystal growth · promote the electrochemical reaction · promote the dehydrochlorination process · promote the intramolecular electron transfer
(38)	illustrate	process, mechanism, effect, structure, result, procedure, evolution, change, variation, distribution, performance, image, diagram, curve, formation	· illustrate the charge transfer process · illustrate the good synergistic effect · illustrate the thermosensitive mechanism · illustrate the layered structure · illustrate the microstructural evolution
(39)	understand	mechanism, effect, process, role, behavior, origin, interaction, property, influence, structure, performance, change, evolution, state, difference, relationship, impact, dynamics	· understand the adsorption mechanism · understand the above-observed synergistic effect · understand the positive role · understand the enhanced long-term cycling behavior · understand the chemical functionalization process
(40)	prepare	film, sample, composite, solution, electrode, layer, membrane, material, gel, device	· prepare either porous or dense films · prepare fully dense titanium samples · prepare different complex composites · prepare a polyaniline thin film

continued

No.	(Phrasal) Verbs	Noun collocates	Examples
(41)	characterize	structure, property, morphology, sample, performance, composition, film, product, change, size, distribution, device, state, material, effect, behavior, interface	· characterize the complex solidification structure · characterize the long-term energy and power performance · characterize the solidification morphology · characterize the disassembly state · characterize all the phosphorized samples
(42)	examine	effect, performance, morphology, stability, property, structure, behavior, distribution, change, role, activity, sample	· examine solely the effect · examine the filtration performance · examine LPS oxidation stability · examine the interaction effects · examine the reinforcement behavior
(43)	support	hypothesis, conclusion, result, formation, finding, mechanism, observation, presence, assumption, fact, model, interpretation, notion	· support the hypothesis · support a preliminary conclusion · support the complex formation · support the contact angle results · support the photophysical mechanism
(44)	suggest	formation, presence, effect, stability, mechanism, existence, interaction, structure, increase, reduction, process, possibility, reaction, change, capability, distribution	· suggest shape-dependent wrinkle formation · suggest a surface-active presence · suggest excellent chemical stability · suggest a resonant enhancement effect · suggest a mesoporous structure
(45)	induce	formation, change, effect, apoptosis, death, charge, transition, response, increase, decrease, shift, growth	· induce efficient capillary formation · induce a drastic increase · induce any noticeable changes · induce the ferroelectric transition · induce a mismatched growth

continued

No.	(Phrasal) Verbs	Noun collocates	Examples
(46)	consider	effect, value, structure, fact, result, property, case, system, range, process, mechanism, density, size, difference, distribution, application	· consider the diffusion effect · consider such a fact · consider the first cycle results · consider the band structure
(47)	fit	datum, curve, peak, spectrum, plot, result, profile, line, equation, point, model	· fit all the available data · fit all the experimental spectra · fit corresponding curves · fit the voltage decay plots
(48)	display	peak, image, curve, structure, morphology, behavior, spectra, value, result, performance, effect, capacity, property, activity, stability, pattern, change, response, dependence, distribution	· display multiple sharp peaks · display a rapid and repeatable response behavior · display the network structure · display a highly fusicellular morphology · display the calculated phonon dispersion curves
(49)	measure	spectra, property, response, change, thickness, conductivity, size, curve, resistance, current, area, value, performance, characteristic, intensity, concentration, signal, potential, voltage	· measure the scattering spectra · measure any noticeable field effect response · measure the temperature change · measure the uncompensated resistance · measure the fluorescence intensity
(50)	determine	value, size, property, structure, concentration, rate, energy, composition, content, performance, number	· determine the chemical structure · determine the catalytic reaction rate · determine the minimum sample size · determine the abrasion performance
(51)	achieve	performance, density, value, efficiency, effect, capacity, property, conductivity, result, conversion, state, separation, balance, rate, temperature, reduction, level, efficacy, control	· achieve the best device performance · achieve a geometric current density · achieve a greater squeezing effect · achieve high energy efficiency · achieve a critical value

continued

No.	(Phrasal) Verbs	Noun collocates	Examples
(52)	reduce	effect, size, density, concentration, resistance, energy, number, rate, value, area, amount, intensity, barrier, loss, thickness, mobility, conductivity, strength	· reduce the charge transfer resistance · reduce the threshold energy · reduce the voids and defects concentration · reduce alloys density · reduce the average crystallite size
(53)	report	result, value, phenomenon, observation, effect, behavior, increase, formation, trend, structure, method, finding, datum	· report much lower values · report a similar increase · report a shape controlling effect · report the initial formation · report diffusion test results
(54)	propose	mechanism, model, method, structure, diagram, strategy, process	· propose a detailed mechanism · propose the following simple model · propose a novel computation method · propose the simplified band diagram · propose a generalized manipulation strategy
(55)	facilitate	transport, formation, transfer, process, diffusion, growth, separation, reaction, adsorption, interaction, extraction, release	· facilitate efficient electron transport · facilitate efficient energy transfer · facilitate Li dendrite formation · facilitate the cationic exchange process · facilitate the orientated epitaxial growth
(56)	deposit	layer, film, electrode, nanoparticles, molecule	· deposit an active layer · deposit 50 nm thick $Al_{0.3}S$ films · deposit more than eight atomic layers · deposit enough photonic energy · deposit active materials
(57)	meet	requirement, condition, demand, criterion, need, value, standard	· meet practical application requirements · meet the engineering heating demands · meet the necessary nutritional needs · meet different measuring conditions · meet the criterion

continued

No.	(Phrasal) Verbs	Noun collocates	Examples
(58)	undergo	change, process, reaction, transition, deformation, cycle, shift, increase, transformation	· undergo a similar prolongedcondition-ing process · undergo a phase transition · undergo dramatic composite changes · undergo an electron transfer reaction · undergo a lower volume change
(59)	possess	structure, property, activity, stability, area, value, performance, ability, potential, conductivity, strength	· possess the hierarchically porous structure · possess the best electrocatalytic activity · possess a comparable stability · possess better initial catalytic performance
(60)	minimize	effect, energy, loss, interaction, number, influence, formation, degradation, damage, cost, risk, error	· minimize elastic energy · minimize concentration polarization effects · minimize the electromagnetic loss · minimize zinc dendrite formation · minimize the device thermal degradation

Task 12: Work in groups to figure out the words and expressions in the following sentences used to highlight relationships between entities and propositions, e.g. cause and effect, similarity and difference, comparison and contrast, (un)correlation, and (in)consistency, which could be reflected by noun phrases, adjectives, adverbs, verbs, conjuncts, etc.

- Cause and effect:

  ____

  ____

- Similarity and difference:

  ____

  ____

- Comparison and contrast:

  ____

  ____

- (Un) correlation:

____

____

- (In) consistency:

____

____

- Others:

____

____

(1) The modulus as determined by nanoindentation is higher than the modulus as determined by macroscopic tensile test (ASTM D638) for polymer-based composites.

(2) The modulus as determined by nanoindentation for neat epoxy was found to be 3.61 GPa which is 1.33 times greater than that measured by the macroscopic tensile test.

(3) The scaled nanoindentation modulus shows the same trend as the modulus from the tensile tests, with a similar level of uncertainty and a similar concave shape.

(4) The experimental data generally matches the simulated data for the 3D random GNP orientation (see Figure 1) for low levels of dispersion, particularly at the larger GNP volume %.

(5) The overall matching of the experimental data with the 3D alignment predicted data is excellent considering the complex and multiscale nature of the modeling technique.

(6) This result signifies that the preloading has a direct effect on the MR elastomer's behavior.

(7) The relative increase in the natural frequency was found to be 2.65 times higher than that of its original natural frequency.

(8) A relatively small change (2.65 times) in the natural frequency resulted in the very high (720 times) change in the stiffness of an MR elastomer when the magnetic field was applied.

(9) Therefore, large-scale defects were resulted and the removal of CTA^+ via ion exchange (with metal cations) then gave rise to mesoscale voids, [17] which is consistent with the observed transformation of the nonporous HKUST-1-R to the mesoporous *M*/HKUST-1-R (Figure 1 and Table 1).

(10) This reduction of the optical band gap (E_{hv}) by ≈0.1 eV from H-DPP to CN-DPP is in accordance with the narrowing of the electrochemical band gap (E_{CV}), as determined by cyclic voltammetry (CV), due to the larger influence of the substituents on the reduction than on the oxidation potential (Figure 1 and Table 1).

(11) The shortest Ti-O distance has increased from 1.71 Å for $y = 0$ (Table 1) to 1.85 Å in $y = 0.30$ (Figure 1), which confirms that the decrease in the tetragonality with La substitution is associated with the decrease in the degree of the covalent character of the Ti-O bond.

(12) The relatively simple charge transfer integral approach used here captures an essential difference between the polymorphs that can explain the variation in measured triplet transfer rates (see Table 1).

(13) This result also shows a good agreement with PL decay measurement where a reduction of surface trap states is observed with increasing shell thickness (Table 1).

(14) The high η and high rate of temperature increase imply that the photothermal conversion capability of M-AuNPs-2 is higher, agreeing well with the simulated and experimentally measured C_{abs} and C_{abs}/C_{sca}, as depicted in Figure 2 and Table 1.

(15) This is supported by Table 1, showing that increasing PEDOT:PSS loading results in a decline in capacitances due to electrochemical inactivity and poor conductivity of PSS.

(16) EDX analysis showed that concentration of K^+ ions is roughly twice as high as concentration of Mg^{2+} ions (Table 1) resulting in ~50% more water per unit cell in Ti_3C_2-K than in Ti_3C_2-Mg (estimated using water submerged QCM as shown in Table 2).

(17) The size increase of PLGA(OVA/CpG) NPs compared with PLGA(CpG) NPs supported the coating of an OVA layer on PLGA(CpG) NPs (Table 1).

(18) After the first two cycles, the higher voltage plateau shown in Figure 6i, j exhibits reversible electrochemistry as indicated in Table 1, whereas the Li^+/Li^0 plateau exhibits low $\eta_{Coulombic}$ similar to experiments without the $LiAl_{0.3}S$ coating (Figure 4).

(19) The combination of increased efficacy in preventing water crystallization and decreased water content with increasing FOSM content (Table 1) leads to a maximum in the total quantity of nf-water in these hydrogels for HF18.

(20) The grain size statistics reported in Table 1 confirm that the maximum grain size increases as the spincoating temperature increases.

(21) As a consequence, this spectrum was fitted with a distribution of quadrupole splittings, ΔE_Q, with the same isomer shift and linewidth (see Table 1).

(22) In addition, the γ factor decreases to 1.076 for the second calcination temperature of 950 °C. On the other hand, it is worth noting that the dielectric performance shares a similar trend, as shown in Table 1.

(23) The aggregation-induced emission enhancement is further supported by the values obtained for the radiative and nonradiative decay rates (Table 1).

(24) Overall these analyses of the impact of materials energetics and nongeminate recombination on device VOC are in reasonable agreement with the 0.3 V increase in VOC measured directly under one sun irradiation (Table 1).

(25) The bi-modal mesopores of NMC-Tyr lead to a higher surface area, a larger pore volume, and a lower micropore surface area as compared to those of NMC-His and NMC-Trp (Table 1).

(26) Moreover, the solubilized Te^{2-} was found to increase with the increase of Na_2S:PbTe molar ratio, correlating to the decrease in Te observed in EDS data of cuboctahedra (Figure 2) and suggesting that the increasing sulfide drives the equilibrium forward, liberating telluride.
(27) The relative increase in the FF may contribute to the reduction in the recombination from the small excess of PbI_2 and enhance the electron quality of the perovskite films, which is consistent with their more uniform morphologies and larger domain sizes compared to those of the perovskite grown on pristine CPTA (Figure 1).
(28) This finding correlates well with the decrease of the centroid of the Raman band $\overline{\nu}_{water}$ with CO_2 dilution for water concentrations between 10.5 and 9.5 mol L^{-1} for the ternary system water/acetone/CO_2 at 308 K and 10 MPa reported in Figure 1.
(29) The morphologies of the stretched nanocomposite films were characterized using LSM, SEM, and TEM, where a change in the alignment of the h-BN particles in the PVA matrix was observed (Figure 2).
(30) All the observations acquired by absorbance measurements in Figure 2 can be ascribed to higher surface area and the porous structure of nanocrystalline TiO_2 due to the DNA modification.
(31) While the device with two layers of CdS (≈65 nm) showed the best performance as also shown in Figure 1, J_{sc} and *FF* start to decrease gradually with more layers.
(32) For these cells, the *W/L* ratio or circumference was similar to the cell length due to the small change in cell width (Figure 2).
(33) However, the significantly smaller increase in the latter testifies to the fact that the lone pair in perovskite-type $PbTiO_3$ is confined in the cuboctahedral voids, while the lone pairs in $SnTiO_3$ form van der Waals gaps and hence can occupy more space (see Figure 2).
(34) PEG chains with higher molecular weight had more hindrance effect and prevented other PEG molecules to react with amino groups of chitosan (Figure 2).
(35) The percent weight loss from 125 to 600 °C (56% by weight) is attributed to the block co-polymer, which is in agreement with previous studies that indicate the decomposition of pAAc at 260 °C and pNIPAAm from 220 to 440 °C.
(36) The film thickness variation, visible in the STEM images of Figure 2 between panels A and C, could result from several mechanisms.
(37) Besides that, g-CCO contains more V_o (18.7%) than that of h-CCO (15.4%), which may contribute to the observed sensitivity enhancement in Figure 1.
(38) In addition, the EDS data of Figure 2 show the adhesive contains significantly less calcium than the shell, as seen in our previous results.
(39) The discernable variation in binding energies (Figure 1) reflects the change of electronic structure of metal and P atoms, acting as hydride-acceptor and proton-acceptor sites, which creates a positive influence on splitting water.

(40) Therefore, the isoenergy surface reflects straightforwardly the reduction of the SBZ due to the molecular overlayer, as previously described for the LEED data presented in Figure 1.

(41) This enhanced process was demonstrated to be reversible with the decrease of pH from 12 to 6 (Figure 1), exhibiting a reversible switching in photoluminescence.

(42) The large absorption band and high surface area of the freestanding SP-Et-CNMO films allow for sensitive detection of metals that create visible color changes upon binding in the films (Figure 1).

(43) As is aforementioned, the spectral response enhancement of SeMT photo-detectors is highly dependent on the kind of metallic NPs with the same density (see Figure 3).

(44) The intensity decrease of the FDC emission is much more relevant on the high temperature range, reaching 65% for 1 at 320 K (Figure 3c) and 80% for 2 at 423 K.

(45) Similar conclusions were drawn from the results shown in Figure 3, where we saw that the higher the pressure is, the more CO_2 is needed for the appearance of the nanostructuraction.

(46) In contrast to Arnold and co-workers [1] , who detected red-shifts for the Soret and Q bands upon Li^+-coordination, no shifts and/or changes in intensity are noted with the addition of the ionic polyelectrolyte at any mass ratio – Figure 1, suggesting the absence of interaction between the two components.

(47) A high magnification figure shows that the single particles of both coatings were well sintered, leaving no pores between the particles (Figure 1, lower row).

(48) PXRD patterns of the composite after 1, 5, and 10 photocatalytic cycles showed no significant variations from that of fresh composite (Figure 3).

(49) At the higher thinning rate of 42.3%, the FE simulation exhibited the penetrating cracks occurring along hoop direction, which was in good agreement with the experiment results, as shown in Figure 28.

(50) Compared to the control microfibers, cross-linked fibers exhibited significantly higher Young's moduli under both dry and hydrated conditions (Figure 2D-E).

(51) This observation is consistent with the trend in the UV–vis–NIR spectra during a reverse voltage sweep, where only minor changes are observed between +4 V and +1 V as well as between −1 V and −4 V (Figure 4a–l).

(52) The nanoplates synthesized at 180 °C showed much higher crystallinity and fewer NPs than samples synthesized at 150 °C.

(53) Li-$LiFePO_4$ cells enable a capacity increase of ~30 mA h/g at 0.5 C for 110 cycles after adding 0.1 M LAF-60 (Figure 1g).

(54) The cells on the CN and HRN showed a striking difference in apical actin anisotropy ratios, where more polarized actin yields a higher anisotropy ratio (Figure 1G).

(55) This could either be related to the desorption of the organic species, as suggested by the nearly full recovery of the surface state (and also the reduction of the C 1s peak), or to the

decomposition into irregular polymeric structures inferred from the STM images obtained above 175 °C (see Figure 3g,h).

(56) In contrast, nanocrystals prepared using heating-up synthesis show residual surface-bound thiolate species at their surface upon ligand exchange (see Figure 1).

(57) In comparison to monocultured models, dual cocultures and tri-coculture 3D-MCTS with and without LbL-MPs, presented increased contraction (Figure 1).

(58) As seen in the relationship between rotation rate and MRR, the higher polishing pressure causes a significant decrease in MRR with extended polishing time [Inset in Figure 7 (b)], which is also associated with wear of the Ce film.

(59) The difference in the increase of two characteristic strains yields a reduction in actuation strain.

(60) Hence, we expect a less steep increase of ion when compared with the hole conductivity, in agreement with the experimental results (Figure 1b).

(61) The PXRD data for cubes (Figure 2) show that there is no significant change in the crystal structure of the NCs after ligand exchange with Na_2S or Na_4SnS_4, consistent with a simple exchange process.

(62) There is also an increase in the value of the magnetization at lower temperatures.

(63) As illustrated in Figure 1a,b, it is obvious that the ε′ and ε″ values are increasing at the same time with the increase of the f-$Ti_3C_2T_X$ contents added in hybrid foams.

(64) When the Co_3O_4 NPs are supported on a Ti membrane, the Co_3O_4 NPs/Ti exhibits a relatively lower catalytic activity and selectivity to benzoic acid in comparison with the Co_3O_4 NWs/Ti under the same conditions in the ECMR.

(65) Compared with top-down strategies, this approach achieved a higher degree of fabrication accuracy, along with improved time and cost efficiency.

(66) Nevertheless, no significant difference in the XRD patterns was observed for the variation in scan speeds.

(67) These morphology changes, in turn, impacted their OPV device performance, and therefore the correlation of structural evolution with an increase of blade coating speed and the addition of the third component to its OPV function must be investigated.

(68) Note that the AgNWs-PDMS TENGs displayed the opposite trend compared with the AgNWs-SU8 TENGs.

(69) This is caused by the high welding speed, leading to the reduction in the heat input.

(70) The liquid flow is induced by the drag (momentum exchange) from adjacent DEM grains.

(71) An increase in ΔE causes greater magnetic shielding and the Si chemical shifts appearing at lower frequencies (Figure 4a).

(72) The increase in roughness factor allows more interaction to the textured surface with the water droplet and thus reduces the contact angles by decreasing drop height and increasing drop radius.

(73) The forming area of inertia effect stage is significantly affected by discharge voltage.

(74) Analysis using a higher intensity X-ray source would improve the yield and resolution of diffraction peaks from such thin alloy films.

(75) The porous structure can facilitate more deeply Li-ion movement and suppresses the mechanical stress induced by the large volume change during cycling (Figure 1b).

(76) This membrane-thinning effect enables more water molecules to be present at each membrane surface because of the lower lipid density.

(77) We note that these profiles in emission intensity are likely influenced by changes in intergrain connectivity.

(78) Therefore, the use of an external base may prevent the carboxylic acid ligand exchange from occurring.

(79) Moreover, smaller crystal size has been reported to be more beneficial thanks to the higher surface area that facilitates additional adsorption of ions and molecules, thus permitting faster regeneration of tissues.

(80) The 17AAG within the micelles can effectively inhibit the p-ERK1/2 level as well, in addition to the photothermal effect.

(81) The tunnel loops have restricted motion in this direction (toward the stem), consistent with structural observations.

(82) It should be noted that the presence of disulfide bonds in CuS limits the rearrangement of the anion sublattice required for the exchange.

(83) This criterion ensures that the growth of the holes favors the reduction in the interfacial area of the system.

(84) In addition, a higher value of surface hydrophobicity can further trigger Aβ aggregation by means of hydrophobic interactions.

(85) When the pre-drying temperature increases, the interfusion between fibers is impeded

(86) The above analysis and modelling approach helps to simplify, benchmark and validate the changes of surface morphology and elemental composition for the CoCr alloy components with complex surface geometry, before and after laser polishing.

Task 13: Discuss in groups whether each of the following statements is hedged (expressing possibility, uncertainty, estimation, etc.) or emphatic (expressing an opinion, idea, etc. in a clear and strong way or in a more factual way).

(1) A possible cause for this difference is the pile-up of material around the point of contact.

(2) This is a strong indication that the fabricated composite specimens contained GNPs that were randomly distributed in 3D space.

(3) The as-received TC307 GNP contained approximately eight layers so processing the composites reduced this to three to four layers.

(4) It can be claimed that both the stiffness and damping capability of the MR elastomer increased with increasing preload because the slope and area under the force-displacement curve represent stiffness and damping capability, respectively.

(5) The relative shift of preload curves seems to be relatively greater than that in the presence of a magnetic field when a preload was applied.

(6) This indicates that the influence of the preload was more prominent when a magnetic field was not applied.

(7) The experimental results revealed that the preload has a significant effect on the response of the MR elastomer under both conditions, without and with the presence of a magnetic field.

(8) The amplitude of the vibration has been significantly reduced in the resonance zone when the magnetic field is applied.

(9) There is no doubt that the magnetic interaction between the magnetic particles would always increase the natural frequency.

(10) As a result of a combined effect, the speedy shifting of the transmissibility curve is observed.

(11) The acceleration amplitude is significantly attenuated in the resonance zone when the magnetic field is applied.

(12) The increase in Raman shift upon cycling suggests a modification in the geometry of the P3 molecules on the surface of CB, and is thus consistent with the modeling and cycling experiments.

(13) Analysis of Scherrer crystallite size from PXRD data provides insight into the mechanism of exchange (Table 1).

(14) Clearly, the increase in the quantity and size of the crystalline regions reported in Table 1 is what is responsible for the observed mobility increase.

(15) As evident from Table 1, A_1 component increases from CS0 to CS5 because, during cation exchange, Cd^{2+} ions primarily reside at surface sites before reaching the lattice sites, thus increasing the surface trap-state contribution.

(16) Our data (Table 1) propose that at 4 mA cm^{-2} (under our experimental conditions), the threshold for the limiting current density may have already been met and the observed nonlinear increases in particle size, film roughness, and the plateau in film thickness could be indications that films formed using a current density of 4 mA cm^{-2} under our experimental conditions are associated with a decrease in film growth efficiency.

(17) The DLS data for the hydrodynamic diameters as obtained for the different NPs shown in Table 1 indicate the surface functionalization of the gold nanoparticles (Au NPs) with ligands by a corresponding increase in size, in particular for the PEGylated NPs.

(18) Measurement of the *g*-factors reveals a 3-fold increase in *g*-factor for CdSe bound by D-(+)-malic acid over L-(+)-lactic acid (Table 1).

(19) The reversibility of the higher voltage plateau was especially obvious in the final two cycles that did not access the Li^+/Li^0 couple, but Table 1 shows this was a general trend for all cycles.

(20) Even though the Ba K-edge energy seems to be insensitive to the coordination number, fitting of EXAFS data revealed a pronounced shift, with a drop in average CN from 11–12 at low x to 8 at $x = 0.55$ (Table 1).

(21) For the deep-blue region, the best sample of PhC2 crystals has a PLQY of 79%, a notable improvement compared with previously reported 2D perovskites (Table 1).

(22) Although this may seem like a modest increase in the CE, Table 1 emphasizes how critical it is to achieve a nearly 100% CE to ensure the long-term cyclability of anode-free batteries, as well as other Li metal batteries.

(23) We note that the significantly increased average value of the particle size distribution for SiNP-Mg (D50 value in Table 1) is due to a higher reaction temperature leading to more pronounced coalescence (see also SEM micrographs in Figure 2a, b).

(24) Therefore, a dramatic improvement in ICE from 36.9% for C-800 to 70.9% for PO/C is obtained (Figure 1b and Table 1).

(25) The mesopore size is much larger than the pore wall thickness (Table 1) of the initial SBA15 template, partially due to a large thermal shrinkage of the template leading to an increased pore wall thickness.

(26) Importantly, the trends we observed here and in our previous work are consistent with local variations of nonradiative recombination and a reduction in these pathways at higher excitation fluences (cf. Figure 2).

(27) The most prominent feature visible in the SDS gels appears in the 24 h serum incubation (Figure 2, channel D), which shows a heavy concentration of proteins in the middle of the shortrun SDS gels, visually indicating a shift in the relative abundance of adsorbed proteins forming the bulk hard corona.

(28) When the stored cells contained K metal, however, significant changes occurred as observed in the photographs of the potassium electrodes, separators, and Sb electrodes (Figure 2).

(29) This effect was observed experimentally when the film folded at room temperature when PSMA was stiff which produced a bigger diameter than that at increased temperature when PSMA was molten and soft (Figure 2).

(30) Surprisingly, this trend was not explainable by an increasing amount of negatively charged carboxyl groups (see XPS spectra in Figure 2).

(31) Thus, the effect of nanopore size on electroosmotic flow explains the most surprising observation in Figure 2, that *fd* and M13 translocate in opposite directions under certain conditions.

(32) The observed size reduction of Ag in heterodimers, when more gallic acid is present (Figure 2), is likely due to the larger gallic acid domain from where sulfidation initiates to form Ag_2S.

(33) A dramatic change in the surface morphology is immediately observed in Figure 3 after annealing the sample to 400 °C for 0.5 h.

(34) This statement is also underlined by the results shown in the context of the pressure variation (Figure 3).

(35) Clearly, introducing the TPBi suppresses the CBP spectral shifts almost completely proving that the shifts observed in Figure 3 are indeed associated with molecular aggregation in the host materials.

(36) It should be pointed out that the shear damage entered the accelerated growth stage under the thinning rate of 32%, beyond which a slight increase of thinning rate may induce ductile fracture in the inner surface of as-spun workpiece, as can be seen in Fig. 1.

(37) The dark-field scanning TEM (STEM) analysis revealed a clear morphology evolution of the film with the increase of film thickness (Figure 1g-i).

(38) The FE_{CO} of pure Pt wire is barely detected, at a level of 0.08% at −1.6 V (vs RHE), suggesting the Pt^0 is not an efficient electrocatalyst for CO_2 reduction (Supplementary Figure1).

(39) Definitely, it is of great significance to achieve high performance in more figure of merits in a single device.

(40) Remarkably, the robot is able to continue traversing the smooth terrain as before without manual intervention, use of external energy sources, redundant electronics or changes to the environmental conditions.

(41) On the basis of the abovementioned results, the possible mechanism for activation and reduction of CO_2 to CH_4 was given in Figure 1.

(42) We illustrated the microstructural changes occurring in the pure and Cu-particle-alloyed magnets (Figure 1).

(43) We indeed find that the photoluminescence intensity of the pristine materials is significantly enhanced following the addition of PS (Figure 1), where the peak shift might be due to variations in the morphology of the materials after blending.

(44) Figure 1 shows that the absolute value of the influence coefficient decreases approximately linearly with the increase in the number of layers.

(45) The increase of TFR with reducing grain size is probably caused by the deferred depolarization due to a large internal compressive stress.

(46) This weight loss was almost negligible, indicating the better thermal stability of 3dd support in itself.

(47) The observed increase of the bandwidth due to the STO buffer layer is possibly the driving force for enhanced conductivity.

(48) However for higher pitch, droplet occupies more untextured portion and mostly stays on the top surface.

(49) With further increasing the sulfidation temperature to 900 °C, the ORR catalytic activity is somewhat enhanced.

(50) So, based on above observations the equilibrium adsorption time was kept at a little higher side for 3 h for all adsorption experiments.

(51) These were followed by a high temperature process starting at 480 °C and completing seemingly at 610 °C.

(52) The day-by-day increment in the CPI clearly shows the increase in cell number in the control sample.

(53) This electrochemical experiment is highly sensitive to changes in Pt interfacial properties, such as faceting.

(54) This contrast is directly manifested as a reduction in the Coulomb integral determining the matrix element in Equation 2.

(55) Nevertheless, it is reasonable to assume that the TiO_x units remain intact and tightly bound.

(56) Conversely, increasing the OH coverage leads to a reduction of the bandgap.

(57) The larger pore size distribution can obviously promote the increase of CO_2 capture amount.

(58) The photoluminescence intensity of the gelatinated QDs was greatly increased by 245% with the increase of pH from 6 to 12.

(59) This will be particularly important in investigating the feasibility of a "layer-by-layer" exchange mechanism

(60) Remarkably, all superlattices are extremely flat.

(61) The titania precursor was apparently fully hydrolyzed before it could reach the water droplets.

(62) Hence, these electrochemical data evidently confirmed the dramatic increase in intrinsic electrochemical properties of the Ti membrane electrode *via* an *in situ* loading of Co_3O_4 NWs.

(63) Comparing to that with celgard separator, the CV curves of NCM-based cell show higher electrochemical stability and distinctly lower polarization, suggesting its fast reaction kinetic.

(64) Noticeably, the shift of binding energy of N 1s of MoC@NCS as compared with NCS suggests that the strong interaction between Mo_2C nanoparticles and NCS.

(65) This less recombination absolutely means a higher density of electrons.

(66) Both samples demonstrate response times longer with the higher concentrations of glucose solutions, which may be possibly ascribed to the concentration dependency of glucose oxidation kinetics in the presence of GO.

(67) Crystallization of drop cast silk fibroin is mostly owed to extended drying time during which silk fibroin has to reorder into its lowest energy conformation, sheets.

(68) This contrast is directly manifested as a reduction in the Coulomb integral determining the matrix element in Equation 2.

Task 14: Discuss in groups the functions of the underlined present and past participial forms or clauses, whether serving as pre- or post-nominal modifiers or adverbials, reflecting the formality and precision of scientific language.

(1) Based on the results of mode II experiments conducted by means of the described test rig, the values of mode II notch fracture toughness for the tested PMMA were obtained for each notch opening angle and tip radius.

(2) Dynamic light scattering (DLS) data reveal that the average sizes of NUS-30, NUS-31, and NUS-32 nanosheets suspended in acetonitrile solutions are around 4.2, 1.3, and 2.1 m, respectively (Figure 5g), matching well with their HR-TEM and AFM results.

(3) Inset in Figure 2b is a simulated HRTEM image based on the MEAM potential optimized structure shown in Fig. 2f, which shows that the relaxation of the surface atoms around the vacancies results in a trapped dislocation residing at the Cu_3Au/Cu(Au) interface, matching well with the in situ TEM observation.

(4) Both MTT and light microscopy images obtained in these experiments were consistent with the aforementioned results that the PGL-DiR MBs could effectively induce cavitation perturbing cell membranes, resulting in the enhanced intracellular uptake of the therapeutic agents.

(5) In particular, calculated from the rotating ring-disk electrode (RRDE) data, the H_2O_2 yield measured for Co-N,B-CSs (highest yield is 2.2%) is much less than that for Pt/C (the highest yield is 5.9%) at all potentials, implying the value of n is between 3.98 and 4.00 at 0.20–0.50 V, well consistent with the results obtained from the K–L plots (Figure a). Taken together, it suggests an efficient oxygen reduction activity *via* a dominant four-electron pathway.

(6) Repeating these measurements at different temperatures results in the data shown in Figure 1, which shows the mean of all the acquired conductivity values plus error bars, which represent the 95% confidence bounds derived from each data set.

(7) The impact of the hardening exponent n on the cutting and tangential forces, shown in Figure 1 (e and f), features a significant difference to the results discussed above, since differences in the dependence of Fc on UCT for various values of n are very small, particularly for small UCT.

(8) Consider the spectral position in the simulated absorption spectra of Figure 1a of a linear trimer that is shown in the curve composed of orange squares is only blue-shifted by 4 nm from a hexamer, shown in the brown curve.

(9) Figure 1d presents the prediction of flow front position (t) solely from electrical resistance data, according to Equation 6, compared with the results extracted from optical images.

(10) The results obtained with such a small domain may be strongly influenced by image forces, leading to important differences in the rate at which the dislocation overcomes the GP zone.

(11) Samples with cryogenic cycling fit well the trend of each curve, indicating that cryogenic cycling prior to bending has not noticeably affected the volume fraction occupied by potential STZs of any type/size.

(12) Employing this analysis, we find the aggregation number of Cy^+ to be 55 (using the values provided in Table 1) in the presence of PSS, which is in agreement with the reported value.

(13) Numerical results obtained using this approach for every frequency value is compared with the closed-form analytical expressions derived in the paper.

(14) Results of several works, in which the phase composition was analyzed using Rietveld method, show a higher fraction of the MgO phase (in the case of stoichiometric Mg_2Si samples) compared to the results presented in this work.

(15) To compare the experimentally measured total thermal conductivity with the model prediction using either L_{surf} or L_0, the in-plane thermal conductivity is calculated using the following equation Equation 2.

(16) The high frequency semicircles of the Nyquist plots were fitted into a two-component equivalent circuit (inset in Figure 1B, the dash lines show the fitted results), representing two interfacial components existing in the Li/LPS/Li symmetric cell.

(17) Additionally, similarly to LSF, there is an opportunity to select more reliable pOT data sets for SF using the results of independent calorimetric studies, especially taking into account the good agreement between the values of the oxidation enthalpy of SF reported by Cheng et al. [1] and Haavik et al. [2]

(18) The HRTEM images (Figure 5c and f) evidently demonstrated the measured neighbouring interplanar spacing of about 0.47 nm, corresponding to the {111} facets of the spinel $LiNi_{0.5}Mn_{1.5}O_4$.

(19) These results are in good qualitative agreement with the experimental indentation curves reported in Figure 1 and 2, demonstrating that only an ultrastiff 2-L film, considerably stiffer than graphene, can lead to the stiffening effect, compared to bare SiC, as reported in Figure 2.

(20) Finally, starting from the observation of the small front shoulder presented at low temperatures in the SRM 1475a peak obtained with classical TREF (Figure 6), we tried to obtain a better separation of the two peaks using dynamic cooling.

(21) To understand these observations, we develop a model describing the interface energy of this system assuming that (i) the interface deformation caused by the nanometer-sized Janus particle is negligible, (ii) and that the particle radius is substantially smaller than the droplet radius.

(22) The results of the most deviating measurements are shown in Figure 1a, obtained with sharp diamond tip (radius of curvature 5 nm) and those shown in Figure 2a, obtained with a blunt Si tip (radius of curvature of 20 nm), both from data gained at room temperature.

(23) The corresponding coarsening rate constant, K_R, is 3.9×10^{-31} m^3 s^{-1}, which is about half the value, 8.95×10^{-31} m^3 s^{-1}, measured by Bocchini et al. at 650 °C for a ternary Co-8.9Al-7.3W at.% alloy.

(24) As shown in Figure 2 all the prepared emulsions are non-Newtonian from the shape of the viscosity-shear curves, i.e. the emulsion systems depict a shear thinning profile, characterized by the decrease of viscosity at higher shear rates preceded by a Newtonian plateau at lower shear rate values.

(25) Furthermore, the simulated isotropic averaged spectrum has a similar peak energy and broadness to the projection spectrum, but with only about 70% of the amplitude, agreeing well with the results shown in Figure 1.

(26) One can note that the OCP variation always followed a similar trend after activation or after recording a CV: it decreased sharply to a minimum value of ~0 V and quickly increased to a more stable value, thus leading to a dip followed by a plateau.

(27) To investigate the critical effect of phonon softening that manifests as a reduced group velocity we replot the measured thermal conductivity as a function of (1−*P*), where *P* is the porosity of each silicon nanowire in Figure 4b, where v_{eff} is proportional to (1−*P*), deduced from the reduction in the Young's modulus as explained above, and the linear trend is very clearly observed.

(28) As in Figure 4, the accuracy of the data was improved with each force-extension curve showing the overstretching transition at 65 pN.

(29) The predicted capillary limit is over an order of magnitude higher than the measured CHF values, which corroborates with the trend shown in Figure 1 indicating that the capillary transport limitation is alleviated by the short transport length of working fluid over the active area in the experiments.

(30) Taken together, these observations support the assertion that $\Delta I/I_0$ was a measure of the molecular volume occluding the pore.

(31) Comparing these spectra with the spectra of the oxide references (see also Figure 1) shows that the Co and Cu atoms are predominantly present in oxidation state +2, agreeing well with XPS results discussed above.

(32) The plot of N_{EL} versus N_{PL} is close to a linear dependence with a slope of one (inset of Figure 1e), validating once again the assessment of average QD occupancies realized in the EL measurements and the conclusion about achieving the regime of population inversion.

(33) These two mechanisms are relative to the same group of mechanisms, i.e. a lack of energy provided to the powder material, resulting in a lack of fusion of the powder.

(34) Therefore, referring to the results obtained via our data set, it seems that none of the qualitative characteristics, separately or jointly employed, could better explain the level of financial sustainability of the WM companies.

(35) The mechanism of collective diffusion identified for the intercalated water and the calculated coefficients of diffusion enables quantitative estimation of the contributions for mass transport through both the pristine and oxidized regions, provided with their areal fraction and spatial patterns identified for the GO membranes.

(36) Intercalation results in the possible formation of a two phase structure consisting of a hard phase (intercalated structure) and a soft phase (polymer matrix alone), making it easy to remove the soft phase (polymer), and, therefore, resulting in an increased wear rate for the intercalated structure, as reported by Jawahar et al..

(37) The magnitude of the response in STO, compared with the response detected in ferroelectric $BaTiO_3$, led to the conclusion that polarity in STO resides in the ferroelastic domain walls.

(38) Cell infiltration results are represented as the percentages obtained for each animal, followed

by average SD.

(39) There are different approaches in the literature to estimate the Hubbard U parameter, including the linear response method, fitting to experimental heats of formation, and using a higher level of ab initio theory where exact exchange is included.

(40) The worse activity of samples obtained with a higher SF/KB weight ratio could be attributed to the lower surface accessibility of active sites induced by the more inactive bulk structure with less developed pores and surface area, demonstrated by SEM images (Figure 1) and N_2 adsorption/desorption results (Figure 2).

(41) This mixture, combined with residual stress caused by uneven welding, results in the microhardness values of the fusion zones (zone 2 and zone 3) between the substrate and sublayer being slightly higher than those of the substrate (276 $HV_{0.2}$) and sublayer (237 $HV_{0.2}$), which were 305 $HV_{0.2}$ and 268 $HV_{0.2}$, respectively.

(42) These new data provide further compelling evidence supporting the translational significance of these compounds.

(43) A similar trend is exhibited for all samples where the stress drop value increases with increasing strain, indicating that more and more energy is necessary to initiate a new shear band accompanying the proceeding of deformation.

(44) These results are consistent with the FTIR data described above showing that the reduced and cycled species are chemically similar.

(45) Large amounts of sputter-coated W may partially block the access offering a larger top surface where Au can be deposited.

(46) Coinciding with these results, the pore size distribution calculated from the N_2 sorption data is in the range of 16-20 nm, confirming the coexistence of mesopores and macropores (Figure 1, Table 1).

(47) The results yielded from bright-field and fluorescent images of the same frame are highly consistent for all tested conditions (Figure 1), suggesting that immunofluorescent staining, which is costly and time-consuming, is unnecessary to evaluate myotube alignment and orientation as a DMD biomarker.

(48) The first step, observed in the first 5 min at a cathodic current density of 5 mA cm^{-2}, results in a dispersion of discrete submicron particles rooted on the surface of the graphite substrate.

(49) It is noteworthy that the XRD patterns of SnO_2/OMCS are relatively broad in comparison with the as-prepared SnO_2 nanoparticles, presumably attributed to the small size of SnO_2 crystals onto OMCS surfaces, which is in good agreement with the results obtained from the TEM images.

(50) As a result, the two series produced with an aluminium flyer presented good quality welds, being in agreement with the window.

Task 15: Read Text Ⅲ carefully, a research article with results and discussion sections presented separately, and discuss in groups the differences in the use of language features between the two sections based on the following questions.

(1) What is the main purpose of this research according to the Abstract?

____

(2) What tests have been conducted as mentioned in the Abstract?

____

(3) What main results are reported in the Abstract and Conclusions Sections?

____

(4) What features of the specimens were found from various CTS tests?

____

(5) What issues related to the specimens were identified based on Tekken tests?

____

(6) Compared to CTS specimens, what behaviors of Tekken welds were observed?

____

(7) For what purpose was Temper Bead Welding (TBW) technique finally applied in wet welding conditions?

____

(8) What typical microstructure and hardness distribution for specimens with selected pitches were reported in TBW applications?

____

(9) What could be concluded from the results of CTS and Tekken weldability tests?

____

(10) Compared to CTS and Tekken weldability tests, what advantages does the TBW method have and what benefits have been experimentally demonstrated in the Discussion section?

____

(11) What could be concluded by comparing the hardness range of the maximum HAZ obtained from different methods?

____

(12) What are the possible reasons for the overall frequent use of past tense in the Results section while present tense in the Discussion section?

____

(13) Any other language features used in the Discussion section (e.g. causative verbs like "enable" and "cause", evaluative adjectives like "high" and "worse", and emphatics like "significant" and "considerably") are clearly distinguishable from those in the Results section, so as to generalize and interpret the research findings in the context of existing literature and engineering applications?

____

Text Ⅲ

Improvement of S355G10+N Steel Weldability in Water Environment by Temper Bead Welding[5]

Abstract: The normalized S355G10 + N steel is used in a variety of applications including the building of offshore structures, which may require repairs in water environment. The main aim of the work was to check susceptibility to cold cracking for fillet welds – Controlled Thermal Severity (CTS) tests and butt welds – Tekken tests and in the next step evaluation of effectiveness of Temper Bead Welding (TBW) application in wet welding conditions for joints welded by covered electrodes. The TBW effectiveness was experimentally verified as a method that may reduce the susceptibility to cold cracking in water environment. An application of the TBW technique led to the maximum hardness of the heat affected zone to fulfill the criterion of EN ISO 15614-1 standard in level of 380 HV10. It was determined that for S355G10 + N steel the beneficial range of the pitch value between beads is 75%–100%.

1. Introduction

...

2. Materials and experimental procedure

...

3. Results

3.1. CTS tests

1 Many welding imperfections were found in the specimens after non-destructive tests. The most frequently found imperfections were lack of fusion, undercuts and metal spattering. Visual and penetration tests showed that the test pieces welded underwater were of much lower quality than those welded in the air environment. After non-destructive tests (NDT), second welds from specimens C1 and C2 were not capable of undergoing further examinations.

2 After macroscopic testing, imperfection was only found in one specimen welded in the water. In specimen C3 there is a lack of deposited metal near the notch. Exemplary examination results are shown in Figure 7 (omitted).

3 A microscopic examination showed that S355G10 + N steel structure consisted of fine-grained ferrite and fine-grained perlite with layers. Examination of the joints showed the presence of microcracks in all specimens made in the water environment. These cracks occurred in the heat affected zones (HAZ) and ran along 15%-25% of the length of the fusion line. A few cracks are also observed in the weld. Specimens welded in the air had no cracks. In Figure 8 (omitted), the results of the microscopic examination of S355G10 + N steel joints are presented.

4 The hardness measurements were performed on cross-sections (both sides) of at least one weld from each CTS specimen. Examples results are shown in Figure 9. In Table 6 (omitted) the maximum HAZ hardness values of the tested joints are presented. In all welds made in the water, hardness values exceeded the critical value.

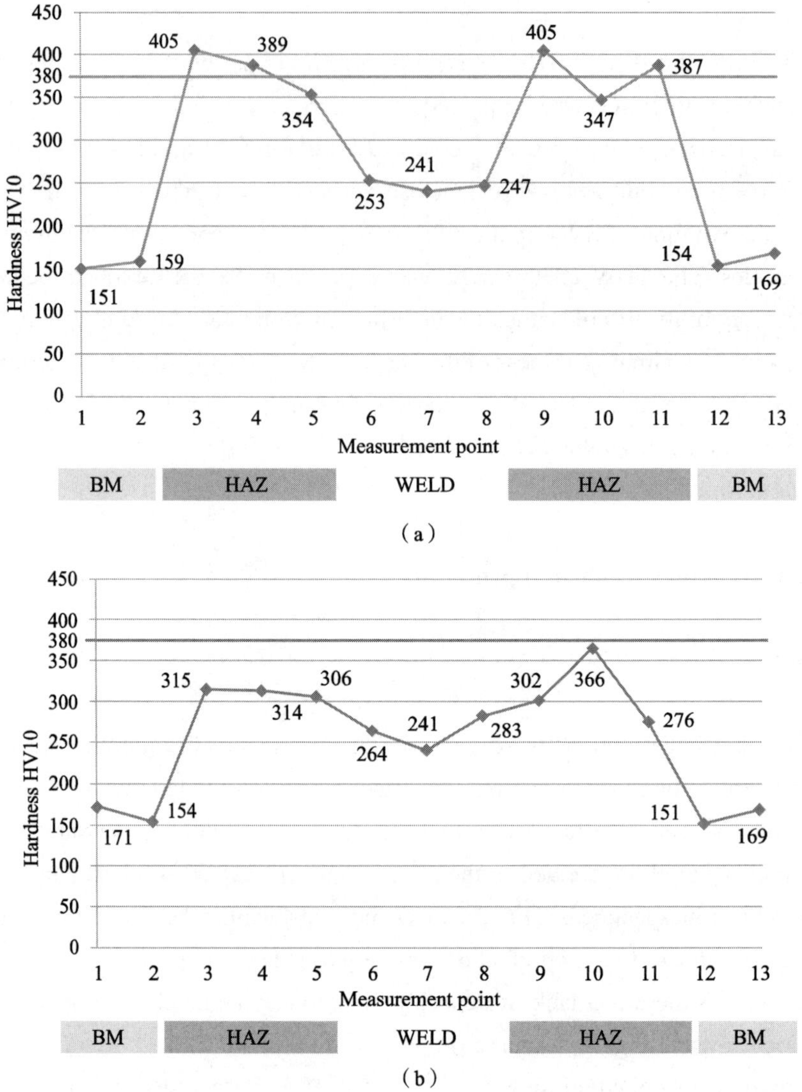

Figure 9 Results of hardness measurements, (a) specimen C3 made in water, (b) specimen C5 made in air

5 The investigation of S355G10 + N steel showed that it has limited weldability in underwater conditions. CTS specimens showed many cold cracks located in the weld metal and HAZ. Specimens welded in the air had characteristically good weldability. To confirm the results, Tekken tests were carried out.

3.2. Tekken tests

6 Non-destructive visual and penetrant tests showed that there are many welding imperfections in the tested joints. The most frequently found issues were lack of fusion, undercuts and metal spattering just like in CTS joints. After NDT testing, the weld from specimen T4 was determined to be unfit to the next examination.

7 Macroscopic examination showed that the only specimen welded in the water that was of good quality was specimen T1, all other specimens had imperfections. The worst of these imperfections were cracks. Cracks went through the whole joint and divided the specimens into two parts. The same situation was found in specimen T5 which was welded in the air. Cracks in the specimens most often ran through the fusion line and sometimes even reached the weld. Exemplary examination results are shown in Figure 10 (omitted).

8 Hardness measurements were performed on a minimum of one cross-section (both sides) of each weld. Exemplary results of the hardness measurements are shown in Figure 11. Table 7 (omitted) presents the maximum HAZ hardness values of the tested joints. In all joints, the maximum HAZ hardness values exceeded the critical value (380 HV10). In specimen T1, which did not break into two pieces, the maximum hardness was close to the critical value (at 387 HV10 level).

9 Tekken tests on S355G10 + N steel joints were shown to have more imperfections than CTS joints. The investigated material had a high susceptibility to cold cracking in underwater conditions. Except for specimen T1, all Tekken welds were broken into two pieces. However, the hardness value of specimen T1 also exceeded the critical value of 380 HV10. The results of the weldability tests on S355G10 + N steel have made it clear that an efficient way to improve the quality of the joints is necessary. Most of the imperfections occurred in specimens with the highest HAZ hardness. It was decided to use a method that would reduce the hardness of the welding process – TBW.

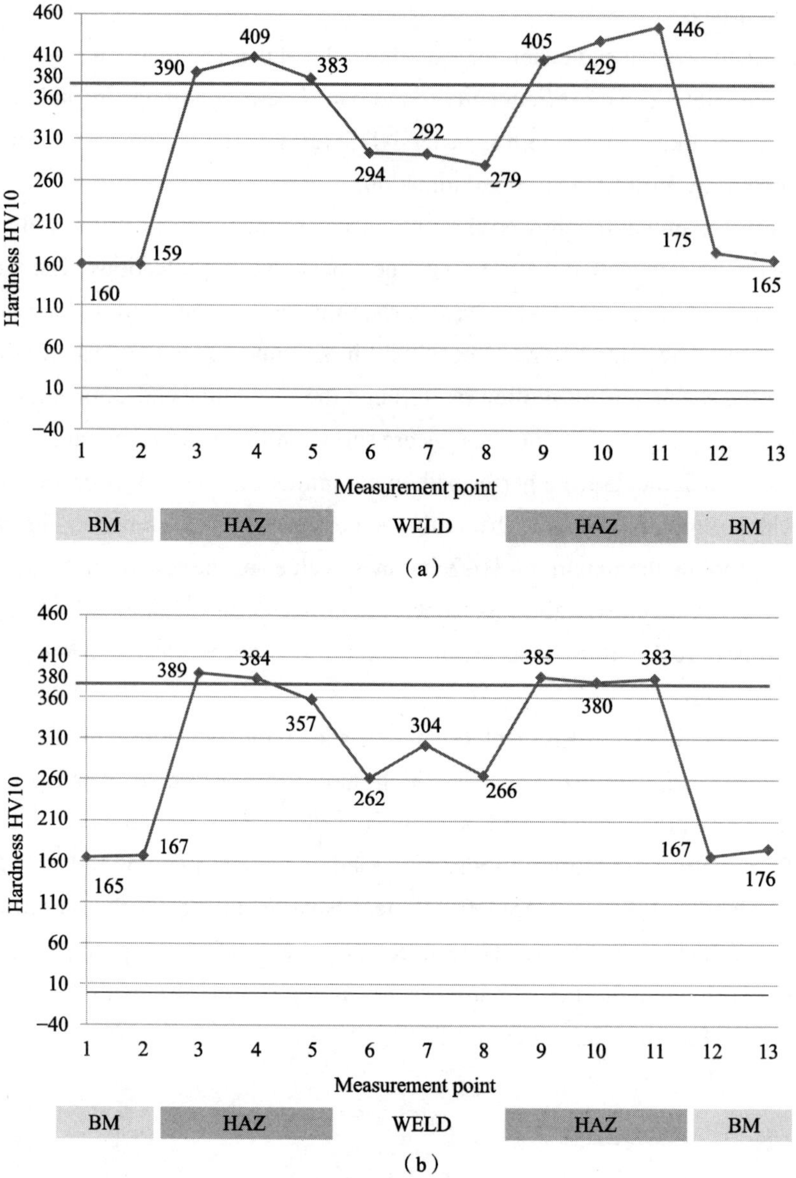

Figure 11 Results of hardness measurements, (a) specimen T2.2 made in water, (b) specimen T5.1 made in air

3.3. Temper bead welding technique

10 Macroscopic examinations were performed to evaluate the pitch. The test welds were cut perpendicular to the axis of the first bead weld. Figure 12 (omitted) shows the exemplary results of the macroscopic examinations.

11 Specimens with pitches of 9%, 21%, 37%, 50%, 75%, 91% and 100% were selected to show microstructures and the hardness distribution of two overlapping beads in a pitch range from 10% to 100%. The welds have a Widmanstätten structure, consisting of ferrite grains in a

columnar arrangement. There are cracks in HAZ (haza and ohaza areas) in the welds. The pitch increases, and the number of cracks decreases. In Figure 13 (omitted) exemplary results of microscopic studies are presented.

12 Hardness measurements were performed on cross-sections with the selected pitches. Exemplary hardness measurement results are shown in Figure 14 (omitted).

13 A HAZ hardness value lower than 380 HV10 was found for specimens with a pitch to 75%. Specimens with higher overlaps had hardness levels below the critical value. In Figure 15 hardness in two specific areas of HAZ (overheating and normalization) of the first tempered weld as a function of pitch between beads is presented.

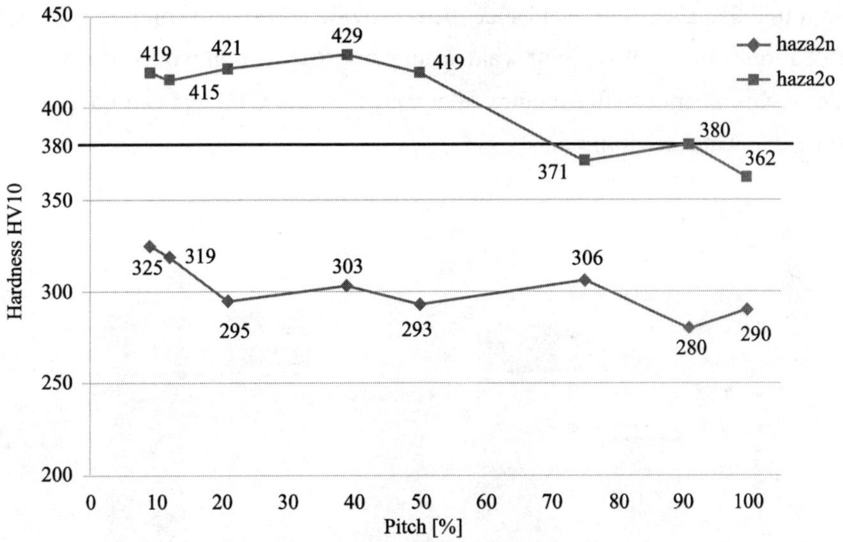

Figure 15 Hardness distribution for specimens with different overlap values at areas haza2o and haza2n of the first (tempered) weld

4. Discussion

14 The results of CTS and Tekken weldability tests showed a high susceptibility to cold cracking of S355G10 + N steel welded in the water by manual metal arc welding (MMA) method. The cracks were observed in HAZ and there are many more cracks in Tekken weldability testing joints. It was decided to use the TBW method to reduce the hardness of HAZ and increase its plasticity. The heat of tempering beads will refine and temper the coarse grained HAZ of the first (tempered) weld. Decreasing of the grain size leads to considerably improving of HAZ toughness, and tempering causes decreasing of joint hardness. Temper bead welding also can cause increasing diffusion of hydrogen from the joint. This technique helps also to reduce residual stresses after welding. All these factors have a positive effect on the weldability of steel. Experiments show that HAZ areas showed a significant reduction in hardness after

welding with temper beads. It was shown that the best when considering the weldability of S355G10+N steel, value of the overlap between two welds is in the range of 75%-100% because measured hardness levels were below 380 HV10. This TBW allows one to obtain joints that meet the hardness requirement of the EN-ISO 15614-1 standard.

15 A comparison of the hardness range of the maximum HAZ from the CTS and Tekken tests and the hardness in the HAZ of the base pad weld after welding and applying the TBW technique are shown in Figure 16. The maximum hardness in HAZ of the base pad weld is given for the hazao area for a pitch not exceeding 20%. The presented results showed that the S355G10+N steel in Tekken tests is characterized by having worse weldability than during CTS tests. This is in line with results given by Kannengiesser and Boellinghaus (2013), who found that CTS fillet weld specimens provide less severe conditions for the cracking of welded joints than Tekken butt weld specimens. Renovation works of constructions used in water environment usually require fillet welds, so the CTS test can be used for weldability testing in a water environment.

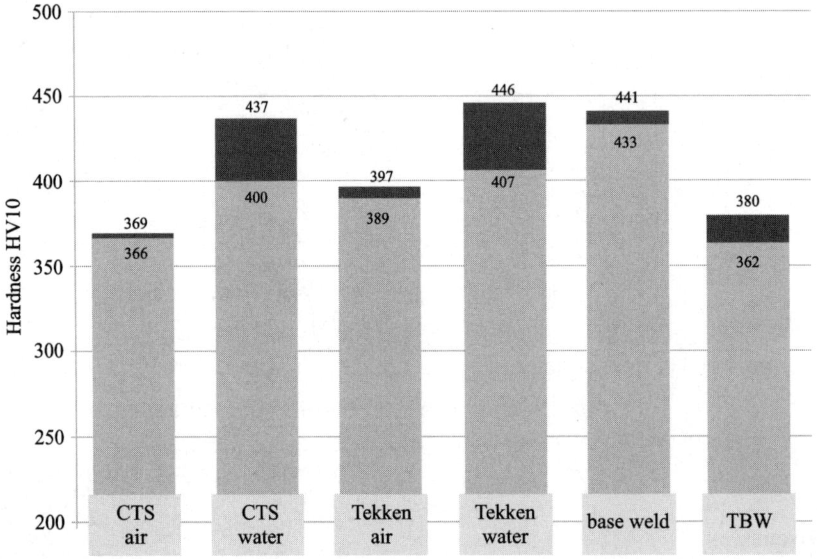

Figure 16 Hardness measurements results for CTS specimens, Tekken specimens and Temper Bead Welding specimens, welded in a water environment. Red color presents range of HV10 values measured in specimens

5. Conclusions

16 From these experiments the following conclusions can be drawn:

1) S355G10+N steel welded in a water environment is characterized by high susceptibility to forming cold cracks. The weldability of the investigated steel in the air is better than in underwater conditions.

2) TBW technique can be an effective method to improve the weldability of steel, such as S355G10 + N welded by covered electrodes in water environment.

3) TBW technique allows to reduce maximum hardness in the HAZ below the critical value of 380 HV10 as stated by the EN-ISO 15614-1 standard.

4) From the viewpoint of weldability it was determined that the best range for the overlap value between subsequent beads is 75%-100%.

References

...

Task 16: Identify the evaluative markers (words and expressions showing the author's attitude to viewpoint on, or feelings towards the entities or propositions being presented) in the following statements, which could be in the form of modal verbs, adverbs, adjectives, nouns, conjuncts, and even sentence patterns.

(1) These undesirable effects become stronger with increasing active layer thickness, as shown in the performance of DSCs with a 5.1 m active layer.

(2) In contrast with carbon black, the HGNs could provide more nanosized triphase regions as active regions for oxygen reduction (Figure 5a-c).

(3) The connection between V and O mainly depends on the pH value of the solution.

(4) As evidenced by XRD patterns (Figure 1a), scanning electron microscopy (SEM) observation, and N_2 adsorption results, the introduction of defects here presents little effect on the morphology and crystal structure of g-C_3N_4.

(5) Especially, the Sn and K element are slightly richer in the precipitate than in the matrix.

(6) Similarly, the phase of WO_3 changes gradually following the microwave heating.

(7) It is generally known that the surface defects such as oxygen vacancies and the large surface areatovolume ratio make ZnO nanoparticles very sensitive to the ambient gases, particularly oxygen.

(8) According to the ICP optical emission spectrometer (ICPOES) analysis, the QDs of sample 3 are typically composed by Cu:Mn:Zn:In of 1.24:0.57:71.03: 27.16 in mol%.

(9) In comparison, the neat ZnO seeds show the red shift of the emission spectra and the significant reduction of the fluorescent intensity over 10 h storage in ethanol solvent.

(10) Differently, the capacitance of PFE is $\approx 0.03\ \mu Fcm^{-2}$ at 1 kHz, and one order of magnitude larger than that of neat PFR.

(11) The hierarchical mesoporous hollow structure with open ends and mesoporous walls significantly shortens the diffusion pathway of MB molecules and promotes their mass transfer and diffusion rate.

(12) Interestingly, brain signaling of the CbV emission presented notably different temporal profiles, depending on the excitation condition.

(13) Notably, all of the surface cells were strongly luminescent, indicating TPE-PyN_3 can penetrate through extracellular matrix and cell membrane and retain strong emission inside living cells.
(14) Importantly, the peak area ratio of V^{5+} and V^{4+} changes greatly in different samples.
(15) Most importantly, this specific core-sheath heterostructure is indeed efficient for enhancing the stability of SEI.
(16) Remarkably, R_{ct} is only 6.8 after 1,000 cycles.
(17) The charge transfer resistance of g-C_3N_4 is remarkably decreased by the continuous p-n band offset in DCN.
(18) Interestingly, we find that the photoresponse increases dramatically when the device is exposed to UV light irradiation for a short time.
(19) Not surprisingly, this phenomenon may be ascribed to the fact that smaller MDCPs have a greater surface-to-volume ratio.
(20) Surprisingly, all DOTAP-RNA mixtures were negative regardless of the DOTAP:RNA molar ratio.
(21) The laser line width, threshold, and cavity quality factor exhibit surprisingly good performance.
(22) These results establish that this new class of DNA cross-linked PIC hydrogels perfectly mimics the mechanical properties of semi-flexible biopolymer networks.
(23) Unfortunately, with poor water solubility, SN38 can not be directly administrated into patients as a drug.
(24) The correlation between water uptake and film swelling, shown in Figure 1, is strikingly nonlinear.
(25) The 74 kg/mol BCP-homopolymer blends generate a similar disordered or poorly ordered morphology at a 90% homopolymer mass fraction.
(26) Moreover, ideally the recovery should be as quick as possible for quick movements such as running, jumping, or other sports and exercises.
(27) More intriguingly, as illustrated in Figure 1, no toxic H_2O_2 remained during UA degradation.
(28) Encouragingly, significantly improved conductivity and electrochemical properties are observed for the treated composite films.
(29) Unexpectedly, soon after the injection many cells outside blood vessels internalize liposomes.
(30) In contrast, the current became very small when a negative voltage was applied.
(31) Stretchability tests showed better efficiency for the PUA substrates than commercially available and highly stretchable organic substrates such as PDMS.
(32) This observation is also fully consistent with previous observations.
(33) However, the pathway is altered completely once ZnO/SiO_2 nanoparticles are present.
(34) This indicates strongly the serious aggregation and quenching of ZnO seeds over time.

(35) Obviously, channeled scaffolds promoted better bone structure compared to the other groups, as more deep pink can be observed.

(36) Indeed, a reduced coordination could result in a decrease in the screening of charges and thus in an effective increase in the concentration of mobile ions.

(37) Instead, these wrinkles greatly obstruct the charge transport pathway in the direction parallel to strain.

(38) The toxicity of such materials, dependent on their contents in biological media, is extremely low.

(39) At first, the Ag NPs are too small to have a significant resonance to affect the reaction.

(40) In fact, no cracks could be observed on the cross-linked film even when stretched to 100% strain for 500 cycles (Figure 1).

(41) The set voltage is only 2.8 V, substantially lower than the initial forming voltage of 7.6 V.

(42) The decrease in capacity upon cycling is solely due to the second reduction as shown by the limited potential window cycling (Figure 1c).

(43) However, the capacitance value of PFE is about two orders of magnitude smaller than those of ionic liquid and ionic gels.

(44) The dye loading amounts are almost the same for both flat and patterned photoelectrodes.

(45) The two CD spectra for each SPPCM without polystyrene microspheres and glass substrate remain nearly unchanged for both front and backside illuminations.

(46) The slightly decreased BET specific surface areas and pore volumes for the composites are probably due to the little amount of MoS_2 or MnO_2 particles trapped into the pores of OMCRs.

(47) This could possibly be explained by an absorption and adsorption phenomena.

(48) The gradient is most likely caused by light leakage under the pattern of the mask during UV exposure.

(49) Such a confinement pyrolysis strategy can potentially be accomplished with the assistance of other chemicals.

(50) This characteristic maybe has great potential as an application in a p-type semiconductive window.

(51) Most likely, the gold precursor diffuses through the pores of the silica.

(52) These results indicate that the open grain boundaries in the perovskite film are detrimental to the device performance.

(53) The π^* resonances of the R6G molecules were highly angularly dependent, indicating that the orientation of the R6G on BN was not random.

(54) Assuming that the forest behaves like an open-cell foam, the collapse stress is proportional to the elastic modulus of the forest.

(55) However, we believe that the intensive improvement would be achieved by adopting a chemical mechanical polishing process to flatten electrodes.

(56) Compared with the micro talc, it seems that the nano talc is more effective in promoting crystallization due to the corresponding higher peak temperature of crystallization.

(57) It appears that the mesopore volume around 2.1 nm is once decreased with reaction time and increased again after reaction for more than 3 d.

(58) As an approximation, we estimate that the transition in the brush conformation responsible for the transition occurs when $z = z_S$.

(59) We supposed that the specific intracellular trafficking of TMCMS might be due to the specific interactions at the nano-bio interface.

(60) It is noteworthy that the surface modification of TiO_2 concretizes particles' borders and minimizes nanoparticle aggregation.

(61) It is well-known that the porous structure is often designed for preparing pressure-sensing materials.

(62) Apart from water as an energy source, it is interesting that ethanol is also able to activate the power generation of g-3D-PPy.

(63) It is notable that the alignment between the transmission and PL peaks in our metasurface systems is very interesting.

(64) It is noticeable that the viscosity of the samples initially decreases sharply and then significantly slower to reach a relatively stable value.

(65) It is somewhat surprising that light-activated ROS production yields such rapid eradication of a multi-drug resistant biofilm-associated infection, since ROS is extremely shortlived.

(66) Thus, it seems reasonable that the carrier recombination emission efficiency is reduced for $Zn_{1-x}Mg_xO$ NPs with a relatively high Mg content.

(67) It is plausible that the characteristic broad amine peaks were preserved in the NPs.

(68) It is important that multiple functionalities integrated into a single device do not interfere with each other.

(69) The observed scattering patterns are difficult to correlate to the specific physical dimensions of the system.

(70) It was possible to precisely guide the hybrid NT along a preplanned trajectory, as can be seen from the time lapse image in Figure 1a.

(71) It is important to note that reflectance spectra can not be well fitted using a coupled oscillator model which does not take account of retardation phase shifts (Figure 1a).

(72) The isotopic distribution is likely to reflect on the characteristic features of this cluster.

(73) These results clearly confirmed that SAW is necessary to process the entire fluid sample.

(74) It is necessary to discuss the effects of the high-energy electron beam used in transmission electron microscopy (TEM) and STEM experiments.

(75) It is quite hard to enlarge the twinned area by just increasing the plunge depth.

(76) For high temperatures (i.e., above 400 °C), the Te loss is accelerated, making it impossible to maintain the initial 2H structure.

(77) It is useful to recall these findings when interpreting the results presented here.

(78) However, it is still crucial to evaluate the safety limits by conducting human-subject experiments with the desired imaging setup.

(79) The pillar height becomes insufficient to prevent water from wetting the gaps.

(80) This is likely because the generations of h^+ and thus •OH are more favorable to occur than that of • O_2^-.

Task 17: Discuss in groups and identify conjuncts (usually in the form of adverbs and prepositional phrases) signifying different relationships between propositions, namely listing/sequencing, summary, apposition/specification, result, inference, and transition.

- Listing/sequencing:
  ____
  ____
- Summary:
  ____
  ____
- Apposition/specification:
  ____
  ____
- Result:
  ____
  ____
- Inference:
  ____
  ____
- Transition:
  ____
  ____

(1) The electrolyte was then changed to 0.5 M H_2SO_4 and the corresponding capacitance was measured (Figure 5).

(2) In addition, depending on the alkyl chain length, especially RK2 and 3, their device performance and trends are quite different.

(3) Moreover, our results show a clear separation between two adjacent MoS_2 arrays.

(4) Finally, combining piezoelectric materials with other tribological strategies (such as those involving 2D materials, nanocomposites, or charged polymer brushes) may extend performance.
(5) Besides, these noble metal nanostructures absorb visible light to generate collective oscillation of valence electrons known as the localized surface plasmon resonance.
(6) Second, the rigid matrix provided by LDHs nanosheets immobilizes and isolates the DCM molecules and suppresses the aggregation caused quenching.
(7) More importantly, the impact of the presence of oxidized and reduced species on the PL features of both porphyrins was investigated by means of spectroelectrochemical assays.
(8) Initially, the coated surfaces were subjected to knife scratch and peel-off tests (Figure 1A, B).
(9) Overall, a rather complex redox behavior is noted.
(10) In general, all DND particle types show photobleaching behavior.
(11) Generally, the molecular weight of polycations significantly affects the transfection efficiency.
(12) Taken together, these results indicated that the PC molecules were successfully coated onto the surfaces of the UCNPs.
(13) Briefly, a solid paraffin pattern was created by a solution-molding method and then immersed into the SF/HFIP/H_2O solution.
(14) In summary, the elastic stretchability increases significantly with decreasing substrate thickness.
(15) Together, these two aspects contributed to the robustness of the LETD and enabled stable contact forming and deforming over 1000 times.
(16) In brief, DFT simulations provide us with energies and geometric structures at the ground state,
(17) In conclusion, the distribution of Ag NPs follows exactly the localized E-field enhancement.
(18) In short, the photocurrent in vacuum is larger than that in ambient while the response rate in vacuum is much slower.
(19) In total, we studied 25 samples using new, randomly selected tips for each sample and generally found similar results.
(20) However, when the sputtering process continues (i.e., 150 s in this study), Au NPs become so dense that they can block the light intake by Se, resulting in a decreasing responsivity.
(21) For example, the proposed structure for hydrated CHT chains is a twofold helix stabilized by intra- and inter-molecular hydrogen bonds.
(22) We synthesized GO using a simple and cheap solution process, that is, a modified Hummers method.
(23) At each excitation wavelength the same set of lifetimes, namely, 3 and 10 ns, are seen.
(24) Specifically, we hypothesized that it could stem either from the core-shell band structure discussed above.

(25) In other words, nanosheets should have better surface adsorption than the corresponding bulk materials even with the same surface area.

(26) This means that our conductor is stretchable and wearable for human beings.

(27) In detail, the synthesized CFNCs show uniform solid nanocubes with an average edge length of 550 nm.

(28) As an example, we show in Figure 1 the analysis of the spinning behavior of the molecular "rotor" over the span of 72 STEM frames.

(29) Therefore, it is not surprising that the fabrication of scaffolds has become to precisely control its construction.

(30) Thus, the mirror symmetry recovered in the STO layer is now transferred into the LSMO layer.

(31) Mesoporous nanorods exhibited an increased ability to accumulate the drug and thus higher retention efficiency.

(32) Hence, hydrosilylation reactions are used to functionalize and stabilize sensitive silicon-based nanomaterials.

(33) We have consequently simulated this configuration (displayed in Figure 1f), and the result of the DOS calculation is presented in Figure 1g.

(34) The BOPP/graphite/paper actuator was fabricated at RH of 40% so that paper could swell at high RH and shrink at low RH.

(35) As such, the role of Ns in particle clearance of all types should not be overlooked.

(36) As a consequence, the ESR signals for 4-oxo-TEMP/$^1O_2$ (Figure 1) were recovered after these treatments.

(37) Ultimately, this new synthesis methodology creates a MF/silica composite as the shell of a hollow particle.

(38) Significant weight loss accompanying an endothermal peak occurred at around 200 °C, possibly implying the decomposition of the Cu lactate.

(39) This can be due to the accumulation of deep traps, probably caused by iodide vacancies at the interface.

(40) Presumably, for this reason, it was difficult to distinguish G from S because each quadromer had the same volume.

(41) Most likely, H and CH_3 radicals produced after the decomposition of H_2 and CH_4 in the presence of Cu catalysts are prerequisites of the reaction.

(42) From this perspective, the least strain on the burr tip should leave the tip less damaged and likely is more preferable for enhancing edge stretchability performance.

(43) In this sense, an electrode made of monolayer black phosphorus (phosphorene) would be ideal.

(44) However, the as-formed gas bubbles adhered to the surface of electrode firmly.

(45) In contrast, the output characteristics of the proposed VFET show clear drain current saturation (Figure 1b).
(46) In contrast, the BOPP film is hydrophobic and is inert to humidity change.
(47) Si-C bonds, by contrast, are stable against oxidation and hydrolysis.
(48) On the other hand, the S signal was weak on the side facing the separator.
(49) The hexagonal crystals are not evident after biomineralization; instead, a layer with mud cracks is evident.
(50) Nevertheless, the bio-mineralized deposits are non-uniform and exhibit globular crystal agglomerates on the surface.
(51) In fact, the amorphous structure persists even after scanning the same area for much longer times (2 h).
(52) On the contrary, VO_2 films irradiated in O_2 atmosphere showed no significant change in the MIT ratio.
(53) Si-C bonds, by contrast, are stable against oxidation and hydrolysis.
(54) For comparison, the carbonate-based electrolytes and other voltage windows are also tested.
(55) The electrode configuration of such type should integrate electroactive materials, with the proper loading amount, into a robust, yet flexible network.
(56) The specific capacitance of HAGFF decreases with increasing current density, but still, it shows good rate performance with retention of 84% and 68% for CG and CA, respectively,
(57) Nonetheless, the component change of the hydrogel could lead to dramatic changes of the peptide fibers.
(58) Here, only the position of the laser focus relative to the synthesis slide has to be controlled.
(59) Herein, the bending performance of the BOPP/graphite/paper actuator when driven by electricity is further investigated.
(60) Recently, photothermal and electrothermal actuators have been widely studied.
(61) Previously, many design approaches have been invented to make charged biomaterials surfaces.
(62) After that, the pressure is reduced back from 30 to 10 MPa, in order to rule out any hysteresis phenomenon.
(63) In the following study, all the experiments have been run using a 98% purity OA (98% assay and 98% primary amine).
(64) The well-constructed nanoplate structure helped to perpetuate the superior mechanical and electrical contacts, and in the meantime afforded a great deal of exposing surface.
(65) Once again, Figure 1b emphasizes the fact that the higher the humidity level, the higher the accuracy of discrimination.
(66) Later, the FTIR was employed to investigate the successful conjugation of Tam (Figure 1b).

Task 18: Fill in the blanks with appropriate verbs, prepositions, and adverbs in the following extracts of research article results and discussion sections.

I

As-welded microstructure[6]

Optical micrographs of cross section of the surfacing sample are (1) ________ in Figure 1. From Figure 1a, the surfacing sample was (2) ________ into seven zones with the micrographs varying along the cross section. The H13 steel substrate (zone 1) exhibited a fine, uniform microstructure (Figure 1b), (3) ________ of α-Fe phase, pearlite and a large number of small carbides dispersed throughout its matrix. This (4) ________ that the substrate is a mixed structure (5) ________ ferrite, pearlite and carbides.

II

Experiments with mixtures of low-density and high-density PE samples[7]

In order to fully investigate these phenomena, we (1) ________ experiments using different cooling rates around 2 °C/min. Figure 1 (2) ________ the elution profile at the two cooling rates where peak areas are close to the expected value for a 50:50 mixture, as well as the thermogram at the critical cooling rate where the entire mixture crystallizes as low-density PE. The thermogram (3) ________ for the cooling rate of 1.85 °C/min (starting cooling program at 85.5 °C) is similar to the expected result for the mixture (4) ________ in Figure 2 at a cooling rate of 0.1 °C/min. The important difference is the cooling time between the two experiments: 11 h in classical TREF (5) ________ with only 30 min under dynamic cooling.

III

Surface characterization of SnO_2/OMCS nanocomposites[8]

A series of SnO_2/OMCS nanocomposites with various Sn/C weight ratios (1) ________ from 15 to 35 wt% ($OMCST_x$, x = 15, 25 and 35) were synthesized by microwave-assisted hydrothermal method. The SEM images of as-prepared OMCS (2) ________ in Figure 1a–d clearly show the uniformly spherical structures with an average diameter of around 90 nm. After modification with Sn precursors, the size of spheres (3) ________ upon increasing the Sn/C ratio and the mean diameters of $OMCST_{15}$, $OMCST_{25}$ and $OMCST_{35}$ are 97, 102 and 124 nm, respectively. The increase in diameter of $OMCST_x$ at various Sn/C ratios is mainly (4) ________ to the coverage of ultra-small SnO_2 nanoparticles on the surface. The TEM image shown in Figure 1e (5) ________ the highly ordered meso-structure of OMCS. After the addition of Sn precursors, a thin and discrete layer of SnO_2 satellite nanoparticles is (6) ________ onto the surface of OMCS core and the thickness increases with the increase in Sn/C ratio from 15 to 35 wt% (Figure 1f–h).

Task 19: Arrange the sentences in each set logically and coherently into a research article results and discussion section.

(1)

(1) Both methods clearly show that the peak of SRM 1476 elutes at a different time as compared with SRM 1475a.

(2) Figure 1 shows a comparison between the thermograms obtained for SRM 1475a and SRM 1476 samples with fast/quench cooling and those given by classical analytical TREF.

(3) As for the first analyzed samples, the peak elution times are similar with both methods.

(4) However, the peaks obtained with the classical method are significantly broader due to the increased column dimensions.

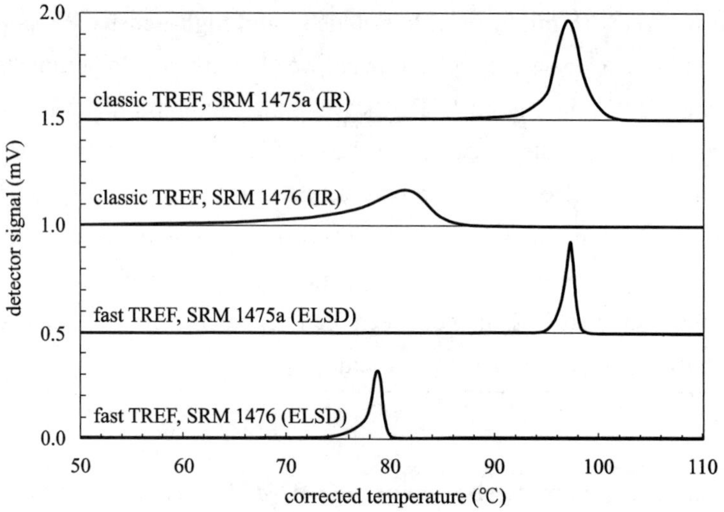

Figure 1 Fast and classical TREF results for SRM 1475a and SRM 1476

(2)

(1) The important difference is the cooling time between the two experiments: 11 h in classical TREF compared with only 30 min under dynamic cooling.

(2) The thermogram obtained for the cooling rate of 1.85 °C/min (starting cooling program at 85.5 °C) is similar to the expected result for the mixture given previously at a cooling rate of 0.1 °C/min.

(3) In order to fully investigate these phenomena, we performed experiments using different cooling rates around 2 °C/min.

(4) Figure 1 shows the elution profile at the two cooling rates where peak areas are close to the expected value for a 50:50 mixture, as well as the thermogram at the critical cooling rate where the entire mixture crystallizes as low-density PE.

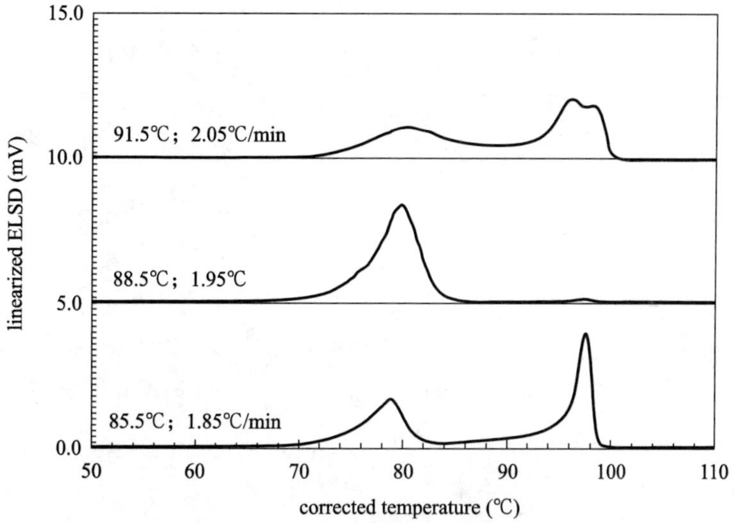

Figure 1 TREF elution profiles of 50/50 mixture of SRM 1475a and SRM 1476 for dynamic cooling rates of 1.85, 1.95 and 2.05 °C/min

(3)

(1) In the presence of the AuSi interlayer, the interfacial resistivity is found to be near 3000 Ω, an order of magnitude larger than the interface in the absence of the AuSi.

(2) Figure 1A shows the first four cycles along with the 20th and 30th CV cycles scanned between −0.1 and +0.1 V, at a scan rate of 0.2 mV/s.

(3) Figure 1B shows the Nyquist plots obtained following cycles 1–4 and cycles 20 and 30.

(4) Quantitatively, the EIS spectra were fit into the two-component circuit model described above (dashed lines overlaid on the Nyquist plots show the fitted results).

(5) These observations suggest that the AuSi interlayer effectively prevented both interfacial decomposition and Li dendrite formation.

(6) The figure shows that the *I*–*V* behavior was relatively constant as a function of cycle number.

(7) The impedance of the same Li/AuSi/LPS/SiAu/Li cell was monitored after each CV cycle.

(8) Figure 1A shows, however, that the current density values near −0.1 and +0.1 V are approximately one order of magnitude lower compared to that seen in the Li/LPS/Li cell.

(9) To prevent LPS material from directly contacting metallic Li, we applied a thin Si interlayer, encapsulated by Au, between LPS and the Li electrode.

(10) The figure shows that no significant impedance change occurs during the CV cycles.

(11) Thus, the AuSi interlayer likely results in a cell impedance increase.

(12) In contrast with the Li/LPS/Li cell, the Li/AuSi/LPS/SiAu/Li cell ran at least 30 scans without shorting, and no obvious Li^+ depletion phenomena were observed.

(13) The R_{SE} and CPE_{SE} values associated with the bulk LPS were fixed, and the interfacial components were obtained from the fit.

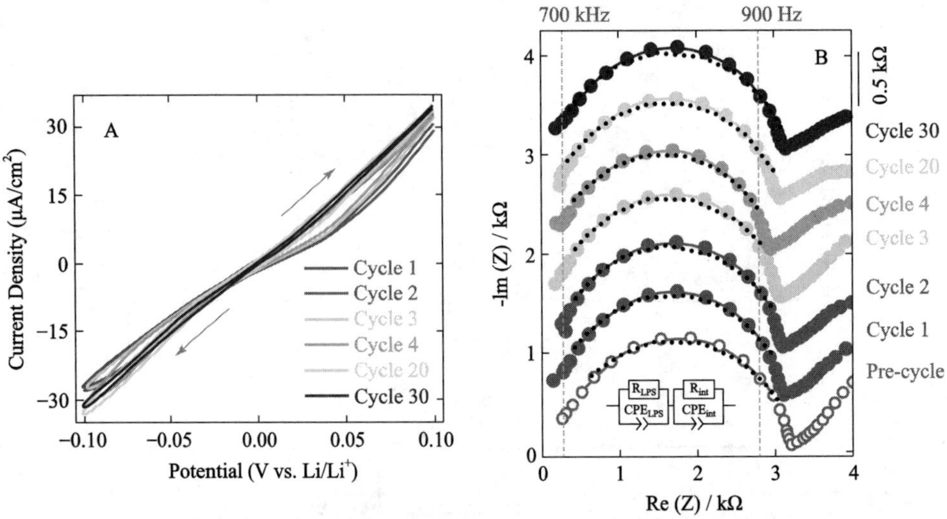

Figure 1 (A) Cyclic voltammetry (CV) of Li/AuSi/LPS/SiAu/Li cell obtained at 0.2 mV/s and (B) the Nyquist plot of the same cell before (red open circle) and after CV cycles

Task 20: Write a short essay to present the main results and discussion of one recent study you have conducted with the aid of tables and figures, and give an oral presentation in groups.

Unit 5
Conclusion Sections

Objectives

- Understand the summarizing and affirming purposes of research article conclusion sections;
- Identify typical structural patterns in research article conclusion sections;
- Figure out the overlapping or echoing structural patterns between conclusion sections and other sections like highlights, abstracts, and introductions;
- Acquaint with generalization strategy used in conclusion sections to synthesize and condense information;
- Identify words and expressions mostly used to conclusively refer to methods and results in previous sections;
- Determine various language features emphasizing the importance, significance, novelty, efficiency, superiority, or applicability of research findings;
- Identify emphatic and evaluative attitudes in presenting key results;
- Identify differences in epistemic modality among conclusion sections from various disciplinary research articles;
- Learn to synthesize and condense ideas based on research details;
- Raise awareness of phrasal and syntactic complexity in research article conclusion sections.

Task 1: Discuss with each other which of the following communicative purposes a typical research article conclusion section needs to fulfill.

Communicative Purposes	Yes/No?
(1) Briefing research purposes of present study	

continued

Communicative Purposes	Yes/No?
(2) Briefing methodology and procedures of present study	
(3) Indicating the effectiveness of present research methodology	
(4) Restating theoretical/methodological hypothesis made in present study	
(5) Describing main difficulties encountered in present study and discussing how the problems were confronted	
(6) Highlighting overall experimental or analytical results of present study	
(7) Presenting key results one by one both numerically and qualitatively	
(8) Condensing research results qualitatively in a logical way	
(9) Referring to tables or figures to report current findings	
(10) Specifying significant results obtained in present study	
(11) Exemplifying significant results obtained in present study	
(12) Interpreting specific research outcomes	
(13) Explaining possible errors of quantitative results	
(14) Explaining satisfactory/unsatisfactory research results	
(15) Relating research results to the critical issue (gap) identified in the introduction section	
(16) Comparing current research results with previous studies, whether citing agreement or disclosing discrepancies	
(17) Relating research results to (inter) national industry standards	
(18) Discussing practical applications of research results with a focus on potential benefits	
(19) Stating values and advantages of present study and results	
(20) Emphasizing the uniqueness of present study and major contributions	
(21) Suggesting the applicability of research results	
(22) Pointing out the limitations of present study	
(23) Reflecting on uncertainty about the correctness of research outcomes in real-world scenarios	

continued

Communicative Purposes	Yes/No?
(24) Indicating research implications	
(25) Making recommendations or stating a practical need for further study	
(26) Indicating potential value of further study	
(27) Providing new insights into the research topics	

Task 2: Read the following versions of a research article conclusion section, in the style of whether an oral presentation, a popular science article, a short essay produced by novice student writers, a formal academic writing, or a commercial advertising discourse, then discuss in groups which version is the most formal, precise, and objective to serve as a research article conclusion section, and finally analyze the differences in language use among the versions.

(1)

Hey, check out what we've done in this paper! We've come up with this cool new method to make these funky spherical structures called hierarchical urchin-like $LiNi_{0.5}Mn_{1.5}O_4$ hollow spheres. They're made up of lots of thin sheets that are all tangled together, with dominant {111} facets exposed. You won't believe it, but these new hollow spheres, or HUL-LNMO as we like to call them, are super fast and stable when it comes to their electrical performance. They can handle high charging and discharging rates without breaking a sweat. The reason behind this awesome performance is the unique hollow structure of the HUL-LNMO, with those {111} exposed facets. This structure lets lithium ions enter and exit really quickly by keeping the distance they have to travel short. It also helps control the expansion of the material during intense discharging.

So, basically, we've made a material that's perfect for electric vehicle batteries. It performs like a champ and has the potential to be used in practical applications. And guess what? This method we used to make these cool hollow spheres can be applied to make all sorts of other materials with a similar structure. That means we can make better batteries for all kinds of high-performance energy storage devices.

(2)

Revolutionary Breakthrough in Energy Storage for Electric Vehicles!

In a groundbreaking study, a team of brilliant materials science researchers has just unveiled an incredible advancement in the field of energy storage. Get ready to be amazed as we dive into the world of futuristic battery technology!

So, what's the buzz all about? Well, these ingenious scientists have developed a never-before-seen method of creating something called hierarchical urchin-like $LiNi_{0.5}Mn_{1.5}O_4$ hollow spheres. It may sound like a mouthful, but trust us, it's mind-blowing!

Imagine these hollow spheres made up of interlocking layers of tiny particles, forming a unique urchin-like shape. And the best part? These spheres have special exposed surfaces, known as {111} facets, that play a vital role in their exceptional performance.

Now, let's talk about what makes these spheres so extraordinary. Brace yourself for some mind-boggling electrochemical kinetics! These spheres demonstrate outstanding capabilities when it comes to charging and discharging at high speeds. It's like having a battery on steroids!

But how do they do it? The secret lies in their structure. You see, these hollow spheres have a hierarchical design that enables the swift insertion and removal of lithium ions—the energy carriers in batteries. By reducing the distance these ions have to travel, these spheres can charge and discharge rapidly, without breaking a sweat.

That's not all—there's another critical advantage. When discharging at high rates, batteries often undergo significant expansion, which can lead to performance issues. However, these clever hollow spheres tackle that problem head-on. They alleviate the volume expansion, ensuring the battery remains stable even during intense usage.

But what does all of this mean for electric vehicles (EVs)? Well, it's a game-changer! With these innovative hollow spheres powering EV batteries, imagine a world where you can charge your electric car in a fraction of the time it takes today. Not only that, but the battery will also maintain its top-notch performance over the long haul. Say goodbye to range anxiety and hello to lightning-fast charging!

The implications of this breakthrough extend far beyond electric vehicles. The method used to create these hollow spheres can be applied to produce a wide range of two-dimensional (2D) nanosheet structures. These structures hold immense potential for developing high-performance energy storage devices, revolutionizing the way we power our lives.

In conclusion, the future of energy storage looks brighter than ever, thanks to the ingenuity of these materials science researchers. With their hierarchical urchin-like hollow spheres paving the way, we're on the cusp of a new era in battery technology. Get ready for faster charging, longer-lasting batteries, and a greener future for us all!

(3)

In this paper, we have successfully developed a novel method to synthesize hierarchical urchin-like $LiNi_{0.5}Mn_{1.5}O_4$ hollow spheres comprising interpenetrating nanosheets with dominate {111} exposed facets. This novel HUL-LNMO exhibits excellent electrochemical kinetics in rate capability and high-rate cyclic stability. The great improvement in rate and cycle performance could be attributed to the hierarchical hollow structure of HUL-LNMO with dominant {111} exposed facets, which could enable fast insertion and removal of lithium ions by reducing the diffusion length and remit the severe volume expansion during the high-rate discharge, demonstrating its potential practical application in LIBs for electric vehicles. Moreover, this method could be easily extended to the fabrication of a wide variety of 2D nanosheet structural cathode materials for high-performance energy storage devices.

(4)

In this paper, we did a really cool thing! We figured out a new way to make these hollow spheres that look like urchins. We used some special sheets that fit together, and they had these surfaces called {111} facets.

These spheres are really good at charging and discharging. They work fast! The reason they're so good is because of their special structure and those {111} surfaces.

The special structure helps lithium ions go in and out really quickly. It's like a shortcut, so they don't have to go far. This helps with fast charging. And even when the battery discharges really fast, these spheres can handle it. They stay stable and don't get too big.

These spheres are really exciting, especially for electric cars. Just imagine being able to charge your car really fast and still have it work great. It would be amazing!

And here's another really cool thing: we can use the same method to make other materials, like flat sheets, for storing energy. They could be really good too!

So, in conclusion, we found a cool way to make these hollow spheres that work really well for charging and discharging. They could be great for electric cars, and we can use the same method to make other useful stuff too.

(5)

Our groundbreaking research has unveiled a revolutionary method to synthesize astonishing hollow spheres resembling majestic urchins. These spheres are composed of interpenetrating nanosheets, and their dominant {111} facets are simply awe-inspiring. They possess unparalleled electrochemical kinetics, showcasing extraordinary capabilities in rate capability and high-rate cyclic stability. The jaw-dropping improvement in both rate and cycle performance can be attributed to the hierarchical hollow structure of these spheres. The dominance of {111} facets plays a pivotal role in enabling the lightning-fast insertion and removal of lithium ions, as it significantly reduces the diffusion length. Brace yourself for the astonishing fact that these spheres even manage to control the volume expansion during high-rate discharge.

The potential practical application of these marvels is mind-blowing. Their exceptional performance paves the way for their seamless integration into lithium-ion batteries (LIBs) for electric vehicles, promising a new era of electrifying transportation.

The implications of this extraordinary method transcend beyond our initial findings. We can easily extend its application to fabricate an extensive array of 2D nanosheet structural cathode materials. Prepare to be dazzled by the unlimited possibilities that lie ahead, as these materials hold the potential to revolutionize the landscape of high-performance energy storage devices.

In conclusion, this paper signifies a monumental milestone in the realm of materials science.

Our exceptional achievement in synthesizing hierarchical urchin-like hollow spheres unveils a promising future for advanced energy storage. Prepare for a paradigm shift as we embark on a journey to reshape the world of electric vehicles and energy storage with these remarkable advancements.

Task 3: Read Text Ⅰ carefully and identify the specific information listed in the given outline.

Text Ⅰ

Formation, Photoluminescence and Ferromagnetic Characterization of Ce Doped Aln Hierarchical Nanostructures[1]

Highlights • Ce was incorporated in AlN forming hierarchical nanostructures. • The Ce doped AlN exhibited an intensive PL emission band at 600 nm. • The AlN:Ce nanostructures exhibited room ferromagnetism.	**Key results**
Abstract: Cerium doped aluminum nitride (AlN:Ce) hierarchical nanostructures were fabricated by a direct reaction of Al and CeO_2 mixed ingot with ammonia using modified arc discharge method. X-ray diffractometry, Raman spectrum and x-ray photoelectron spectroscopy analysis obviously indicated that Ce^{3+} ions incorporated inside the AlN nanostructures. The prepared AlN:Ce exhibited a strong red-orange emission at 600 nm and showed room temperature ferromagnetism. These results suggest that AlN:Ce hierarchical nanostructures have potential as a light-emission nanodevice and diluted magnetic semiconductor.	**Purpose & methods** **Methods & results** **Discussion**
1. Introduction As an important III-nitride semiconductor, aluminum nitride (AlN) possesses many excellent properties such as high thermal conductivity, superior mechanical strength, excellent thermal and chemical stability and low dielectric constant [1]. These excellent properties make it promising for electronic, optoelectronic and field emission devices [2-4].	**Important research field**

continued

For enhancing the application of AlN, doping is regarded as an effective method. First of all, doping can improve electronic properties of AlN by increasing the number of carriers [5-6]. Secondly, as the largest band gap (6.2 eV) semiconductors, AlN can be modified wide band from UV to infrared wavelengths through dopants [7-9]. Finally, the diluted magnetic semiconductors (DMS) can be produced by introducing appropriate dopants in AlN [10-13]. Therefore, it is important to synthesize AlN doped with transition metal or rare earth (RE) elements for their novel electronic, optical and magnetic properties, which are used as optoelectronic and spintronics devices.	**Important research topic**
As an important RE metal dopant in semiconductors, Cerium (Ce) is the most commonly used activators, because its luminescence properties are attributed to fully permissible electric dipole 5d-4f transitions, which lead to a large absorption cross section in UV–visible range [14-15]. Because Ce^{3+} ionic radius is significantly larger than the Al^{3+} ions, it has been difficult to stably dope Ce^{3+} ion into the AlN host lattice. Up to now, there are few reports about Ce doped AlN.	**Literature review**
For example, Si^{4+} and Ce^{3+} ions co-doped AlN polycrystalline powders which emit blue color have been produced through gas pressure sintering method [16]. Ce^{3+} doped single-crystal AlN bulk with intense pink-colored emission at 600 nm has been synthesized by an optimized high pressure and high temperature flux method [17]. The high quality AlN:Ce ceramic with viable white emission has been produced by processing procedure based on current activated pressure assisted densification [18]. The AlN:Ce thin films with strong blue emission were synthesized using radio-frequency reactive sputtering [19].	**Typical studies**
All above reports are focused on AlN:Ce bulk materials or thin film and their optical properties. It is well known that hierarchical nanostructures with structural complexity and greater functionality have attracted increasing interest for their application on nanodevices [20-21].	**Synthesis of previous research**
To our knowledge, there is no report on AlN:Ce nanostructures. In addition, previous theoretical calculations predicted that Ce doping can induce the room temperature ferromagnetism in AlN using density functional theory [22-23]. However, the experimental study of magnetic property of AlN:Ce has not yet to be reported.	**Research gap in the literature**

continued

In this work, we present that AlN:Ce hierarchical nanostructures were synthesized using an improved direct current (DC) arc discharge method. Strong red-orange emission and ferromagnetism have been obtained from AlN:Ce hierarchical nanostructure. These results suggest AlN:Ce hierarchical nanostructures have potential applications in light-emission and spintronic nanodevices.	**Purpose of present study** **Major findings**
2. Experimental	
The synthesis was carried out in a modified DC arc discharge apparatus described previously [24-25]. A tungsten rod was used as the cathode. Al (purity: 99.999%) and CeO_2 (purity: 99.99%) powders mixed with a molar ratio of 100:1 were pressed into ingot. The ingot was tightly inserted into the water-cooled graphite crucible serve as anode. When the plasma arc is kindled, the input current was set at 90 A and the voltage was a little higher than 40 V. The synthesis process was maintained for 30 min in pure NH_3 (purity: 99.999%) with a pressure of 30 kPa. Finally, the yellow products were collected at the water-cooled collecting wall, after passivation in Ar for 5 h.	**General description of methods** **Procedures**
The crystal structures of AlN:Ce were characterized by X-ray diffractometry (XRD, Rigaku D/Max γA, Cu-Kα target, $\lambda = 0.154178$ nm). The morphology and chemical composition of these samples were characterized by field-emission scanning electron microscopy (SEM, FEI MAGELLAN-400) equipped with an energy dispersive spectrometer (EDS). The bonding states of the elements were studied via X-ray photoelectron spectroscopy (XPS) on an EASY ESCA spectrometer (VG ESCA LAB MKII). Micro-Raman and photoluminescence (PL) spectra were obtained by JY-T800 Raman spectrometer excited with an Ar^+ line at 514 nm and He-Cd line at 325 nm, respectively. The magnetic measurement was investigated using a vibrating sample magnetometer (VSM). All measurements were performed at room temperature.	**Data collection & analysis**
3. Results and discussion	
The crystal structure of as-prepared sample was characterized by XRD, shown in Fig. 1a. All diffraction peaks can be assigned to pure wurtzite AlN structure with space group: $P6_3mc$ (186) (DPF Card No. 08-0262). As shown in Figure 1b, the (100), (002), and (101) peaks in AlN:Ce display an obvious shift to the lower angles compared with undoped AlN, indicating the expansion of lattice constants after doping is due to the	**Referring to figures** **Highlighting significant results**

continued

substitution of larger Ce^{3+} (1.34 Å) for Al^{3+} (0.54 Å). This phenomenon agrees well with previous reports about Sc or Y doped AlN [11-13].

Interpreting & discussing results

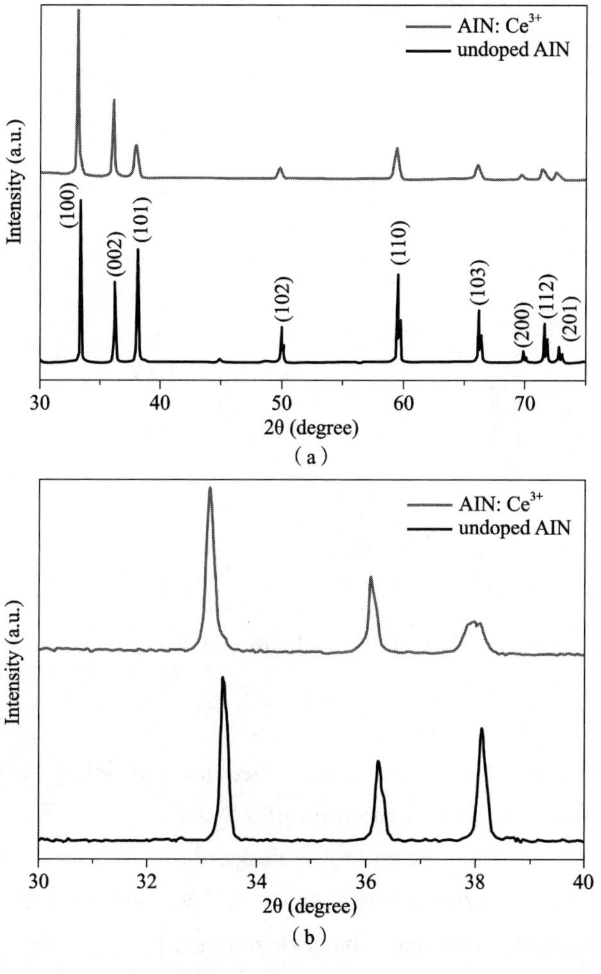

Figure 1 (a) XRD patterns of AlN and AlN:Ce hierarchical nanostructures. (b) The enlarged (100), (002), and (101) peaks

The Raman spectra of undoped and Ce doped AlN were used to further study the crystal structure and the results are shown in Figure 2. The spectrum of undoped AlN reveals the four Raman active modes around 247.3, 611.2, 655.9 and 668.2 cm^{-1}, corresponding to vibrational modes of E_2(low), A_1(TO), E_2(high) and E_1(TO), respectively. The weak and broad peak at 904.2 cm^{-1} is indexed to the overlap of the modes A_1(LO) and E_1(LO) [26]. These locations display a wurtzite AlN structure. Compared with undoped sample, Raman peaks of A_1(TO) and E_2(high) in

Briefing methods

Referring to figures

Presenting main results

continued

<table>
<tr>
<td>AlN:Ce broaden and shift to lower frequencies with increasing tensile strain, may be ascribe to the disorder of the crystals as a result of the substitutional doping of Ce ions in the AlN lattice. This result agrees well with the observed tensile strain in XRD. Similar phenomena were also observed in Sc, Mg and Co doped AlN [12], [27-28].

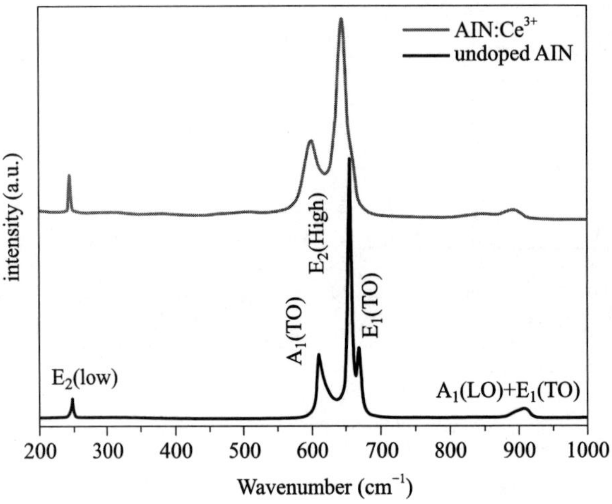

Figure 2 Typical Raman spectra of AlN and AlN:Ce hierarchical nanostructures</td>
<td>Interpreting & discussing results</td>
</tr>
<tr>
<td>XPS measurement is an effective tool to characterize electrical structure and composition. The survey scan XPS spectrum of AlN:Ce sample is shown in Figure 3a. Besides the C 1s and O 1s peaks, the spectrum exhibits strong peaks for Al 2p, Al 2s and N 1s as expected. The Al 2p peak at 73.8 eV (Fig. 3b) is assigned to aluminum bound to nitrogen in wurtzite AlN, while the N 1s peak at 397.0 eV (Figure 3c) represents binding energy of nitrogen in AlN [29]. Complex XPS spectrum of Ce 3d is presented in Figure 3d. Two obvious peaks at 885.6 eV and 904.5 eV, and two shoulder peak at 880.7 eV and 898.0 eV corresponding to two pairs of spin-orbit doublets is identified as trivalent Ce. The weak peak with a binding energy of 916.1 eV is attributed to the $3d_{3/2}4f^0$ component of Ce, indicating mixed valence of Ce. The XPS spectrum of Ce 3d matches with the previously reported value of Ce-N bonding in CeN [30-31]. On the basis of XPS analysis, it can be confirmed the successfully incorporation of Ce^{3+} ions into AlN lattice.</td>
<td>Briefing methods

Referring to figures

Highlighting main results

Interpreting & discussing results</td>
</tr>
</table>

continued

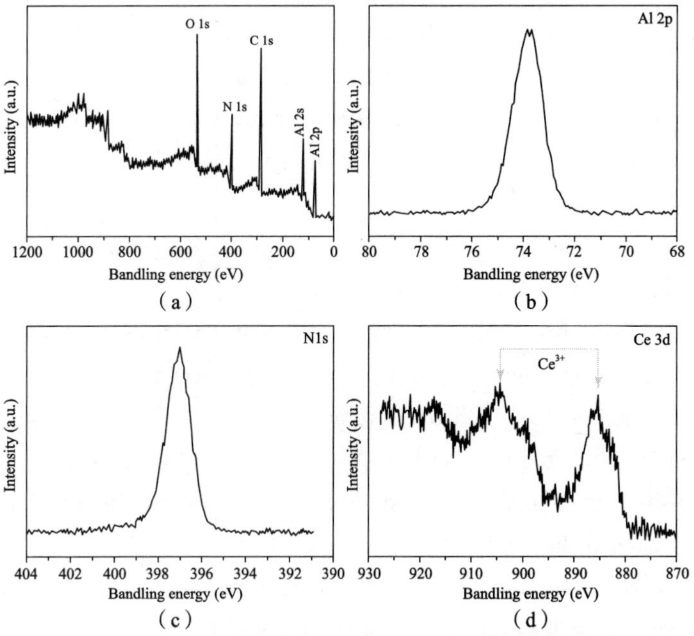

Figure 3 XPS spectra of AlN:Ce hierarchical nanostructures: (a) survey scan, (b) Al 2p, (c) N 1s, and (d) Ce 3d core levels, respectively

The morphology of the AlN:Ce sample was examined by SEM. The low-magnification SEM image in Fig. 4a shows the bulk densely packs with nanowires or nanobelts which constitute complex hierarchical nanostructures. Figure 4b and c are low- and high-magnification images of detailed fringe bulk, they display many branches irregular radiate from the main trunk of the crystals forming hierarchical nanostructures. The morphologies of branches are nanowires or nanobelts. The chemical composition of AlN:Ce hierarchical nanostructures was determined from EDS spectrum (Figure 4d). Quantitative analysis reveals that the dopants of Ce ions in AlN correspond to a concentration close to 1.09%.	**Briefing methods** **Referring to figures** **Highlighting main results** **Interpreting & discussing results**
(Figure 4 Omitted)	
The hierarchical nanostructures are easily formed under non-equilibrium and supersaturated growth condition. In previous studies, the AlN hierarchical nanostructures were also observed under high temperature and high N_2 or NH_3 pressure accompanied with non-equilibrium growth process [32-33]. Arc discharge plasma with high energy supplies a sharp temperature gradient and heat convection in	**Discussing results**

continued

reaction chamber, which can effectively synthesize AlN:Ce hierarchical nanostructures. In addition, Ce ions incorporate into AlN lattice, leading to high concentration of Al and N vacancies. These imperfections can act as second new nucleation sites for growing branches. Therefore, AlN:Ce hierarchical nanostructures can be easily prepared through arc discharge method.	
Figure 5 (a) shows the room temperature PL spectrum of the AlN:Ce hierarchical nanostructures excited by 325 nm from He–Cd laser, demonstrating a broad emission ranging from 550 to 750 nm with a peak at about 600 nm (2.06 eV) in red-orange light range. This red-orange emission can be clearly seen by the naked eye [Figure 5 (c and d)]. In previous studies, PL spectrum of AlN:Ce have a blue emission [16], [19] or red emission band [17]. It is well known that nephelauxetic effect and crystal-field splitting of the 5d states of luminescent ions influence final PL emission [34]. The blue emission is always observed in Ce doped AlN with oxygen impurities. However, our synthesis of AlN:Ce with ammonia can effectively reduce oxygen impurities. Therefore, the emission peak shape and location are exactly the same as previously reported in AlN:Ce single crystals [17]. Ishikawa et al. reported aluminum vacancies (V_{Al}) connecting with nearby Ce_{Al} (Ce substitutions on the Al site) can form complex and stable defect structure (Ce_{Al}-V_{Al}) in AlN:Ce single crystals, and experimentally confirmed the luminescent center in AlN:Ce is neutral Ce_{Al} [17], [35]. As a result, the nonsymmetrical PL emission of AlN:Ce can be assigned to the 4f-5d electron transition of Ce^{3+} dopants in AlN.	**Referring to figures** **Highlighting main results** **Interpreting & discussing results**
(Figure 5 Omitted)	
The reasonable studies on light-emitting properties of AlN:Ce are hot topics in literature but ferromagnetism is seldom explored. Recently, Dar et al. pointed out the AlN:Ce has ferromagnetism for DMS applications through first principles calculations [22]. Later, Majid et al. predicted the favorable formation of Ce_{Al}–V_N complex in AlN:Ce introduces ferromagnetism [23]. However, the experiment on the magnetic properties of AlN:Ce has not been reported. Figure 6 shows typical magnetization curves of AlN:Ce hierarchical nanostructures versus field were measured with a VSM at room temperature. It is evident that AlN:Ce hierarchical nanostructures exhibit a well defined hysteresis loop and show ferromagnetic behavior at room temperature. The saturation magnetization	**Emphasizing research gap in literature** **literature review and existing gap** **Referring to figures** **Highlighting main results**

continued

and remanent magnetization of the AlN:Ce hierarchical nanostructures is 0.038 emu/g and 0.01 emu/g, respectively, while its coercive field is 284 Oe. The results from the XRD, Raman and XPS measurements in AlN:Ce samples demonstrated that the Ce^{3+} ion was incorporated into the host lattice without changing its structure.	**Specifying main results**
The ferromagnetism of AlN:Ce hierarchical nanostructures should be intrinsic and come from Ce doped AlN lattice. According to bound magnetic polaron model, the magnetic behavior in AlN:Ce is sensitive to defects, such as anion or cation vacancies and interstitials. The stable Ce_{Al}-V_{Al} complex has been confirmed by experiment and theoretical calculation [17], [35]. Wu et al. found that Al vacancies in AlN can introduce spin polarization to N atoms around the defect site resulting in ferromagnetism [36]. Lei et al. reported introduction of rare-earth element Y into AlN leads to high Al vacancies responsible for the ferromagnetism [11]. Given the above, the observation of ferromagnetism in AlN:Ce nanostructures may be attribute to Al vacancies which are introduced by incorporating of Ce^{3+} ions into AlN lattice. 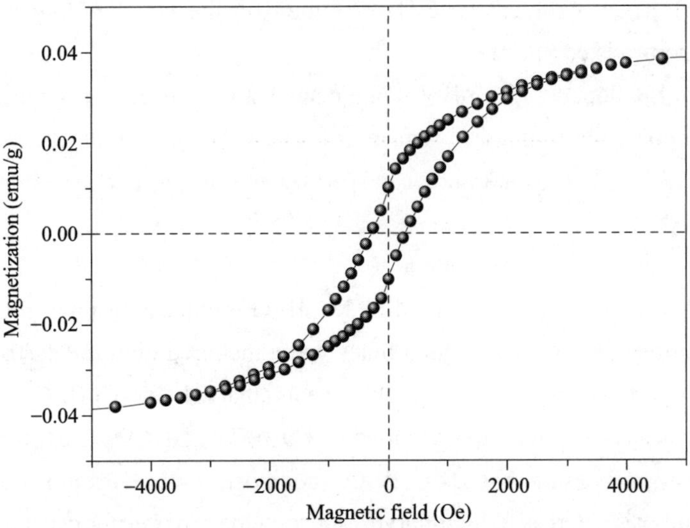 Figure 6 Magnetization hysteresis loops of AlN:Ce hierarchical nanostructures measured at room temperature	**Interpreting & discussing results**
4. Conclusions Through a simple arc discharge method, AlN:Ce hierarchical nanostructures have been successfully fabricated. The synthesis approach provides an opportunity for doping large size rare-earth elements into AlN	**Highlighting overall results** **Stating research significance**

continued

materials. AlN:Ce hierarchical nanostructures exhibit strong red-orange emission and room temperature ferromagnetism, which have applications in optoelectronic and spintronic nanodevices. **References** …	**Discussing practical applications**

Task 4: Read Highlights, Abstract, and Conclusion sections of Text Ⅰ again and discuss in groups the similarities and differences with respect to types of information presented among the three sections.

Task 5: Identify the sentences in the following research article conclusion sections that closely match the corresponding article title and research purpose extracted from the Introduction section.

Ⅰ

Title: Enhanced visible-light-driven photocatalytic H_2 evolution on the novel nitrogen-doped carbon dots/$CuBi_2O_4$ microrods composite

Purpose: In this study, a novel N-doped CDs/$CuBi_2O_4$ photocatalyst was successfully fabricated by loading N-CDs nanoparticles onto the surfaces of $CuBi_2O_4$ micro-rods by a facile hydrothermal method, and applied to the photocatalytic water splitting under visible light irradiation.

Conclusion

A novel nitrogen doped carbon dots (N-CDs) modified $CuBi_2O_4$ photocatalyst was prepared for photocatalytic water splitting. The N-CDs nanoparticles were anchored onto the surfaces of $CuBi_2O_4$ micro-rods. The hydrogen production result reveals that N-CDs/$CuBi_2O_4$ shows enhancing photocatalytic activity compared to the pristine $CuBi_2O_4$ and CDs/$CuBi_2O_4$. The excellent photocatalytic performance of N-CDs/$CuBi_2O_4$ composite was attributed to the introduction of N-CDs in composite increased light harvesting and electron transfer capacity.

Ⅱ

Title: Wafer-Scale Growth and Transfer of Highly-Oriented Monolayer MoS_2 Continuous Films

Purpose: In this work, we report the growth of highly oriented continuous ML-MoS_2 films on 2 in sapphire wafers with only two mirror-symmetric domain orientations present.

Conclusions

In summary, we present the wafer-scale growth of highly oriented ML-MoS_2 continuous films. As produced films are uniform and of high quality, which was demonstrated by optical and electrical characterizations. We also demonstrated the film can be transferred cleanly and innocuously; meanwhile, the substrate was reusable for subsequent growth. We believe that this simple, efficient, reproducible, and low-cost method can shed light on the practical applications of wafer-scale ML-MoS_2.

Ⅲ

Title: Reduced Graphene Oxide-Based Artificial Synapse Yarns for Wearable Textile Device Applications

Purpose: In this study, we fabricated fiber-based artificial synapses that use reduced graphene oxide (RGO) on yarns.

Conclusions

We fabricated artificial synapses that used RGO-coated conductive yarns. RGO was directly formed on the yarn by electrochemical reduction. The device emulated synaptic functions including EPSC and PPF. Also, we confirmed that the device can be transformed from STP to LTP by controlling the amplitude of applied pulses. Furthermore, the device was not degraded noticeably during bending. We demonstrated the feasibility of these synapses in the cross-point structure for integrated textile device applications. We believe that the RGO-coated yarn-based artificial synapses can be a good candidate for future wearable neuromorphic systems.

Ⅳ

Title: VO_x@MoO_3 Nanorod Composite for High-Performance Supercapacitors

Purpose: Herein, we propose an effective strategy to simultaneously improve the capacitance and cycling stability of VO_x-based electrode.

Conclusion

In summary, we demonstrate a facile electrochemical method to obtain the heterogeneous VO_x@MoO_3 composite. The thin layer of MoO_3 effectively interacts with VO_x and modifies its chemical environment and electronic structure. This largely boosts the electrochemical performance of the latter. The composite electrode shows a high capacitance of 1980 mF cm^{-2} at 2 mA cm^{-2} and still maintains 1166 mF cm^{-2} at the high current density of 100 mA cm^{-2}. It also allows a stable long-term cycling, with 94% capacitance retained after 10 000 cycles. The strategy to introduce effective interactions between heterogeneous components presented here can also be applied to other pseudocapacitive type electrodes to promote their electrochemical behavior.

Ⅴ

Title: Al_2O_3/reduced graphene oxide double-layer radiative coating for efficient heat dissipation

Purpose: Motivated by such considerations, in this work, the radiative heat dissipation performance of Al_2O_3/rGO double-layer radiative coating was revealed theoretically and experimentally.

Conclusion

A double-layer radiative coating was proposed, and the radiative heat dissipation performance was investigated theoretically and experimentally. Specifically, the double-layer structural coating fabricated by assembling rGO on Al_2O_3 bottom layer promoted the infrared emissivity value from 0.17 (bare aluminum radiator) to 0.57 at 3–8 μm and 0.83 at 8–20 μm waveband range, respectively. The Al_2O_3/rGO duplex coating with higher emissivity value improved the efficient heat dissipation performance of LED. It can enable a significant LED temperature drop by 5.6 °C (1 W LED chip), which was coincident with the result by simulation (1 W). More remarkably, the temperature drop further decreases to 14.0 °C for 5 W LED chip. Furthermore, the Al_2O_3/rGO duplex coating shows superior hydrophobic and anti-corrosion properties, which expects to be a promising candidate for heat dissipation of electronic products in oceans and humid ambient.

Ⅵ

Title: Characterizing Temporal Heterogeneity by Quantifying Nanoscale Fluctuations in Amorphous Fe-Ge Magnetic Films

Purpose: In this article, we study the phase transition pathway from a helical to a paramagnetic phase in amorphous Fe_xGe_{1-x} magnetic thin films in the limit where the sample is a bad metal and magnetic interactions are weak.

Conclusion

In summary, our study provides a new framework for describing and understanding phase transitions in materials on a microscopic scale. We showed that the existence of a fluctuating domain fraction that follows a specific scaling law can drive a phase transition. It is conceivable that controlling and tailoring the fluctuating fraction will change phase transition pathway thereby arriving at a new phase. It is conceivable that fluctuating population and effect of its growth on phase-stability can be determined for other magnetic systems exhibiting fluctuation. Our study of the question of fraction of domain fluctuation growth at the local level and how that influences global properties and phase transition can find similarity and use-case in a variety of scientific fields, for example, in biological and social sciences where growth models are often used and is an important area of research.

Task 6: Transform the following high-frequency verb-noun collocations used in research article conclusion sections into passive voice. Determine what verb-noun collocations are typically used for each of the listed communicative purposes.

1) Briefing research purpose:

____

2) Briefing research methods:

____

3) Summarizing or highlighting overall results:

____

4) Interpreting or evaluating research results:

____

5) Indicating implications and/or discussing applications of main results:

____

6) Making recommendations or stating a practical need for further study:

____

No.	(Phrasal) Verbs	Noun collocates	Examples
(1)	play	role, function	· play a highly significant role · play an important protective role · play multiple functions
(2)	have	effect, potential, influence, impact, value, advantage, application, property, structure, implication, stability, size, ability, density	· have a favorable photothermal effect · have higher corrosion potential · have a serious and extensive influence · have a direct negative impact · have significantly higher flow stress values · have great potential applications · have the following advantages
(3)	use	method, model, technique, approach, process, simulation, system, film, measurement, array, combination, strategy, microscopy, electrode, condition, calculation, device, datum	· use straightforward, low-cost fabrication methods · use 3D thermal and deformation models · use a scalable, facile, and technologically mature high-energy ball-mill technique · use a patterned photocrosslinking approach · use the electric field alignment process · use Finite Element simulation · use the coaxial powder laser system
(4)	improve	performance, property, stability, efficiency, strength, resistance, conductivity, quality, activity, ability, compatibility, toughness, sensitivity, interaction, effect, density	· improve the electrocatalytic hydrogen evolution performance · improve the charge transfer efficiency · improve the operational device stability · improve the high temperature strength · improve the laser damage resistance

continued

No.	(Phrasal) Verbs	Noun collocates	Examples
(5)	provide	insight, strategy, approach, way, platform, route, method, opportunity, evidence, avenue, understanding, path, basis, possibility, guidance, solution, framework, tool, pathway, potential, example, proof, idea	· provide critical and imperative scientific insights · provide a promising and simple strategy · provide a simple environmentally friendly approach · provide a computationally efficient way · provide an efficient sensing platform · provide an effective and accurate method
(6)	exhibit	property, performance, stability, activity, capacity, behavior, response, increase, strength, effect, value, conductivity, structure, efficiency, density, capacitance, capability, rate, morphology, mode, sensitivity, resistance, durability, ability	· exhibit good photocatalytic degradation performance · exhibit excellent long cycle stability · exhibit superior electrocatalytic activity · exhibit an excellent reusability and oil retention capacity · exhibit maximum piezoelectric response · exhibit strongly nonlinear and internal dynamic behaviors
(7)	show	performance, property, stability, potential, effect, increase, activity, capacity, behavior, value, improvement, change, agreement, rate, efficiency, trend, strength, distribution, decrease	· show excellent sodium storage performances · show a vastly improved thermal stability · show unprecedented application potential · show a good magnetic separation effect · show high and fast adsorption capacity
(8)	pave	way, road, pathway, path	· pave a broad and novel way · pave the road · pave a solid path
(9)	develop	method, model, strategy, material, approach, system, device, process, technique, sensor, membrane, composite, therapy, battery	· develop a novel inspection method · develop high performance dielectric materials · develop a simple but robust strategy · develop a new thermoresponsive shape-changing system

continued

No.	(Phrasal) Verbs	Noun collocates	Examples
(10)	deliver	capacity, density, performance, property	· deliver a stable discharge capacity · deliver an extremely high volumetric energy density · deliver much better electrochemical performance
(11)	investigate	effect, property, mechanism, behavior, influence, structure, role, performance, evolution, relationship, parameter	· investigate the piezotronic coupling effect · investigate the electrochemical performance and lithium storage mechanism · investigate the influence · investigate the strain sensing behaviors
(12)	demonstrate	performance, approach, potential, method, strategy, ability, feasibility, effect, property, application, synthesis, advantage, utility, importance, control, capability, possibility, mechanism	· demonstrate excellent sensing performance · demonstrate a simple and effective approach · demonstrate a new synthetic method · demonstrate a facile and efficient strategy · demonstrate a strong commercial potential
(13)	draw	conclusion, attention, guideline	· draw more general conclusions · draw wide attention · draw some design guidelines
(14)	enhance	performance, property, stability, activity, efficiency, effect, rate, interaction, conductivity, transport, resistance, strength	· enhance specific fluorescence imaging performance · enhance the flame retardant and thermal stability · enhance power conversion efficiency · enhance the intervalley scattering rate
(15)	achieve	performance, efficiency, effect, density, value, property, improvement, conductivity	· achieve enhanced electrochemical performances · achieve the highest transporting efficiency · achieve enhanced photocurrent density · achieve an outstanding shear thickening effect

continued

No.	(Phrasal) Verbs	Noun collocates	Examples
(16)	increase	density, concentration, rate, strength, number, efficiency, temperature, conductivity, capacity	· increase the electron density · increase the nutrient concentration · increase the recycling rates · increase the coordination number
(17)	affect	property, behavior, morphology, formation, structure, stability, size	· affect the global macroscopic behavior · affect local oxygen vacancy formation · affect the electronic structure · affect distinct microscopic processes
(18)	form	structure, layer, bond, phase, network, crystal, particle, interface, contact, composite	· form a stable hybrid structure · form a very slippery and conductive outer layer · form strong hydrogen bonds · form perfect striped phases
(19)	promote	formation, development, growth, transfer, kinetics, activity, process	· promote the conductive network formation · promote the development · promote the good growth · promote rapid mass transfer
(20)	propose	model, mechanism, method, approach, strategy, framework, fixture, design	· propose a novel method · propose a new 3D growth model · propose a promising synthetic approach · propose a novel oxygen management strategy · propose a plausible two-step mechanism
(21)	present	strategy, method, approach, potential, study, value, technique, system, synthesis, effect	· present a simple and cost-effective strategy · present a facile yet efficient method · present a novel room-temperature approach · present a great potential · present higher values
(22)	open (up)	avenue, door, way, opportunity, route, possibility, pathway, window, range, venue, perspective, channel	· open up new avenues · open up a new possibility · open a new door · open an efficient way · open up a simple and effective route

continued

No.	(Phrasal) Verbs	Noun collocates	Examples
(23)	inhibit	growth, proliferation, formation	· inhibit cavity growth · inhibit secondary structure formation · inhibit the immune response
(24)	fabricate	device, composite, structure, film, sensor, electrode, array, system, material	· fabricate light-emitting devices · fabricate 3D carbon hybrid structures · fabricate macroporous films · fabricate an ultrasensitive NO_2 sensor
(25)	report	method, synthesis, system, strategy, approach, study, surface, material, fabrication, effect, design	· report a promising method · report a flexible sensor system · report the facile and scalable synthesis · report a feasible design strategy
(26)	prepare	composite, film, material, electrode, series, particle, fiber, electrocatalyst, device	· prepare novel resin-based composites · prepare macroscopic solid supramolecular plastic films · prepare desired materials · prepare an extremely thick electrode
(27)	overcome	limitation, problem, drawback, barrier, limit, hypoxia, challenge, resistance, loss	· overcome the analysis limitations · overcome the stubborn problem · overcome the drawbacks · overcome the electron transfer barrier
(28)	address	issue, problem, challenge, question, need	· address the safety issues · address the pertinent problem · address future challenges · address relevant biological questions
(29)	offer	advantage, approach, possibility, opportunity, insight, platform, strategy, potential, route, prospect, pathway, method, framework, solution, understanding, tool	· offer great advantages · offer a more promising approach · offer valuable insights · offer an efficient strategy · offer an effective platform
(30)	obtain	result, value, film, conclusion, property, efficiency, composite, strength, density, structure, parameter, material, layer	· obtain a resistant mechanical film · obtain more accurate calculation results · obtain qualitative pI values · obtain the gradient emission

continued

No.	(Phrasal) Verbs	Noun collocates	Examples
(31)	reduce	resistance, cost, size, time, energy, strength, rate, effect, number, loss, emission, density, temperature, property	· reduce the charge transfer resistance · reduce manufacturing costs · reduce the adsorption energy · reduce the fabrication time
(32)	observe	effect, increase, change, behavior, trend, mode, reduction, property, improvement	· observe unusual diffraction effects · observe significant morphological changes · observe a significant increase
(33)	hold	potential, promise	· hold great application potential · hold great promise · hold the prospect
(34)	induce	response, formation, change, growth, differentiation, charge, transition, stress, effect, transformation	· induce potent immune memory response · induce gel formation · induce surface charge · induce enhanced grain growth
(35)	highlight	importance, potential, role, versatility, point, opportunity, efficiency, difference, advantage	· highlight the importance · highlight the important engineering potential · highlight the prominent role · highlight the need
(36)	design	material, system, device, structure, composite, setup, sequence, platform	· design new electrode materials · design improved optical devices · design a synergistic antibacterial system · design composite structures
(37)	indicate	potential, formation, effect, transition, presence, possibility, interaction, change, ability	· indicate a great potential · indicate single phase formation · indicate the effectiveness · indicate the presence
(38)	solve	problem, issue	· solve the cracking problem · solve the environmental issues
(39)	harvest	energy, layer	· harvest higher frequency ocean wave energy · harvest the surface oxide layer

continued

No.	(Phrasal) Verbs	Noun collocates	Examples
(40)	suppress	growth, formation, recombination, response, reaction	· suppress the continuing growth · suppress the surface recombination · suppress Li dendrite formation · suppress the reverse reactions
(41)	summarize	result, finding, conclusion, remark	· summarize our PRP critical findings · summarize the analyses results · summarize the conclusion from this study
(42)	reveal	mechanism, structure, role, effect, potential, performance, stability, response, presence, distribution, change	· reveal stable atomic structures · reveal the crucial mechanistic role · reveal the distinguished electrochemical performance · reveal a rather complex mechanism
(43)	require	study, investigation, research, use, voltage, step, procedure, optimization, change	· require the use · require further investigation · require simple processing · require a slight change
(44)	confirm	presence, formation, result, validity, structure, stability, role, effectiveness, ability	· confirm the presence · confirm the combined effects · confirm the important role · confirm the deformation mechanisms
(45)	synthesize	composite, material, nanostructures, particle, sheets, film, structure, sample, polymer, graphene, compound	· synthesize highly dense porous carbon materials · synthesize such hybrid nanostructures · synthesize biocompatible amphiphilic hydrogel-solid particles
(46)	predict	property, behavior, value, structure, stress, strength, generation, effect, trend, shape, result, recovery, reaction	· predict tensile mechanical behaviors · predict accurately the values · predict the exact crystal structure · predict the tensile stress · predict the effective thermal conductivity

continued

No.	(Phrasal) Verbs	Noun collocates	Examples
(47)	perform	study, test, measurement, simulation, experiment, analysis, task, calculation, reaction, investigation	· perform optical measurements · perform a different sensing task · perform an in-depth experimental study · perform wet chemical reactions
(48)	lay (down)	foundation	· lay down the foundation · lay a solid foundation
(49)	employ	method, strategy, approach, technique, electrolyte, process, model, layer, film	· employ different standard test methods · employ microfluidic or 3D printing techniques · employ solid-state or ionic liquid electrolytes
(50)	influence	property, behavior, performance, quality, value, structure, stability, process, reaction, rate, formation, distribution, development	· influence the nanocomposite structure · influence data quality · influence the electrochemical performance · influence lubricating behaviors

Task 7: Identify the tenses, voices, and sentence patterns mostly used in the following sentences to brief research purposes and/or methods in research article conclusion sections.

(1) An in-depth study was performed to investigate the effects of aggregation behavior on the interaction between D-A polymers and F4TCNQ dopants to achieve efficient molecular doping in the solution state.

(2) We prepared a 3D ZnTe@Zn electrode by simple electrodeposition and in situ vapor–solid reaction method for rechargeable AZIBs.

(3) In summary, we proposed a novel strategy of laser plasmonic fabrication to assemble metal nanocrystals into interconnected network using glue molecule modulation.

(4) In this work, we have proposed a rainwater-based L–L TENG by investigating L–L CE in various L–L systems.

(5) To sum up, we carried out a detailed study on ten hybrid metal-halide crystals grown by solution-based methods using a series of transition metals (Cu^{2+}, Mn^{2+}, and Co^{2+}), while varying both the organic cation nature (alkyl and aryl) and the perovskite phase (RP and DJ).

(6) In summary, we report a thiourea competitive crystallization strategy to manipulate top-down growth process of FA-based perovskite in two-step sequential deposition method.

(7) Here, a multi-functional p-type 8TPAEPC with high carrier mobility of 3.12×10^{-3} cm^2 V^{-1} s^{-1} was synthesized and incorporated into the perovskite.

(8) Sodium local environments and dynamics in hydrated PEDOT:PSS mixed ionic-electronic conducting polymer films have been characterized under operando conditions by solid-state NMR spectroscopy.

(9) The liquid triboelectric properties of the test liquid were precisely measured using a validated liquid triboelectric properties measurement platform.

(10) We have developed an effective and scalable method for building low-density and site-controlled quantum emitters by utilizing FIB-induced luminescence quenching.

(11) The atomic arrangements of carbonates are clearly visualized in a (Cu,C)-1234 oxycarbonate superconductor by iDPC-STEM technique.

(12) We have provided a succinct but comprehensive overview of the properties of GDYs and their applications in semiconductor devices. The GDYs own natural bandgaps and high carrier mobilities.

(13) We have reported the photolithographic fabrication of oriented MOF micropatterns from oriented MOF films.

(14) In the Janus-IS hydrogel evaporator, we carefully design a stacking cation-selective permeable hydrogel with an anion-selective hydrogel, which removes the cations and anions, respectively, from the evaporation surfaces and establishes a built-in electrical field within the evaporator.

(15) By copolymerizing the monomers, the CF morphology was modulated to cluster-based CF with gradual conductance modulation, and a volatile nature was achieved by utilizing a high-density Al_2O_3 layer.

(16) The isolated manganese sites are primarily generated during synthesis via a simple impregnation, while in situ redispersion of few scattered MnO particles results in full atomic dispersion.

(17) In this study, an asymmetric Janus hydrogel bioadhesive was prepared via one-step fabrication by adjusting the interfacial distribution of free–COOH groups on the top/bottom surface.

(18) The asymmetric structure was systematically investigated through SEM, 3D contour, XPS, and optical microscopy.

(19) A solvent-induced method was used to fabricate a widely used polymer PCDTPT to a chiral structure.

(20) Owing to the broadband CD spectrum and excellent electrical conductivity, we designed chiroptical detectors for CPL detection.

(21) The wood was chemically pretreated to remove lignin and filled with PPy as a photothermal material. A Ni complex hygroscopic gel and a PI layer were introduced as hygroscopic and hydrophobic layers, respectively.

(22) We engineered a glucose-responsive nanosugar and demonstrated its potential as a highly biocompatible glucose-responsive insulin delivery platform for diabetes treatment using two distinct insulin-deficient diabetic mouse models.

(23) In this work, we report on a continuous flow process for the synthesis of functional core cross-linked polymeric micelles (CCPMs) allowing for precise control over the molecular properties of polymeric nanomedicines. The CCPMs are produced by self-assembly and cross-linking in two consecutive slit interdigital micromixers using chemo-selective disulfide bond formation of thiol-reactive polypept(o)ides (pSar-*b*-pCys(SO_2Et)) with functional cross-linkers.

(24) We investigated the realistic and statistical contact scaling behavior of MoS_2 FETs by proposing and using ACMs.

(25) In conclusion, we proposed a biomimetic structural design by incorporating gradient structure into Bouligand structure to improve the impact resistance of ceramic-polymer composites.

(26) In summary, we propose controlled isotropic canalization as a distinct top-down hierarchical strategy for microsized silicon-based lithium storage.

(27) In conclusion, in this work we screened for two-dimensional monolayers that can be exfoliated from experimentally known stoichiometric materials, including a new database of experimental structures (MPDS).

(28) AZO, a solution-based TCO, was used as an oxygen source for WSe_2 to change the chemical environment by systematically filling Se vacancies with oxygen. This process overturns the limited performance by eliminating the trap states acting as a scattering factor that hinders the electrical properties of FETs.

(29) We systematically investigated the catalytic abilities of a SnS_2 sheet on the electrochemical reduction of CO_2 to formate to elucidate the catalytically active sites and their catalytic mechanism.

(30) In summary, we used a droplet-microfluidic device as a tool to form protein nanoparticles with control over size and uniformity.

(31) Our combined experimental and theoretical study investigates the differences and correlations between the plasmon-enhanced ASPL and SHG processes under various ultrashort pulsed laser excitation conditions, including incident fluence, wavelength, and polarization.

(32) The successful synthesis was based on the decalcification–intercalation mechanism in a nonaqueous environment, revealing high-quality and high-uniformity 2D SiGe nanosheets with low-level hydrogenation.

(33) In summary, we describe here a controllable sustained feedstock release growth strategy that has successfully achieved SCBP film growth.

(34) In summary, in situ electrochemical BCDI was employed to probe the potential-dependent strain evolution of a single Pt nanoparticle.

Task 8: Discuss in groups the tenses, voices, and sentence patterns typically used in the following sentences to summarize the main results in research article conclusion sections.

(1) Through a simple arc discharge method, AlN:Ce hierarchical nanostructures have been successfully fabricated.

(2) By taking advantage of high-resolution cryo-TEM imaging, we were able to directly observe the corona and crystalline core of nanofibres in a vitrified solution.

(3) The findings presented here show that both the ammonium carbonate diffusion technique and the method of mixing $CaCl_2$ with $NaHCO_3$ to prepare $CaCO_3$ crystals in the presence of certain organic additives can eventually produce a morphology previously attributed to "mesocrystal" formation.

(4) Capitalizing on the incremental negative torsional stiffness induced by magnetic interactions in a rotational magnet pair, we have developed a motif for rotational bi/multistability.

(5) In summary, we have developed a solar-driven co-production of H_2 and value-add conductive polymer PANI that is 10 to 1000 times more valuable than the O_2 that is normally generated in conventional water splitting.

(6) We have demonstrated a phototransistor based on graphene/BHJ configuration that can perceive weak light intensity signals of less 10 nW cm^{-2} at visible to NIR region.

(7) We demonstrated a retina-inspired vision chip prepared by the phototransistor capable of in-chip processing visual images for light intensity range over six orders of magnitude ($10–10^7$ nW cm^{-2}).

(8) This work reveals that the key strategy for preventing charge injection, which generates noise currents in perovskite PDs, lies in the use of non-toxic Pb-free Sn-based perovskites.

(9) The PEEK/TiO_2/HA hybrid composite demonstrated significantly greater mechanical strength than the conventional PEEK owing to its complementary TiO_2 reinforcement and synergistic HA coating.

(10) In summary, we designed a novel and versatile shape memory ink system suitable for 4D printing at the macro- and microscale using DLP and DLW printing, respectively.

(11) To conclude, we have shown that turbulent and diffusive mass transfer regimes can be exploited in combination with stochastically distributed defects in the surface of robust engineering alloys to allow the electrochemical engraving of dendritic structures under the application of an electrical field.

(12) We have shown that logical data strings can be extracted from these patterns using computationally inexpensive feature extraction methods, and these strings can be rapidly cross correlated to discriminate one pattern from the next.

(13) We confirmed their successful synthesis and layered nature by XRD analysis.

(14) The intelligent incorporation of TzBI into PM6 skeleton as an A_2 unit for constructing terpolymers is found successful in improving the blend morphology and device performance of the PSCs.

(15) In summary, we present how peripheral fluorination on FePc molecules (FeF_8Pc and $FeF_{16}Pc$) impacts the structure and electronic properties of the commensurate monolayers they form atop Ag(110).

(16) We have, for the first time, successfully synthesized diketopyrrolopyrrole(DPP)-EDOT-based copolymers that exhibit all-electronic conductivities as high as $\approx$300 and $\approx$200 S cm^{-1} on being doped by $B(C_6F_5)_3$ and its Wheland complex.

(17) 2D metal carbides contacts demonstrate great improvements for photodetection as shown in MXene to semiconductor devices.

(18) It is demonstrated that this strategy is access to high-quality perovskite polycrystalline films with larger grains and fewer defects.

(19) It is concluded that the embedded thiourea will heal the interface lattice mismatch by forming 2D perovskite buffer layer in SnO_2/perovskite heterojunction.

(20) In conclusion, we have developed a stable, mixed shell polymeric micelle with cross-linked, in-shell EGCG and core-loaded with ciprofloxacin.

(21) The depth-dependent laser-induced conversion of organic precursor coatings has successfully been utilized to fabricate a complete flexible sensor architecture for selective CO_2 sensing at room temperature.

(22) Enabled by the unique set of advantages of the BC-CPH, we 3D print monolithic all-hydrogel bioelectronic interfaces capable of long-term high-efficacy electrophysiological stimulation and recording of diverse tissues and organs in rat models.

(23) Our work suggests that the electro-thermal approach could be harnessed to yield ferroelectric polymer nanocomposites featuring exceptionally high actuation performance in terms of strain, mechanical energy density, electromechanical coupling factor, actuation strength and fatigue durability.

(24) As a proof of concept, we demonstrate cell-nascent protein interactions in three dimensions as an area of growing interest, especially with respect to the dynamic reciprocity of cell-matrix signalling and its effects on mechanosensing and differentiation.

(25) In summary, by taking advantage of high-resolution cryo-TEM imaging, we were able to directly observe the corona and crystalline core of nanofibres in a vitrified solution.

(26) In summary, using the polymer-thermal-decomposition-assisted melting strategy methodology, we report the simple preparation of a type of glass foam, a_{gf}ZIF-62, combining the structural advantages of MOF glass and foam.

(27) In summary, a material that contained Ru sites with a low oxidation state successfully stabilized in a solid catalyst is developed, enabling CO_2 activation under mild conditions.

(28) We have experimentally demonstrated that gate voltages affect both the IX PL intensity and

its lifetime, indicating non-radiative decay rate modulation by a correlated electronic phase in a moiré superlattice.

(29) We show that the WSe_2/MoS_2 heterostructure exhibits a rich layer-dependent correlated electronic phase, including single-layer, layer-selective, interlayer-excitonic and layer-hybridized phases.

(30) In conclusion, we show prominent effects of electronic interactions in Γ-valley moiré bands over a wide range of parameter space in $tdWSe_2$.

(31) Through the design and synthesis of new polymers comprising linear alkyl linkers and TADF units, we have imparted stretchability onto a light-emitting polymer that can break the 5% EQE limit only from singlet emission, thereby achieving a record EL efficiency for intrinsically stretchable emitters.

(32) By using a higher NA objective, better detectors and optimized samples, both the sensitivity and spatial resolution can be improved several times.

(33) In conclusion, we realized in situ magnetic detection of magnetic materials using an implanted V_{Si} defect ensemble in SiC-based anvil cells under high pressure.

(34) We demonstrate that the sensitivity of single-atom vibrational spectroscopy in a STEM can reach the chemical-bond level.

(35) Our combined experimental and theoretical results unambiguously demonstrate that the defect-induced local perturbations of the pristine-graphene vibrational modes are extremely sensitive to the chemical-bonding configurations and atomic mass.

(36) We have demonstrated STM-induced excitonic luminescence nanoscopy of an atomically resolved vdW heterostructure featuring short-period moiré superlattices.

(37) We directly reveal how the nanoscale environment influences the luminescence characteristics, leading to sizeable excitonic energy shifts and the emergence of emission from charged and localized excitons on areas separated only by a few nanometres.

(38) Halide ions/vacancies and their subsequent migration in polycrystalline MHPs have been shown to be the most consequential ion migration in MHPs.

(39) Our results show that small volume and large GB diffusivities are a necessary combination for achieving stable and hysteresis-free MHP devices, requiring strong GBs that confine fast ion motion to the GB.

(40) A liquid triboelectric series was first established based on liquid-solid contact electrification.

(41) Our results demonstrate that these gains are obtained by the NPM effect without degradation in the electrical properties of membranes.

(42) We have presented an ion-transfer engineering via a Janus-IS hydrogel solar evaporator.

(43) We developed a compact and energy-efficient memristor-based ReLU activation neuron device that can solve the chronic vanishing gradient problem in advanced DNN.

(44) In summary, low-valent manganese atoms stabilized on CeO_2 are revealed as the first stable catalyst for NH_3 oxidation to N_2O, rivaling state-of-the-art Au/CeO_2 in terms of selectivity

and reaching a two-fold higher N_2O productivity.

(45) The complexity and organization of mSPs are similar or identical to those observed in viruses, which led us to conclude that the nanoscale compartments can effectively protect the cargo against heat and other environmental factors representing one of the biggest bottlenecks for deployment of vaccines.

(46) We have introduced the concept of the SPIRIT printing strategy based on the MB bioink, which was demonstrated to serve as both an excellent bioink and a suspension medium simultaneously for embedded bioprinting.

(47) In summary, we have designed and assembled a hygrothermic wood actuator.

(48) In summary, the PULSED microparticle fabrication method, in combination with an adapted syringe pump filling method, enables the scalable production of small, fully biodegradable microparticles that exhibit pulsatile release after a customizable, material-dependent delay.

(49) In this work, PET incorporated with bioderived diphenolic TPA counits was successfully synthesized by a two-step polycondensation reaction.

(50) FTIR analysis revealed that trend of DHTE loading in the copolymers was consistent with the change in ratio of absorption intensity, and the phenolic group reacted with hydroxy ester group of the monomer and polymer chain to form the permanent covalent network.

(51) Overall, the highly biocompatible $PG_{EDA\text{-}FPBA}$ NPs provide an excellent platform for the assembly and transport of insulin molecules by several mechanisms, enabling control over the pharmacokinetic properties of the released insulin.

(52) We have developed a strategy for bulk RNA labeling and SG dynamics imaging in live cells.

(53) We discovered that the thermodynamically immiscible water/EmimFSI system would experience spontaneous miscibility transition to form a homogeneous IL-AE with the introduction of $Zn(OTf)_2$.

(54) The experimental observation confirms that contacts with small contact lengths can have similar contact resistances to large contact lengths.

(55) The GB composite exhibits a compact ceramic front layer to directly resist impact force and a tough polymer-dominated basal layer to maximally absorb residual impact energy.

(56) This work demonstrates a sophisticated biomimetic model design of superhigh-sulfur-content cathode material with a unique topology structure at the molecular level, which is of great significance to developing Li–S cells.

(57) In summary, we have developed a colorless transparent PU elastomer with extremely high strength and toughness, the unprecedented crack tolerance, and biodegradability.

(58) In summary, a feasible in situ cation-exchange approach of PQDs was demonstrated to tailor the optoelectronic properties and stoichiometries of $FA_xCs_{1-x}PbI_3$ PQDs for high-performing solar cells.

(59) We revealed that PBI coordinated to the surface of ZnS NC is displaced quasi-reversibly by visible-light irradiation.

(60) We have demonstrated an approach for designing the supercool materials that cut off NIR light from the sunlight to achieve thermal shielding.

Task 9: Skim through the Abstract and Conclusion sections of Text Ⅱ and identify their overlapping and distinctive types of information so that to understand the more affirming and reflective nature of the Conclusion section.

Text Ⅱ

High-Resolution Cryo-Electron Microscopy Structure of Block Copolymer Nanofibres with a Crystalline Core[2]

Abstract: Seeded growth of crystallizable block copolymers and π-stacking molecular amphiphiles in solution using living crystallization-driven self-assembly is an emerging route to fabricate uniform one-dimensional and two-dimensional core–shell micellar nanoparticles of controlled size with a range of potential applications. Although experimental evidence indicates that the crystalline core of these nanomaterials is highly ordered, a direct observation of their crystal lattice has not been successful. Here we report the high-resolution cryo-transmission electron microscopy studies of vitrified solutions of nanofibres made from a crystalline core of poly(ferrocenyldimethylsilane) (PFS) and a corona of polysiloxane grafted with 4-vinylpyridine groups. These studies show that poly(ferrocenyldimethylsilane) chains pack in an 8-nm-diameter core lattice with two-dimensional pseudo-hexagonal symmetry that is coated by a 27 nm 4-vinylpyridine corona with a 3.5 nm distance between each 4-vinylpyridine strand. We combine this structural information with a molecular modelling analysis to propose a detailed molecular model for solvated poly(ferrocenyldimethylsilane)-*b*-4-vinylpyridine nanofibres.

INTRODUCTION

1 The solution self-assembly of crystallizable polymeric or π-stacking molecular amphiphiles is attracting increased attention as a route to fabricate one-dimensional nanofibres and two-dimensional (2D) platelet micelles and more complex morphologies with a range of applications that include therapeutics[1-4], catalysis[5-6] and nanoelectronics[7-10]. In particular, the ambient-temperature seeded-growth approach termed living crystallization-driven self-assembly (CDSA) allows dimensional control and access to low-dispersity samples of one-dimensional and 2D core–shell nanoparticles with predetermined dimensions and analogous processes can be applied to supramolecular polymers[11-23]. Studies of the growth mechanism

have indicated that the formation of a highly ordered crystalline core occurs by the epitaxial deposition of the amphiphile in a molecularly dissolved unimer state at the seed termini[23]. Although scattering methods have been used to evidence core crystallinity in the resulting nanoparticles[21-22], including field-aligned samples[24], direct observation of the crystalline core using high-resolution microscopy has not been achieved.

2 A theoretical framework for understanding the morphology of block copolymer micelles with a crystalline core was originally proposed by Vilgis and Halperin[25]. Their model was centred on the well-known lamellar morphology for 2D polymer single crystals, and for one-dimensional fibre-like micelles, a rectangular cross section was proposed in which chain folding of the core-forming block influences the spacing between coronal chains. However, although the presence of a rectangular cross section has been evidenced for block copolymer nanofibres with crystalline polycarbonate[26], polyfluorene[27] and polythiophene[28] cores, for the case of poly(ferrocenyldimethylsilane) (PFS), studies have indicated an oval-shaped cross section at low-to-moderate degrees of polymerization based on transmission electron microscopy (TEM) cross-sectional analysis[29-30].

3 In this work, we have focused on nanofibres with a crystalline PFS core as the presence of highly scattering electron-rich iron centres provides excellent TEM contrast[31]. We previously reported studies of an electric-field-aligned sample of low-dispersity PFS-*b*-polyisoprene fibre-like micelles in solution, which were grown from PFS-*b*-polydimethylsiloxane seeds[24]. Alignment was essential to study the orientation of the PFS polymer chains in the crystalline core with respect to the direction of the nanofibre long axis. Synchrotron small-angle and wide-angle X-ray scattering provided data that suggested single-crystal-like order in the PFS core with folded chains perpendicular to the nanofibre long axis. However, a direct observation of the crystalline core was not possible. Moreover, although the swollen corona chains play a crucial role in providing colloidal stability and also in the fibre fragmentation phenomena[32-33], limited structural information on their arrangement has been obtained to date.

4 TEM has taken an increasingly prominent role in materials science with the development and widespread adoption of cryogenic transmission electron microscopy (cryo-TEM), where the solution-phase sample is imaged in a vitrified near-native state and is therefore free from drying artifacts[34-36]. Cryo-TEM offers direct physical characterization and provides a real-space image or description of the specific physical form of individual nanostructures in the sample. This method has been previously used to detect the relatively large lattice spacings for nanofibres with smectic liquid crystalline cores (>4.2 nm)[37-38]. In the present study, we have investigated nanofibres with a crystalline PFS core by cryo-TEM in a vitrified solution. Crucially, the high-resolution cryo-TEM study allowed for the direct observation of both solvated corona chains and crystalline nanofibre core (*d* spacing, around 0.65 nm) in solution without the interference of an adhering surface. We use the phrase 'high-resolution cryo-

TEM' to emphasize the clear observation of the core lattice and the in situ morphology of the corona chains of the PFS_{24}-*b*-$P4VP_{192}$ nanofibres. This was achieved by the combination of a field-emission-gun cryo-TEM with a direct electron detector and was not possible using a conventional thermionic-source-based TEM with a charged-coupled device camera in the dry state or cryo-TEM in solution.

Nanofibre preparation and characterization

5 The amphiphilic block copolymer used in this work contained a crystallizable PFS core-forming block and a polysiloxane-based corona-forming block with 4-vinylpyridine (4VP) groups grafted to each repeat unit and is represented for simplicity as PFS_{24}-*b*-$P4VP_{192}$ (Figure 1a). This material was prepared by the previously reported sequential anionic polymerization methods followed by a thiol–ene 'click' reaction to attach the 4VP groups[6]. The PFS_{24} and $P4VP_{173}$ homopolymers were also prepared as controls according to reported methods[6], [39]. Nanofibre seeds of PFS_{24}-*b*-$P4VP_{192}$ (number-averaged length $L_n = 29$ nm and length polydispersity $Đ_L = 1.09$ obtained by TEM) were prepared by the self-assembly of the PFS_{24}-*b*-$P4VP_{192}$ block copolymer in isopropanol under homogeneous nucleation conditions to afford long polydisperse nanofibres with a crystalline PFS core and a P4VP corona followed by sonication according to previously reported methods[6]. Using the seeded-growth living CDSA process (Figure 1b), low-length-dispersity nanofibres were formed at ambient temperature with predictable and controllable lengths ranging from 199 to 1,096 nm (by TEM analysis), which were dependent on the ratio of the added PFS_{24}-*b*-$P4VP_{192}$ to the seeds. Low- and high-magnification TEM images of the PFS_{24}-*b*-$P4VP_{192}$ nanofibres ($L_n = 759$ nm, $Đ_L = 1.03$) are shown in Fig. 1c–f. In the dry state, only the high-contrast PFS core of the nanofibres was detected by TEM as the low-contrast corona could not be distinguished from the carbon film support (Figure 1c,d). We further studied the cryo-TEM of the nanofibres in a vitrified solution of isopropanol at ~90 K (Figure 1e,f). The core and corona could be readily distinguished by high-magnification cryo-TEM of a vitrified solution without the interference of the supported carbon film as a substrate (Figure 1f). The height (~8 nm) and overall width (~83 nm) of the nanofibres were determined by atomic force microscopy height images. In the former case, the major contributor is the core, whereas in the latter case, both nanofibre core and corona contribute to the measurement. We also used cryo-TEM to analyse the cross-sectional shape of the PFS core by imaging the PFS_{24}-*b*-$P4VP_{192}$ seed nanofibres. The examination of examples that were oriented along the axis of the electron beam indicated the presence of an oval-shaped cross section for the core domain with a major axis of around 9 nm and a minor axis of around 8 nm.

High-resolution observation of the nanofibre corona by cryo-TEM

6 In this study, we utilized a Talos Arctica cryo-TEM (FEI) operating at 200 kV and equipped with a K2 direct electron detector and BioQuantum energy filter for the high-resolution cryo-TEM images. Sample images were energy filtered ($\Delta E = 20$ eV) and recorded at a total dose

of 13 e^{-} Å^{-2}. The PFS$_{24}$-*b*-P4VP$_{192}$ nanofibres ($L_n = 759$ nm, $Đ_L = 1.03$) in an isopropanol solution were studied. Compared with other block copolymer nanofibre systems such as worm-like micelles that generally have packing reminiscent of spherical micelles with high curvature at their termini, the PFS$_{24}$-*b*-P4VP$_{192}$ nanofibre termini displayed a very different morphology. The relatively exposed and apparently planar crystalline PFS core at the terminus presumably facilitates the living CDSA behaviour in solution where unimer addition is relatively unimpeded.

7 The substantial contrast between the PFS core and P4VP corona under cryo-TEM conditions allowed the core–corona interface to be readily distinguished (Figure 2a). A model was built to demonstrate the side view of the corona and core domains of the nanofibre of PFS$_{24}$-*b*-P4VP$_{192}$. In a typical measurement, the width between the two edges of the corona domains was found to be around 62 nm, and the width of the core domain was found to be around 8 nm. This gave a maximum extended length of 27 nm for the corona chains attached to the core domain in the vitrified solution (Figure 2a). We created a model to simulate this length in Materials Studio 7.0 and the simulated length of 27 nm for the corona chains showed excellent matching with the average experimental length (27 ± 4 nm; Fig. 2b,c). The modelling method resulted in a helical structure for the corona chains (Figure 2d).

8 To provide further insights, we studied the arrangement of the corona chains of PFS$_{24}$-*b*-P4VP$_{192}$ nanofibres in more detail (Figure 2e). As the intrinsic contrast of the corona domain was relatively low, a fast Fourier transform (FFT) was used to convert the image into an FFT pattern (Figure 2e, inset), which was filtered by applying a spot mask in the DigitalMicrograph software (v. 3.43). A pattern for the selected area was generated by an inverse FFT (Figure 2f). A profile of an intensity analysis of the inverse FFT image is shown in Fig. 2g. The average distance between the two strands of the P4VP corona chains was determined to be 3.5 nm (Figure 2c).

Characterization of the nanofibre core

9 An analysis of a solution of the PFS$_{24}$-*b*-P4VP$_{192}$ nanofibres ($L_n = 759$ nm, $Đ_L = 1.03$) in an isopropanol solution (80 mg ml^{-1}) by wide-angle X-ray diffraction revealed a strong peak located at $2\theta = 13.7°$ (corresponding to a lattice spacing of $d = 0.65$ nm), which represents the d spacing of the lattice of the PFS core domain[24]. We found that the solid-state XRD analysis of a nanofibre film cast from isopropanol on a silicon wafer substrate showed a similar peak. In addition, a new broad peak was present at $2\theta = 6.64°$ ($d = 1.33$ nm), which is presumably associated with the adjacent pyridine-to-pyridine distance along the helix of corona chains, as shown in the modelling of the P4VP corona. This broad peak was observable at a relatively high concentration (400 mg ml^{-1}) of PFS$_{24}$-*b*-P4VP$_{192}$ nanofibres and was not detectable at a low concentration (80 mg ml^{-1}), presumably due to the low scattering signal of the helical structure of the P4VP segment. The wide-angle X-ray diffraction profile of a powdered sample of PFS$_{24}$-*b*-P4VP$_{192}$ also showed the presence of three peaks but with a much weaker

peak intensity at 13.7°, indicative of a lower degree of crystallinity of PFS in the bulk state. A very broad but more intense peak was also observed at $2\theta = 19.3°$ ($d = 0.46$ nm), which was assigned to the π–π-stacking distance of pyridine rings of the P4VP corona. The wide-angle X-ray diffraction profiles of the individual PFS_{24} and $P4VP_{173}$ homopolymers showed a sharp peak at $2\theta = 13.9°$ ($d = 0.65$ nm) and a broad peak at $2\theta = 20.1°$ ($d = 0.44$ nm), respectively, consistent with the proposed assignments.

10 In high-resolution cryo-TEM experiments, the PFS_{24}-*b*-$P4VP_{192}$ nanofibres ($L_n = 759$ nm, $Đ_L = 1.03$) within the vitrified isopropanol solvent possessed random orientations relative to the incident electron beam. In a small percentage (around 5%) of the cases, high-resolution cryo-TEM revealed a clear hexagonal lattice pattern for the crystalline PFS core domain (Figure 3a). However, the core lattice was not observed using a conventional cryo-TEM technique. The boundary of the PFS core domain was distinct as the electron density contrast with the corona domain was substantial[31]. The lattice spacings for the PFS core were determined to be 6.5–6.6 Å. The angle between the two lattice planes was measured to be around 60°, which indicated that high symmetry was present for the crystalline PFS core of the PFS_{24}-*b*-$P4VP_{192}$ nanofibres (Figure 3b). The lateral displacement of the core-lattice pattern outside the crystalline region is caused by a delocalization effect from the spherical aberration of the objective lens or the inelastic scattering of incident electrons that can be eliminated by an aberration correction, as shown in other studies[42-44]. By taking advantage of the high-resolution images (Figure 3a,b), the FFT of the data (Figure 3c) revealed that the PFS chains pack in a lattice with 2D pseudo-hexagonal symmetry perpendicular to the long axis of the nanofibre core with *d* spacings of 6.5–6.6 Å.

11 A particularly useful study to facilitate the interpretation of these results involves packing within the crystal structure of the linear PFS pentamer, which also takes place in a pseudo-hexagonal fashion with three assigned *d* spacings of 5.90, 6.28 and 6.58 Å for planes with Miller indices of [1–2 0], [0 1 1] and [1–1 1], respectively[45]. The *d* spacings observed for the PFS_{24}-*b*-$P4VP_{192}$ nanofibres are similar to those detected for the linear pentamer[45] as well as for the previously studied polyisoprene-*b*-PFS nanofibres (6.29 and 6.43 Å)[24]. However, the structure of the PFS_{24}-*b*-$P4VP_{192}$ nanofibre core possesses relatively higher symmetry with almost identical *d* spacings (6.5–6.6 Å), which can be assigned to interchain packing distances. The intermolecular distance between the adjacent PFS core-forming chains was calculated to be 7.6 Å and the maximum axial length of the lattice inside the PFS core domain was measured to be around 9 nm (Figure 3b). This corresponds to around 12 PFS chains between the two edges of the crystalline core domain. The FFT image in Figure 3b shows a clear hexagonal pattern (Figure 3c), which corresponds to a hexagonal lattice packing of the core-forming PFS polymer, as detected by X-ray scattering[24]. The highly symmetric pattern shown in Figure 3c also reflected the higher-symmetry packing in the PFS core compared with that observed in the pentamer of ferrocenyldimethylsilane[45]. A simulated pattern of the

core domain of PFS_{24}-*b*-$P4VP_{192}$ was generated using an inverse FFT from the selected FFT pattern, which was filtered by applying an array mask. Based on the parameters obtained from the cryo-TEM studies, we were able to propose a detailed structural model for the cross section perpendicular to the long axis of the PFS_{24}-*b*-$P4VP_{192}$ nanofibres (Figure 3d,e). The adjacent strand-to-strand distance of the corona (3.5 nm) and the distance between the adjacent PFS core-forming chains (7.6 Å) are labelled in the proposed model. This indicated that one P4VP corona chain emanates for every 4–5 parallel PFS chains in the core perpendicular to the nanofibre growth direction (Figure 3e).

12 The extended chain length of the PFS_{24} core chains is around 16 nm as the intrachain distance between the Fe centres of neighbouring ferrocenes is 0.69 nm (ref. 45). Based on the nanofibre height of around 8 nm based on atomic force microscopy and core cross-section measurements by cryo-TEM, one chain fold per PFS_{24} chain would be anticipated. We, therefore, proposed a molecular model to demonstrate the core chain folding present in the nanofibres (Figure 4). Based on this model in which coronal chains can emerge from either of the two faces on opposite sides of the core perpendicular to the nanofibre long axis, a brush density of one P4VP corona strand for every four parallel PFS chains is predicted. This is consistent with that calculated based on the corona chain modelling spacing (3.5 nm) and PFS chain separation (7.6 Å) (Figure 3e). Furthermore, the unsymmetrical nature of the PFS core in terms of dimensions (height, ~9 nm; width, ~8 nm; Figure 4) together with the presence of both chain folds and emanating coronal chains at the two opposite faces would be expected to contribute to the observed non-rectangular, oval-shaped core cross-section[29-30].

Conclusion

13 In summary, by taking advantage of high-resolution cryo-TEM imaging, we were able to directly observe the corona and crystalline core of nanofibres in a vitrified solution. We determined that the PFS chains in the core pack in a lattice with 2D pseudo-hexagonal symmetry perpendicular to the long axis of the nanofibre core. With the knowledge of the corona and core structures, we were able to propose a detailed model for the PFS_{24}-*b*-$P4VP_{192}$ nanofibres at the molecular level where coronal chains emanate from opposite faces with a brush density of around one P4VP strand for every four parallel PFS chains. These results provide a detailed insight into the structure of the nanofibres with crystalline cores for the first time, to the best of our knowledge. The high structural order present in the core revealed here by cryo-TEM is likely to be replicated in analogous fibre-like assemblies accessible via living CDSA as a result of the epitaxial growth mechanism. For example, block copolymer nanofibres with crystalline π-conjugated polyfluorene or polythiophene cores exhibit exceptional exciton diffusion lengths, which has been postulated to be the result of the structural order associated with their preparation via living CDSA[8], [27]. Future work will aim to extend analogous studies to other nanofibre systems to provide structural comparisons and key insights into the properties underlying their potential applications.

(Figures omitted)

Methods

...

References

...

Task 10: Read Text Ⅱ carefully and answer the following questions with specifics and examples provided in the text.

(1) Why has increasing attention been paid to the solution self-assembly of crystallizable block copolymers or π-stacking molecular amphiphiles along with the growth mechanism?

____

(2) What research concerning the crystalline core has not been successfully conducted according to paragraph 1?

____

(3) What efforts have been made in exploring the morphology of block copolymer micelles with a crystalline core, and what research gap still exists in this field?

____

(4) Why did this study focus on nanofibres with a crystalline PFS core and adopt cryo-TEM?

____

(5) What is the specific purpose and what is the overall design of this research according to the Introduction section?

____

(6) What steps have been taken to prepare and characterize the PFS_{24}-*b*-$P4VP_{192}$ nanofibres?

____

(7) What has been revealed by the cross-sectional shape analysis of the PFS core using cryo-TEM?

____

(8) What has made it possible to clearly identify the interface between the PFS core and P4VP corona under cryo-TEM conditions?

____

(9) What could be learned about the width and length of the corona and core domains of the PFS_{24}-*b*-$P4VP_{192}$ nanofibres by the simulated modelling?

____

(10) What could be observed and presumed from the wide-angle X-ray diffraction profiles of a solution of the PFS_{24}-*b*-$P4VP_{192}$ nanofibres, a powdered sample of PFS_{24}-*b*-$P4VP_{192}$, and the individual PFS_{24} and $P4VP_{173}$ homopolymers?

____

(11) What property did the PFS_{24}-*b*-$P4VP_{192}$ nanofibres exhibit in high-resolution cryo-TEM experiments?

____

(12) What study was further conducted to help interpret the results obtained by the wide-angle X-ray diffraction and high-resolution cryo-TEM experiments?

____

(13) What structural model was proposed for the cross section perpendicular to the long axis of the PFS_{24}-*b*-$P4VP_{192}$ nanofibres?

____

(14) What molecular model was put forward to demonstrate the core chain folding present in the nanofibres?

____

(15) What specific purpose does each citation in paragraphs 5-12 serve?

____

(16) According to the Conclusion section, did this study successfully address the research gap identified in the Introduction section?

____

(17) Based on specifics in which paragraphs are the 2nd and 3rd sentences in the Conclusion section claimed?

____

(18) How do the authors evaluate the results and what implications could be drawn according to the Conclusion section?

____

(19) What directions for future research are suggested at last?

____

(20) What are the possible reasons for the tense shifting in the Conclusion section?

____

Task 11: Read the following excerpt of the Conclusion section from Text II, in which complex noun phrases are already underlined together with node words bolded, and adverbials are placed in parentheses in either main or subordinate clauses. Discuss with each other the phrasal complexity, typically reflected in the use of pre- and post-modifiers in noun phrases, and syntactic complexity, typically manifested by adverbials in various forms in the other research article conclusion sections below.

Excerpt

(In summary), (by taking advantage of high-resolution cryo-TEM imaging), we were able to (directly) observe the corona and crystalline **core** of nanofibres in a vitrified solution. We determined that the PFS **chains** in the core pack (in a **lattice** with 2D pseudo-hexagonal symmetry perpendicular to the long axis of the nanofibre core). (With the knowledge of the corona and core structures), we were able to propose a detailed **model** for the PFS_{24}-*b*-$P4VP_{192}$ nanofibres at the molecular level where coronal **chains** emanate (from opposite faces) (with a brush density of around one P4VP strand) (for every four parallel PFS chains). These results provide a detailed **insight** into the structure of the nanofibres with crystalline cores (for the first time), (to the best of our knowledge). The high structural **order** present in the core revealed (here) (by cryo-TEM) is likely to be replicated (in analogous fibre-like assemblies accessible via living CDSA) (as a result of the epitaxial growth mechanism).

I

Low-oxidation-state Ru sites stabilized in carbon-doped RuO_2 with low-temperature CO_2 activation to yield methane[3]

Conclusion

In summary, a material that contained Ru sites with a low oxidation state successfully stabilized in a solid catalyst is developed, enabling CO_2 activation under mild conditions. This is achieved by doping RuO_2 with C atoms at interstitial positions of the crystal lattice, using a green and mild hydrothermal synthesis strategy. This work opens up alternative perspectives for the design of catalysts based on transition metal oxides by manipulating interstitial atoms in the metal oxide structure. The application of this C-doped RuO_2 material to energy-demanding processes, such as CO_2 hydrogenation to form methane at a low temperature as presented here, opens up a promising sustainable alternative to the current processes that operate at higher temperatures.

II

High-efficiency stretchable light-emitting polymers from thermally activated delayed fluorescence[4]

Conclusion

Through the design and synthesis of new polymers comprising linear alkyl linkers and TADF units, we have imparted stretchability onto a light-emitting polymer that can break the 5% EQE limit only from singlet emission, thereby achieving a record EL efficiency for intrinsically stretchable emitters. Our systematic experimental characterization and computational simulations show that longer alkyl linkers (in the range of ten carbon units) provide more effective strain dissipation capabilities than short linkers and thus higher stretchability, without sacrificing EL performance. As demonstrated in the fully stretchable OLED devices presented here, we anticipate that the successful development of stretchable TADF polymers offers considerable potential for applications and a technologically feasible path for stretchable optoelectronic devices for use in human-interactive applications.

Ⅲ

Directional self-assembly of facet-aligned organic hierarchical super-heterostructures for spatially resolved photonic barcodes[5]

Conclusion

In summary, a versatile and straightforward strategy for the directional self-assembly of organic epitaxial architectures without complex surface functionalization has been reported, where well-defined super-heterostructures were assembled effectively via directional attraction between different building particles. These super-heterostructures featuring spatially resolved emissive colors and identifiable graphical patterns constitute the distinctive photonic barcodes. A library of super-heterostructures is further acquired through modulation of building blocks with specific color and morphological characteristics, allowing for construction of abundant barcodes with enlarged encoding capacity. These results provide inspiration for the controllable construction of hierarchical microstructures toward anticounterfeiting applications.

Task 12: Familiarize yourself with the two exemplary research article conclusion sections, in which bolded noun phrases typically refer to present or previous studies, underlined italic noun phrases refer to research methods while underlined bold noun phrases to research results in present or previous studies. Identify the noun phrases in the following conclusion excerpts fulfilling each of the abovementioned three functions and proving a link with additional information inside or outside the texts.

Example (1)

The findings presented here show that both *the ammonium carbonate diffusion technique* and *the method* of mixing $CaCl_2$ with $NaHCO_3$ to prepare $CaCO_3$ crystals in the presence of certain organic additives can eventually produce a morphology previously attributed to "mesocrystal" formation, in agreement with **other recent work**. Here, by comparing **the outcomes** from *both methods* with those from *the in situ AFM observations*, we have directly translated **the resulting morphology** from the nanoscale to the microscale and shown how it arises through **completely classical growth mechanisms**. Although we have demonstrated this using the specific case of PSS-modified calcite, **the results** may serve as a more general basis for extending **classical growth mechanisms** to the formation of other complex crystal morphologies. In addition, they re-emphasize that proposals of nonclassical pathways based solely on *observations of* crystal morphology and surface roughness must be carefully considered.

Example (2)

In **this study**, **a coupled model** involving **electromagnetic, thermal, transformation and mechanical model** for the entire induction heat treatment was investigated. Induction hardening and subsequent tempering of a shaft part was simulated based on **the proposed models**. *The corresponding experiments* were conducted to validate **the simulation results**. In addition, **the mechanism** of tempering relaxation was studied by *simulation*. **The following conclusions** of **this study** were drawn. 1) A uniformly hardened layer was obtained on the shaft part after **the proposed induction hardening process**. The quenched residual stress of the surface was compressive and the value of the radial residual stress much less than in the axial and tangential. 2) Residual stress decreased with increased tempering temperature. For SAE 4140H steel, C-atom segregation and precipitation of metastable ε-carbide had little effect on volumetric shrinkage. The quenched residual stress was found to be maintained after 200 °C tempering, but reduced by 13.9 and 92.2% after 400 and 600 °C tempering, respectively. 3) Both volumetric shrinkage, due to phase transformation, and creep have effects on tempering stress relaxation. With increased tempering temperature, the influence of transformation on stress relaxation decreased and creep became **the major relaxation mechanism**. 4) Considering the good agreement observed here of **the simulation results** with **the experimental results**, **the conclusion** was that **the proposed model** could be used to predict temperature, microstructure, and stress evolution during induction hardening and tempering.

I

Solar-driven co-production of hydrogen and value-add conductive polyaniline polymer6[6]

Conclusion

In summary, we have developed a solar-driven co-production of H_2 and value-add conductive polymer PANI that is 10 to 1000 times more valuable than the O_2 that is normally generated in conventional water splitting. The lower operational bias means one single junction perovskite solar cell can be used for H_2 production while inhibiting any oxygen evolution reaction at the anode. The product PANI is solid that is easily harvestable and can be electroplated onto surface of interest. Therefore, the process does not require the use of a separation membrane. The polymerization being a proton releasing process eliminates the need for acid replenishment. In addition, the process is auto-catalytic and self-regulates when driven by solar. All these benefits reduce both the cost and complexity of the process significantly. The ANI polymerization process has a high-power utilization ratio of 93%, while the Faradaic efficiency for H_2 production is 98.6 ± 3.9%. Last, the use of low-cost earth-abundant CoP-Ni foam replacing Pt as counter electrode for the process was demonstrated. This work is an advancement for low-cost solar-driven production of H_2 incorporating simultaneous high value product co-generation in one process accelerating development for rapid transformation to clean H_2 economy.

II

Lithographically patterned functional polymer–graphene hybrids for nanoscale electronics[7]

Conclusion

In conclusion, we have demonstrated a scalable and precise approach for fabricating hybrid polymer–graphene nanoscale devices. Beyond graphene, the ability to dope other 2D materials—including semiconductors such as transition-metal dichalcogenides and phosphorene, among others—in a controlled manner can be pivotal for the development of nanoscale optoelectronic devices. The patterning and synthetic methods developed in this work can be extended more generally to other 2D materials and, in conjunction with polymer dielectric substrates, could offer a path towards low-power, flexible 2D-materials-based electronics.

Ⅲ

Modeling and characterization of crystallization during rapid heat cycle molding[8]

Conclusion

This paper presented a novel method for determining crystallization information during the RHCM process, and the influence of the temperature and shear rate on crystallization were both considered in this method. The non-isothermal temperature field was first converted into several isothermal processes using mathematical ideas, and then the isothermal crystallization parameters were obtained through experiments. Finally, WXRD measurement results were employed to verify the calculated crystallization results. According to the results obtained in this study, the following conclusions can be drawn. (1) The non-isothermal RHCM process can be assumed to be a superposition of finite isothermal processes. (2) A greater degree of supercooling of polymers results in easier nucleation, but spherulite size and growth rate both increase with increasing temperature. (3) The crystallization process in the skin layer is much faster than that in the core layer. At each measurement location, a higher mold wall temperature induces the slowest crystallization speed, yet results in the highest degree of crystallinity in the end. (4) The crystallization in the skin and shear layers during the RHCM process is a typical non-isothermal shear crystallization behavior, and the shear effect cannot be ignored. The core layer is dominated by quiescent non-isothermal crystallization. (5) The proposed method can determine crystallization information during the RHCM process, and its calculated results agree well with polarizing microscope results and WXRD results. In general, the proposed method is accurate and effective. It is a potential candidate technology for determining crystallization evolution information during the RHCM process and thereby optimizing the molding process for manufacturing high-quality products.

Ⅳ

Influence of pulse energy on machining characteristics in laser induced plasma micro-machining[9]

Conclusion

In this paper, a simulation and experimental study was conducted to investigate the influence of pulse energy on machining characteristics during the LIPMM process. An axisymmetric model combining the effects of cascade and multiphoton ionization, and diffusion and recombination losses was developed to simulate the plasma's spatial and temporal density and distribution at various pulse energies. Laser interaction and propagation through the plasma were also taken into account in the developed numerical model. Thereafter, micro-channels on stainless steel using LIPMM with various pulse energies were generated, and the plasma focusing process, depth and width of the machined channels were investigated. The experimental results presented a very close correlation between the simulated plasma evolution and density distribution. The conclusions can be highlighted as follows. 1) An energy threshold of approximately 2.92 μJ is determined to induce optical breakdown resulting in plasma formation. The plasma begins to form near the laser focal plane with a super-threshold pulse energy, and then shifts toward the direction of the incident laser due to the plasma's shielding property. Subsequently, the plasma tends to shirk with further lapse of time resulting from the decreasing laser irradiance. With the increase of pulse energy, a similar plasma evolution process was observed. However, the plasma forms at an earlier stage and plasma size is larger. 2) The machined depth increases with an increase in focusing distance up to $z = 250$ μm for different pulse energies. Simulation results reveal that free electron density peaks occur around 150–180 μm above the laser focal plane, which is approximately 100 μm under the plasma focal plane. The maximal electron density at the plasma front moves toward the direction of the incoming laser beam after the appearance of the peak density, leading to a gap. It is concluded that pulse energy has no significant effect on the variation of the focusing distance during the LIPMM focusing process within the test range.

In terms of the advantages of LIPMM, the results presented here can be used for theoretical understanding and further exploiting its potential. However, this work only correlates plasma evolution with the focusing process and geometric features of LIPMM; the interaction between the plasma and the material being processed was not simulated. Therefore, relevant studies should be carried out to get direct machining results in future works.

Task 13: Identify the structural pattern or types of information in the following two research article conclusion sections, then discuss in groups their varying degrees of writer-reader interaction and involvement conveyed by the underlined expressions, and finally analyze the possible reasons for their overall difference in adopting the past/present tense and the active/passive to present main results.

I

Inner- and outer-wall sorting of double-walled carbon nanotubes[10]

Conclusion

Despite the surfactant or polymer used for separation residing entirely on the outer wall, this work suggests that the two walls of a DWCNT are strongly coupled and that it is possible to develop a solution process sensitive to this interaction. In this work, the four different DWCNT types have been identified and separated. Gel chromatography was shown to be capable of enriching the different DWCNT types, and provided an exquisite example of the sensitivity of the surfactant shell to the electronic properties of a nanotube. Inter-wall coupling was similarly verified with the use of PFO–BPy in toluene, and residual metallic species were removed from the S@S fraction. Resonance Raman maps of sorted DWCNTs were measured, and contributions from the inner- and outer-wall species were identified and were found to be shifted relative to SWCNTs. Access to pure DWCNT samples will allow for the identification of a new set of RBMs associated with DWCNTs, and extended studies of the spectral shift induced upon placing one carbon wall within another will be possible. In particular it is expected that it will be possible to separate DWCNTs not only based on their inner- and outer-wall electronic type, but also based on chirality, and this will enable the investigation of single chirality species with different surrounding inner or outer walls. To achieve this it is proposed that DWCNTs must first be separated into the four electronic classes presented in this work and then separated by employing slight changes in solution pH or redox conditions. Central to the success of this approach will be the ability to perform a diameter-based fractionation of DWCNTs. In the current work and as shown in Supplementary Figure 1, a weak diameter fractionation is observed between fractions, but improving the resolution of this separation will now form the focus of future research.

II

Energy dissipation in fluid coupled nanoresonators: The effect of phonon-fluid coupling[11]

Conclusion

In summary, we examine the effect of phonon-fluid coupling on dissipation of a fluid coupled nanoresonator system driven at gigahertz frequencies. Our MD simulations on frequency and density scaling of total dissipation incorporate the phonon-fluid coupling effect directly. However, to analyze the consequences arising solely due to the coupling effect, we resorted to theoretical formulation of each dissipative process in the system. We identify that the total dissipation is contributed by intrinsic dissipation (D_{int}^{f}) governed by Akhiezer mechanism and fluid dissipation (D_{flu}). We formulate the fluid dissipation in terms of hydrodynamic force on the resonator and observe that its magnitude is either very low or comparable to the intrinsic dissipation. As experimental observations on fluidic channel based resonator [10-11] suggests, the relatively low fluid dissipation is a consequence of confining the fluid inside the CNT instead of placing it outside. Another consequence is nonmonotonic scaling of fluid dissipation with the density of the fluid, which we observe at higher excitation frequencies. Besides that, the fluid dissipation is found to increase monotonically with excitation frequency for a given density and with density for low excitation frequencies. We confirm our formulation of intrinsic dissipation due to Akhiezer mechanism by considering the case of empty CNT in vacuum. The formulation reveals that the strength of the Akhiezer dissipation depends on the metric $\Omega\tau$, where Ω is the excitation frequency and τ is the phonon relaxation time. Since τ mostly varies between 1 and 100 ps for a CNT at room temperature, Akhiezer mechanism becomes dominant at gigahertz excitations. We find these relaxation times are affected by the phonon-fluid coupling and is manifested in the Akhiezer dissipation. This leads to a change in the intrinsic dissipation of the resonator in the presence of fluid at a given density when compared with that of vacuum (D_{int}^{v}). This change ($\Delta D_{int}=D_{int}^{f}-D_{int}^{v}$) has been ignored in all previous studies of fluid coupled resonator systems. In our case, accounting for the change becomes particularly crucial since the order of $\Delta D_{int} \sim D_{int}^{v} \sim D_{flu}$. Further, we find that the phonon-fluid coupling strength and consequently the phonon relaxation times can be manipulated by changing the fluid density. We show that an important consequence of the coupling on the total dissipation is its inverse scaling with the density of the fluid at some frequencies. Although, the consequence of phonon-fluid coupling is presented, in this study, in the context of intrinsic dissipation due to Akhiezer mechanism, the coupling effect is general and extends to calculation of any material parameter that depends on phonon relaxation times, such as thermal conductivity, thermal diffusion time, etc., in the presence of interfacial fluidic interaction.

Task 14: Skim through Text Ⅲ and identify the main idea of each paragraph. Determine the main purpose and major results of this research, and discuss in groups based on which paragraphs each sentence in the Conclusion section is stated.

Text Ⅲ

A Mesocrystal-Like Morphology Formed by Classical Polymer-Mediated Crystal Growth[12]

Abstract: Growth by oriented assembly of nanoparticles is a widely reported phenomenon for many crystal systems. While often deduced through morphological analyses, direct evidence for this assembly behavior is limited and, in the calcium carbonate ($CaCO_3$) system, has recently been disputed. However, in the absence of a particle-based pathway, the mechanism responsible for the creation of the striking morphologies that appear to consist of subparticles is unclear. Therefore, in situ atomic force microscopy is used to investigate the growth of calcite crystals in solutions containing a polymer additive known for its ability to generate crystal morphologies associated with mesocrystal formation. It is shown that classical growth processes that begin with impurity pinning of atomic steps, leading to stabilization of new step directions, creation of pseudo-facets, and extreme surface roughening, can produce a microscale morphology previously attributed to nonclassical processes of crystal growth by particle assembly.

1 Introduction

1 Assembly of nanoparticles, molecular clusters, or other solution species that are more complex than simple ions, is now recognized as a common mechanism for crystal growth in a wide range of materials.[1] Dissolved macromolecules have been reported to drastically alter these pathways and their associated rates of formation. Examples—particularly with the use of acidic (poly)peptides[2] and proteins[3]—include the stabilization of amorphous precursor particles[4] (either solid[5] or liquid-like[6]), and the promotion of nanoparticle assembly with crystallographic coalignment to form so-called "mesocrystals"[1], [7] with exotic morphologies.[8]

2 The concept of a mesocrystal was initially introduced to describe the coaligned assembly of calcite nanocrystals into a 3D kinetically stabilized superstructure in the presence of the organic additive polystyrene sulfonate (PSS).[7] Since then many examples of mesocrystals have been reported.[7], [9] However, a recent critical reinvestigation of calcium carbonate mesocrystals grown with polymer additives similar to PSS showed there was no compelling evidence for the involvement of crystalline precursor particles in the formation of the apparent mesocrystal structure.[10] In fact, although observations of characteristic mesocrystal morphologies, rough nanoparticulate surfaces, high surface areas, and line broadening of powder X-ray diffraction patterns have commonly been considered indicators of a mesocrystal structure, the detailed analyses from that study led to the conclusion that the investigated calcite/PSS-MA (calcite/poly(4-styrene sulfonate *co*-maleic acid)) crystals did not fit the definition of a mesocrystal. Moreover, despite their outer mesocrystalline morphology, the crystals were demonstrated by transmission electron microscopy (TEM) imaging to have internal single crystal character. A strikingly similar morphology was also achieved through bulk overgrowth experiments on calcite single crystals.[10-11] Comparable exotic morphologies reminiscent of mesocrystals were obtained for calcium carbonate in the absence of any additives though classical crystallization pathways.[12] Consequently, in spite of the numerous examples of mesocrystal formation via assembly of particles, both amorphous and crystalline, from biological as well as abiotic origin,[1] the growth mechanism responsible for the creation of the exotic morphologies commonly attributed to mesocrystal structures in the absence of particle assembly is now unclear. Therefore, in this work we have used in situ atomic force microscopy (AFM) to investigate the overgrowth mechanism of calcite crystals utilizing PSS as the growth-modifying polymer. We show that the introduction of PSS does indeed lead to the formation of the exotic structures and morphologies, previously attributed to mesocrystal formation, but does so through stabilization of new step directions, coupled with step pinning that generates extreme roughening of the surface and broadening of crystal edges into pseudo-facets. Neither in situ AFM, cryo-TEM, nor light scattering data provides evidence for the attachment of nanoparticles of either amorphous or crystalline calcium carbonate to the growing crystal.

Hence, we conclude that the final morphology arises through completely classical growth mechanisms.

2. Results

2.1 Bulk Diffusion Experiments

3 In our control benchtop experiments without PSS, after 1 day (1 d) of CO_2 diffusion from the solid $(NH_4)_2CO_3$ source into a solution of 1.25×10^{-3} M $CaCl_2$ within a desiccator, rhombohedral calcite crystals grew on a Si_3N_4 substrate (Figure 1a) alongside a population of vaterite crystals.[13] In contrast, when PSS was introduced to the mineralizing solution, after 1 d of diffusion the typical rhombohedral morphology was modified through the flattening of the (001) face[7], [14] and the three adjacent crystal edges, resulting in a triangular (001) facet with a triradiate pattern of (018) facets (Figure 1b). The observed smaller size of the calcite-PSS crystals (14 ± 4 μm vs 48 ± 9 μm; average ± s.d. of the distribution) after 1 d of diffusion shows that calcite formation was significantly inhibited by the introduction of PSS (Figure S2, Supporting Information). By placing the Si_3N_4 substrate perpendicular to the sedimentation direction in the diffusion setup, we ruled out sedimentation of the crystals from bulk solution and confirmed that the crystals nucleated and grew directly on the substrate (Figure S1, Supporting Information).

4 In all of the experiments, vaterite started to form at early diffusion time and, after longer time periods, spherical vaterite was still present next to the obtained calcite-PSS crystals as determined by Raman spectroscopy (Figure S3, Supporting Information). The initial formation of vaterite in the bulk diffusion experiments fixes the supersaturation for the calcite growth at the value given by the vaterite solubility product ($K_{sp,\ vat} = 1.2 \times 10^{-8}$ M^2). A depletion region was observed surrounding the calcite-PSS crystals (Figure S5, Supporting Information) indicating that the vaterite likely dissolved and reprecipitated as the thermodynamically more stable calcite seeds grew. Despite the morphological changes and rough appearance of the (001) and (018) faces, at this early stage of growth the (104) facets appeared surprisingly smooth (Figure S3, Supporting Information), and etching experiments in deionized water revealed dislocation etch pits on the (104) facets characteristic of single-crystal calcite (Figure S3, Supporting Information). However, the (001) plane dissolved more rapidly, and also the triradiate (018) faces etched more deeply than the (104) facets. We attribute these effects to: (1) a locally increased polymer content at these sites and[10], [14] (2) the higher surface energy of the (001) and (018) planes compared to the (104). Upon increasing the diffusion time (Figure 1c–e), the size of the (001) and (018) facets increased while the (104) facets became roughened, producing the overall shape and morphology previously considered to be an indicator of mesocrystal formation.[7], [14]

5 The resemblance of the smooth (104) facets to those of single-crystal calcite motivated us to perform overgrowth experiments on single-crystal calcite seeds using the same diffusion method (similar to the experiments of Kim et al.,[10] also see the Experimental Section). After

diffusion times similar to those used in the nucleation and growth experiments, we found a remarkable resemblance in size, orientation, and morphology of the resulting crystals, with the (001) facet and adjacent (108) facets again expressed and the (104) faces highly roughened (Figure 1f). Confocal Raman spectroscopy confirmed that in the overgrowth experiments the end product was also calcite (Figure S3c,e, Supporting Information).

2.2 In Situ AFM Observation of Calcite Growth Modification by PSS at the Nanoscale

6 Although the experiments described above clearly demonstrate that the typical calcite-PSS crystal morphology observed in our nucleation and growth experiments can be obtained through overgrowth on preformed single-crystal calcite seeds (Figure 1), the final crystal morphology provides little information about the growth mechanism.[10] Therefore, we used in situ AFM to study the overgrowth on the (104) face of a cleaved single crystal of calcite in a PSS-containing solution generated by mixing PSS-$CaCl_2$ and $NaHCO_3$ solutions at controlled pH (Experimental Section). The crystals exhibited growth hillocks formed at screw dislocations with discrete obtuse and acute steps typical for (104) calcite faces (Figure 2a). Step heights were determined to be 6.4 ± 0.6 Å (mean ± s.d. of the distribution), in good agreement with the expected value for a screw dislocation having a Burgers vector $m = 2$ (6.2 Å, see ref. 15). When PSS was introduced into the growth solution, obtuse and acute steps were progressively modified over time by step pinning, presumably due to the poisoning of kink sites (Figure 2b,c). Moreover, when comparing the growth hillock after PSS inflow (Figure 2b) with the one before (Figure 2a), we observe significant changes in the morphologies of both the obtuse and the acute steps, in terms of lateral dimensions, spacing, and height (Figure S8, Supporting Information). Most significantly, step pinning by PSS led to a vastly roughened expression of the acute steps with the protrusions of the roughened steps directed toward the facet boundaries. In addition, while the angle before PSS inflow between the two obtuse steps ($[\bar{4}41]$ and $[48\bar{1}]$) had a value of ≈103° (vs theoretical value of 101.9°, see ref. 16), after its introduction, rounding of the obtuse steps led to a progressive increase in this angle to produce pseudo-[001] steps, again resulting in protrusions toward the facet boundaries.

7 We were also able to observe the step poisoning process in more detail (Figure 3a–c). In the presence of PSS, the measured 6.3 ± 0.5 Å height of the steps (Figure 3d–f) is in excellent agreement with the step height of 6.4 ± 0.6 Å determined without the added polyelectrolyte (Figure S8, Supporting Information). However, on the terraces between the steps, the fluctuations in height in the presence of the polymer are 7.0 ± 1.7 Å as compared to 2.4 ± 0.6 Å for the calcite surface in pure solution (Figure S8, Supporting Information), with the difference attributable to adsorption of PSS molecules to the crystal surface[13], [17] (Figure 3d–f). The individual growth layer also exhibited the rounding of the obtuse steps and extreme roughening of the acute steps observed overall for the entire growth hillock.

2.3 Role of Amorphous Calcium Carbonate (ACC) in Calcite-PSS Crystal Formation

8 Given the extensive literature reporting the involvement of ACC during growth of calcite, an obvious question is whether there was direct involvement of ACC in the growth of the calcite-PSS crystals reported here. In the in situ AFM overgrowth experiments, the supersaturation with respect to ACC (σ_{ACC}) was calculated by determining the concentration of free Ca^{2+} in solution ($c(Ca^{2+})_{free}$) using a calcium-ion-selective electrode (Ca-ISE). Here, the PSS partially complexes Ca^{2+}-ions in solution due to electrostatic interaction with the SO^{3-} groups.[13] We find that, on average, 0.25 ± 0.05 Ca^{2+} are bound per SO_3^-, effectively lowering $c(Ca^{2+})_{free}$. However, the largest part of the total Ca^{2+} added is present in form of ions in the bulk solution (≈92%) (Figure S6a, Supporting Information). Consequently, we find that $\sigma_{ACC} = -3.11$, that is, the solution is heavily undersaturated with respect to ACC. This is in line with our AFM measurements, in which we do not observe growth on the calcite surface by addition of detectable ACC particles. Additionally, when investigating the growth solution by cryo-TEM and dynamic light scattering (DLS) we could not detect any ACC particles besides the presence of a minor quantity of small ≈6–10 nm structures, which are presumably smaller Ca-PSS globules formed through complexing of Ca^{2+} with negatively charged sulfonate groups of the PSS, in line with our previous findings[13] (Figure S6b,c, Supporting Information). Moreover, these particles show a low electron scattering contrast in brightfield TEM, as opposed to ACC.[18] Therefore, we conclude that the calcite crystals formed in the presence of PSS do not grow through the addition of amorphous $CaCO_3$.

3 Discussion

9 Comparing the in situ AFM images to the SEM images of crystals from the benchtop diffusion experiments, we are now able to translate the modifications of growth hillock morphology from the nanoscale to the microscale, where the bulk calcite-PSS crystals exhibit roughened (104) faces bounded by roughened, threefold (001) faces and elongated (018) faces (Figure 1f). The data clearly show that the roughened (104) faces arise from pinning of the individual atomic steps and the threefold (001) faces are, in fact, pseudo-facets formed by pseudo-[001] steps created through polymer-poisoning of the obtuse steps. Because the facet boundaries lie in the six equivalent (018) planes, the convergence of the highly roughened steps from the hillocks on adjacent crystal faces produces a triradiate pattern of rough (018) pseudo-facets (Figure 2c,e,f). This effect is analogous to the broadening of the growth hillock boundaries created by convergence of the acute steps from adjacent hillock sectors (Figure 2b). The width of that boundary is of the same order of magnitude as the lateral undulations of the step fronts in the direction of the facet edges (Figure 2 b,d–f). In the early stage of this process captured in the AFM experiments, these step undulations are of order 1 μm, as is also the case for the (018) pseudo-facets during the early stages of formation in the diffusion experiments (compare Figure 1b and Figure 2b). The resulting crystal pseudo-facets lie at an angle of ≈27° with respect to the (001) plane, which, as

expected, matches that of the (018) planes for which this angle has a theoretical value of 26.23° (see ref. 19). Indeed, regarding external faces, we find a great resemblance of the calcite-PSS crystal with a simulated single crystal of calcite cleaved along the (001) and (018) planes.

10 Although the early stage of formation of the calcite-PSS crystals in the benchtop diffusion experiments differs from the AFM flow cell experiments, because there is no involvement of either ACC or vaterite in the latter, our study confirms that both the ammonium carbonate diffusion technique (including benchtop overgrowth experiments) and the mixing method produce a morphology previously attributed to "mesocrystal" formation. Moreover, the similarity of the outcome despite the differences in growth regimes further emphasizes the impurity effect of PSS on step growth. In the benchtop experiments, which rely on diffusion of carbonate into a solution with a fixed initial supply of Ca^{2+}, the supersaturation drops as the crystals grow. At early times the supersaturation is high enough to drive nucleation of new crystals or 2D nucleation of islands on existing crystals, but in this regime the crystals are simple rhombohedra with smooth surfaces. In other words, there are no observed impurity effects despite the rapid growth. As the crystals grow and consume Ca^{2+}, the supersaturation drops toward the regime explored by AFM, where the crystals grow more slowly (Figure S2, Supporting Information) and the "mesocrystal-like" morphology emerges. This is exactly what is expected from impurity pinning of steps, whether in a rough-step limit where the key factor determining the extent of pinning is the ratio of the critical step curvature to supersaturation, or in a smooth-step limit where the key factor determining the extent of pinning is the relative rates of polymer and solute binding to kinks.[20]

11 The growth mechanism revealed by these experiments clearly demonstrates the dominance of classical growth processes in producing the exotic crystal morphology with roughened patterns previously attributed to mesocrystal formation. However, the results do not directly reveal the atomic-scale mechanism that causes the acute steps to exhibit greater roughening and bunching than the obtuse steps, which indicates there is a greater degree of binding of PSS to the acute steps. There are two potential reasons for this difference. The first is simply that the configuration of the acute step results in better binding to the sulfonate groups of PSS. This mechanism is illustrated by the case of aspartic acid (Asp) on calcite for which even a switch from left-handed to right-handed Asp led to change in binding energy due to differences in the relative geometry of the Asp molecule and the mineral step edge.[21] These structure-dependent effects led to enantiomer-selective modification of the two acute steps and better overall binding of both enantiomers to the acute steps than to the obtuse steps. However, for larger molecules, a second mechanism associated with the energy removing waters of hydration can play a significant role in determining differences in binding. For the peptide polyaspartate (Asp_n), previous work found that molecules with three or more residues ($n \geqslant 3$) preferentially bound to the obtuse steps, because fewer water molecules needed to

be removed. Thus, the dehydration energy was smaller for binding to the obtuse than the acute steps.[22] The number of water molecules released depends of course on the specific configuration of lowest energy and thus the effect should vary significantly from system to system. Thus, the same reasoning would imply that while PSS binds to both the obtuse and acute steps, the dehydration energy at the acute step is smaller than at the obtuse step.

12 To understand the generation of the pseudo-[001] steps that define the (001) plane, we propose two possibilities. First, PSS exhibits a different binding affinity to the left versus the right facing kinks (Figure 2a), leading to a change in step shape. Maruyama et al.[23] found that the addition of L-aspartic acid during calcite growth led to different lengths for the originally symmetric $[\bar{4}41]$ and $[48\bar{1}]$ obtuse steps. They proposed that the effect could result from a difference in the resistance to incorporation into the two distinct kink-types—designated here as +/+ and +/− to indicate kinks moving toward the obtuse/obtuse and obtuse/acute corners of the growth hillock—due to kink blocking by the aspartic acid. A preferential poisoning of the +/+ kinks will reduce the speed of the step toward that corner and thus lead to a flattening along the [001]. The second mechanism—perhaps acting in combination with the first—is related to the calcium density along the [001] step direction: As the shape of a newly formed step fluctuates during its advance, deformations toward the [001] direction will present an increased fraction of exposed Ca^{2+} sites and a net positive charge (see Figure 4). Therefore, complexing of the negatively charged sulfonate groups of PSS will be favored, thus promoting the stabilization of the (001) pseudo-steps.

13 Undoubtedly, the locally positively charged nature of an (001) pseudo-plane terminated by rows of calcium (Figure 4) will additionally contribute to its stabilization due to the electrostatic binding of PSS, as proposed in numerous other studies.[7], [9], [14], [25] Molecular dynamics simulations of styrene sulfonate (SS) and PSS oligomer adsorption onto calcite surfaces[25] showed a preference for the (001) facet, where the sulfonate interaction takes place through direct and solvent-mediated binding, for which both the PSS-containing molecules showed an approximately perpendicular orientation toward the surface. The solvent-mediated binding refers to the sulfonate group residing in the second layer of high water density which forms very strong hydrogen bonds with water molecules in the (strongly polarized) first and second solvation layers. In contrast, adsorption onto the less favorable (104) facet—where binding of the sulfonate to the surface is solvent-mediated by one or two layers of water molecules—can take place with the molecules oriented more parallel to the mineral surface. Because the structures of a [001] step and a {001} face are closely related, these two phenomena are likely to be related to one another and, as such, formation of the (001) pseudo-facets should reinforce PSS binding to the (001) plane and vice versa.

4 Conclusion

14 The findings presented here show that both the ammonium carbonate diffusion technique and

the method of mixing $CaCl_2$ with $NaHCO_3$ to prepare $CaCO_3$ crystals in the presence of certain organic additives can eventually produce a morphology previously attributed to "mesocrystal" formation, in agreement with other recent work.[10], [11] Here, by comparing the outcomes from both methods with those from the in situ AFM observations, we have directly translated the resulting morphology from the nanoscale to the microscale and shown how it arises through completely classical growth mechanisms. Although we have demonstrated this using the specific case of PSS-modified calcite, the results may serve as a more general basis for extending classical growth mechanisms to the formation of other complex crystal morphologies. In addition, they re-emphasize that proposals of nonclassical pathways based solely on observations of crystal morphology and surface roughness must be carefully considered.

5 Experimental Section

Detailed information on the following methods can be found in the Supporting Information: $CaCO_3$ benchtop diffusion experiments in the presence or in the absence of PSS, $CaCO_3$ overgrowth bench-top diffusion experiments, confocal Raman microscopy, atomic force microscopy overgrowth experiments on calcite, diffusion experiments in the AFM fluid cell, ion-selective electrode experiments, SEM, and cryo-TEM analysis.

References

...

Task 15: Determine typical expressions and sentence patterns in the following sentences used to indicate implications and/or discuss applications of main results in research article conclusion sections.

(1) The synthesis approach provides an opportunity for doping large size rare-earth elements into AlN materials.

(2) AlN:Ce hierarchical nanostructures exhibit strong red-orange emission and room temperature ferromagnetism, which have applications in optoelectronic and spintronic nanodevices.

(3) These results provide a detailed insight into the structure of the nanofibres with crystalline cores for the first time, to the best of our knowledge.

(4) Although we have demonstrated this using the specific case of PSS-modified calcite, the results may serve as a more general basis for extending classical growth mechanisms to the formation of other complex crystal morphologies.

(5) Taking into account the great tunability offered by magnetic interactions, we deem rotational magnet pairs (as negative stiffness elements) can play a fundamental role in realizing reusable rotational energy absorbers and dynamic dampers, the properties of which can be easily tuned in situ and without the need to refabricate the material.

(6) The discoveries made on this benchmark material can undoubtedly be transferred to many

other persistent luminescent compounds which are operating on a very similar principle.

(7) From these points of view, research investigation on the 2D materials at nanoscale level simplify the future industrial relevant applications.

(8) This work is an advancement for low-cost solar-driven production of H_2 incorporating simultaneous high value product co-generation in one process accelerating development for rapid transformation to clean H_2 economy.

(9) This retina-inspired device with real-time sensing and processing, combined with human health monitoring features, opens up a new path for the future development of multi-functional intelligent terminals.

(10) Collectively, we believe that this work provides great insights for understanding the aggregation effects of D-A polymers on molecular doping in the solution state, enabling the engineering of side chains in the polymer suitable for future applications in TEs and other electronic devices.

(11) Based on the demonstration of the proposed L–L TENG as a rainfall alarming system, the sensitive response of the electrical output to various parameters can further extend its application in passive sensors, overturning the perceived dominance of traditional solid material-based electronic devices.

(12) Overall, the TiO_2 nanoparticle-reinforced and HA-coated bioimplant with high strength and biocompatibility, which was developed using an effective additive manufacturing process, represents a great advancement in the field of composite prostheses.

(13) We believe that the presented approach will open new opportunities in different application fields due to the versatility and accessibility of the materials.

(14) We envision large potential of the system at the microscale in microrobotics and biomedicine due to the remarkable printing performance as well as the possible actuation at low temperatures, which to the best of our knowledge is one of the lowest reported in the literature so far.

(15) As such, we have demonstrated the principle of rapidly direct-writing unclonable marks that can be applied toward the authentication of high-value parts and consumer goods that are vulnerable to counterfeiting.

(16) These remarkable electrochemical properties endow this 3D ZnTe@Zn candidate as a promising next-generation advanced anode for AZIBs that need excellent stability, durability, and low voltage hysteresis.

(17) To conclude, our comprehensive study demonstrates that the composition and structural flexibility of hybrid organic-inorganic metal-halides can offer valuable and interesting magnetic properties, opening new pathways toward the design and pursuit of novel layered magnetic materials for optoelectronic and spintronic applications.

(18) Our work presents a novel but effective strategy for resolving the tradeoff between intermolecular interaction and backbone disorder enabled by high electron deficiency and

high molecular dipole segment for improving blend morphology and photovoltaic performance of polymer solar cells.

(19) In the highly oriented molecular p–n heterojunction here designed, the hole injection barrier is effectively reduced, paving the way for its implementation in solar cells and/or light emitting diode (LED-type) devices.

(20) Our study provides further insights to tailor novel polymer compositions with consistent and reproducible electrical parameters and long-term storage and thermal stabilities while retaining the excellent thermoelectric parameters of PEDOT:PSS and its derivatives.

(21) This study leads to new opportunities to explore innovative dopant structures from the myriad of already existing, efficient dopants to fine-tune thermoelectric parameters.

(22) Our work leads to new directions for the exploration of a myriad of Wheland complexes based on the choice of the appropriate oligomer.

(23) Through this scalable fabrication process, MXene-Semiconductor-MXene devices could enable wide usage in high-performance optoelectronic and photonic integrated systems.

(24) This work elucidates the modified mechanism of thiourea in FA-based perovskite crystallization process, which is inspiring for improving the perovskite device with sulfur additives.

(25) The proposed concept provides a materials synthesis strategy for chemical sensors integratable into sensor array technologies which may lead to application in wearable, easy-to-operate, and real-time sensing devices.

(26) This contribution proposes a new strategy to enhance the photovoltaic performance of PSCs by designing narrow bandgap-conjugated molecules extending the photovoltaic response, which is applicable in any device structure.

(27) By addressing the lingering challenges in conductive hydrogels, the BC-CPH provides a promising material for tissue-like bioelectronic interfaces.

(28) This work may offer a versatile tool and platform not only for a vision of hydrogel bioelectronics to achieve better electrical interfacing between machines and biological systems but also for broader applications of conducting polymer hydrogels in tissue engineering and regenerative medicine.

(29) This strategy enriches the design of the ferroelectric phase transition and creates an opportunity for optimizing a variety of functionalities, for example, electrocaloric ferroelectric materials for flexible and wearable devices. Besides ferroelectric polymers, we anticipate that electro-thermal actuation is also applicable to other polymer systems with large volume changes associated with the thermally induced phase transition.

(30) This study provides a practical approach for the preparation of self-supported high-performance membranes.

(31) This work opens up alternative perspectives for the design of catalysts based on transition metal oxides by manipulating interstitial atoms in the metal oxide structure.

(32) The application of this C-doped RuO_2 material to energy-demanding processes, such as CO_2 hydrogenation to form methane at a low temperature as presented here, opens up a promising sustainable alternative to the current processes that operate at higher temperatures.

(33) The operando NMR method reported here opens up new routes to obtain quantitative insights into alkaline-ion dynamics and ion uptake on electrical gating in a wide range of organic mixed ionic-electronic conductors. The approach can be applied to understand the mechanisms, capacity retention and degradation of a wide range of mixed ionic-electronic conducting polymers.

(34) Assuming the MoS_2 electron is itinerant in the observed layer-selective Mott insulator phase, the WSe_2/MoS_2 heterobilayer could be a suitable platform for studying the Kondo lattice model, which can lead to a deeper understanding of strongly correlated systems.

(35) A modest displacement field is sufficient to tune the valley character of the lowest TMD moiré band from Γ to K, providing a way to electrically switch between dramatically different moiré band structures within a single device.

(36) As demonstrated in the fully stretchable OLED devices presented here, we anticipate that the successful development of stretchable TADF polymers offers considerable potential for applications and a technologically feasible path for stretchable optoelectronic devices for use in human-interactive applications.

(37) Silicon-defect-based in situ magnetic detection technologies could provide several immediate research opportunities in materials science.

(38) The experimental capabilities demonstrated in this work allow for direct correlation of the local vibrational properties with defect atomic configurations and for exploring the tuning of phonon-mediated functionalities in graphene by point defects.

(39) STML offers exciting opportunities to unveil near-field charge and energy transfer and proximity effects in vdW heterostructures with unprecedented accuracy, offering outcomes in photonics, optoelectronics, and nano-electronics.

(40) The GB strength model shows how material composition and passivation methods work and provides a guideline for designing strong GBs for the development of ultrastable and hysteresis-free MHP devices.

(41) The multiscale diffusion framework will contribute to elaborating accurate models of ion migration in MHPs across a wide range of scenarios, including devices under external stresses and across different applications.

(42) Our study is the first report on the liquid triboelectric series not only paves a way to understand the liquid-solid contact electrification phenomenon but also provides opportunities to enrich and expand the applications of the liquid triboelectric effect.

(43) In our work, by using the NFP process, we can provide arrays of selected QDs at any position regarding the premade markers. Hence, we can create large-scale multiple QDs integrated photonic circuits by using this technique.

(44) Our results provide a direct and thorough insight into the atomic configuration of carbonates in (Cu,C)-1234 and unveil the underlying influencing mechanism of carbonates on superconductivity.

(45) Together, these results point to a long-sought solution to the problem of maximizing TE material performance by breaking the coupling between the thermal and electrical properties.

(46) We envisage that the here proposed proximity lithography protocol, which combines precisely oriented MOF films with micropatterning, will progress the integration of MOF components with anisotropic functional properties into miniaturized devices.

(47) It is thought that this work provides a promising strategy to create a wide range of functional plasmonic nanocrystals under the laser-material interactions at the nanoscale for boosting wide applications.

(48) In short, the Janus-IS hydrogel evaporator exhibits a breakthrough salt-resistant route, effective salt removal, a record-high evaporation rate, and reproducible evaporator fabrication.

(49) We believe that the proposed mReLU can reduce the energy and area overload on the peripheral circuit for implementing the activation function, and solve the chronic vanishing gradient problem in DNN, thereby providing a solution for energy-efficient hardware neuromorphic systems.

(50) This work sets an important milestone in the design and understanding of catalysts for NH_3 oxidation to N_2O, establishing CeO_2-supported SACs as a highly promising class of materials for this application, as well as a range of other selective oxidation reactions.

(51) Given the unique asymmetric adhesive performance and facile preparation of the MAH hydrogel, it opens a novel and effective method to fabricate a completely different Janus hydrogel bioadhesive for simultaneous in vivo wound healing and anti-postoperative adhesion.

(52) Based on the simplicity of the mSP assembly-disassembly mechanism, we expect the possibility of using mSPs for gene therapies and other areas of medicine as well as biotechnology.

(53) Overall, this work provides a new strategy for the development of organic CPL detectors, which will be beneficial for the future development of sensing.

(54) This SPIRIT approach offers an unparalleled bioprinting capability wherein complex tissue constructs with an embedded internal structure such as a freeform vascular network can be easily created with much less printing time.

(55) Looking forward, we can leverage these SPIRIT bioprinting capabilities to engineer complex tissues and organs and accelerate important breakthroughs in biofabrication and tissue engineering for therapeutic applications.

(56) In summary, this work provides a new strategy for PET chemical recycling while potentially maintaining full functional equivalency and compatibility with legacy PET.

(57) Our technology provides a strong framework for the formulation of a glucose-responsive insulin delivery system for potential clinical translation.

(58) Taken together, this work is of novelty both from fundamental and practical research points of view.

(59) Our research paves a new pathway for the rational design of CDs with high fluorescence brightness and superior RNA selectivity via the spatial organization of recognition motifs and emission centers, which offer guidance for the development of various multifunctional nanomaterials with facile integration and full modularity.

(60) It is the first report of utilizing CDs as bulk RNA labeling agents for high-resolution spatiotemporal dynamics imaging of granules, which uncovers the extensive heterogeneity involved in the SG assembly process and provides a powerful toolbox for illustrating how cells reshape RNA dynamics in response to stress.

(61) Based on this attempt, we believe that ionic liquids showing poor solvating ability to Zn salt are very promising alternatives to the molecular co-solvent in enabling highly reversible, safe, and environment-adaptable AZMBs and beyond.

(62) Characterizing devices with ACMs at lower temperatures (<300 K) could open a new window for exploring metal-semiconductor contacts.

(63) The biomimetic GB structure proposed here effectively inherits the superiority of ingenious natural wisdom, which can be extended to the development of more advanced impact-resistant materials for protective applications.

(64) This work provides a new and systematic approach towards rational design and development of Li-mediated PEC NRR systems for fast, cost-effective, and efficient green synthesis of NH_3.

(65) In terms of the robust performance, this elastomer has great potentials for applications in the defense industry, flexible electronics, cushioning, energy absorption, etc.

(66) Our work sheds light on the critical importance of controlling isotropy of the configuration as well as bonding density of the interface, offering a new horizon for the rational design and manufacture of hierarchized and hybridized silicon granules with practical lithium storage features, and providing insight into full implementation of high-capacity electrode materials at high rates and high loadings for energy technologies.

(67) This work provides a facile approach for delicate regulating the stoichiometries of PQDs to tune their optoelectronic properties and tolerance factor for high-performing and stable solar cells or other optoelectronic devices.

(68) Our findings represent a significant step toward understanding and applying ASPL and SHG upconversion in broadband multiresonant metallic nanocavities for multimodal or wavelength-multiplexed operations in bioimaging, sensing, interfacial monitoring, and integrated photonics.

Task 16: Identify in groups typical expressions and sentence patterns used below to make recommendations or state a practical need for further study in research article conclusion sections.

(1) Future work will aim to extend analogous studies to other nanofibre systems to provide structural comparisons and key insights into the properties underlying their potential applications.

(2) However, some of the details of the microscopic mechanism are still unknown and, as the benchmarking study suggests, this stands in the way of further improvements of the overall performance. Hence there is still ample work left to be done.

(3) Moreover, it is recommended to perform multiple independent thermoluminescence, afterglow and fading experiments that are fitted simultaneously to the same set of parameters such as trap depths to ensure maximal predictability of the resulting fitted model.

(4) The key drawback of the HER process via electrocatalysis is the deterioration of the catalysis during long term catalysis, owing to that more efficient and less corrosive catalysts need to be explored in the near future.

(5) The corresponding HER mechanism of the 2D materials is still not clear and needs further investigation.

(6) Therefore, novel strategies are required to design new polymers for solution mixing.

(7) The conceptual development for the simultaneous read-out during flexible operation is the subject of future feasibility studies.

(8) However, less is known about the nucleus in three dimensions, and studies to date have been limited by resolution. Of further note, since ExM is compatible with RNA imaging, PhotoExM might be tailored in the future to accurately characterize the subcellular and cellular heterogeneities of transcriptional states in 3D microenvironments, which remain largely unknown.

(9) Future efforts will focus on achieving other large-area ultrathin self-supported MOF glass membranes based on their self-supported characteristics, or by using MOF glasses as a continuous phase.

(10) Recently, a great challenge still exists in the stability and far-infrared detection of CPL detectors.

(11) In order to prepare high-performance CPL detectors, it is necessary to further explore new chiral materials, using a reasonable device structure to fully exploit the potential of chiral materials.

(12) Further studies including toxicology, techno-economic and life-cycle assessment, and gas barrier aspects, will be needed to provide a more comprehensive understanding of these materials before they are ready for end-use applications like food packaging.

(13) Looking forward, additional studies are merited to reduce the performance variability of

scaled contacts and to further understand the impact of substrates and contact metals.

(14) Because some contacts in mature transistor technology nodes are not gated, it is essential to explore the effects of contact gating on the scalability of contacts in future studies.

(15) Future device scaling studies are needed that combine scaled contact length with scaled channel length.

(16) Further improvement in the efficiency of photoinduced ligand desorption will help achieve advanced photocatalysts that possess both colloidal stability and high reactivity, as well as photopatterning capability of conductive circuits using NC solid films.

(17) Further exploration of organic-inorganic nanohybrid systems that show similar photoinduced ligand dissociations is crucial for various fields of materials science.

(18) Further investigation is needed to examine the dependence of ASPL and SHG processes on plasmonic metal types (e.g., Ag, Au, Cu, Al) both with and without interband photoluminescence pathways.

Task 17: Identify emphatic and evaluative markers in the following sentences, which could be in the form of an attributive adjective, a predicative adjective, an adverb, a prepositional phrase, a noun phrase, or even a verb, as underlined and bolded in the first ten sentences.

(1) The findings in this work **provide** **great** potential for armouring individual grains with cationic surfactants without a loss of charge transport properties for **highly efficient and stable** optoelectronic devices such as solar cells and light emitting diodes.

(2) It **exhibited** several **significant** advantages, including **high** separation efficiency and **ultrafast** labeling speed, **high** purity of the product, and **much lower applicable** carbohydrate concentration.

(3) The WEL lattices also **open** **meaningful and valuable** opportunities as smart slim films in phone cameras, where enhancing imaging functionality and optimizing flexibility are **critical** parameters for next-generation devices.

(4) It **displays** an **excellent** bidirectional bending performance, which is a **significant** improvement compared with reported photothermal or electrothermal actuators.

(5) The **very high** quality of the ED was **confirmed** by the **good** agreement between theory and experiment in the description of the intralayer Ti-S interactions.

(6) The diffuse critical point does not only **have** a **significant** impact on the piezoelectric and electrocaloric response but also **expands** the temperature range of enhanced properties.

(7) The solar cell **features** **high** thermal stability and **fully** reversible colour and performance, which are **key** requirements for **successful** integration into diverse applications.

(8) The **successful** integration of 2D semiconductors and transparent conducting oxides, demonstrated here **for first time**, **has** **huge** implication for **fully** transparent 2D electronics

and photonics.

(9) Benefiting from the combination of **enhanced** conductivity, **highly** porous structure, and protective RGO, $Mn_xCo_{1-x}CO_3$/RGO composites **shows** **great** promise as an anode material for high-performance LIBs.

(10) This work **provides** not only an **ideal** electrode for commercial EDLCs, but also a **new** chemistry to synthesize **highly** dense porous carbon materials that would have **promising** applications for **various** energy devices.

(11) Overall, owing to the virtues of simple design, high output, efficient energy conversion, and convenient operation, the combined effect generator is very promising in practical application for powering small electronics.

(12) The amount of N and P codoping in NPCNs also has a significant effect on the electrochemical properties.

(13) Systematic electrochemical measurements suggested that the flexible electrodes delivered much better electrochemical performance compared with the traditional electrodes.

(14) Direct harvesting of PEs is a highly efficient way to collect solar generated bioelectricity.

(15) The unique structure of the nanopore array provided a large surface area and facilitated the generation of smaller bubbles, which led to a highly efficient supply of electrolytes.

(16) To conclude, some key points can be summed as follows: The bevel plays a very important role in deciding the crushing behaviours of the composite sandwich panels.

(17) This work not only represents a novel strategy for constructing 2D-based functional nanomaterials but also provides a new avenue for the rational design of next-generation solar steam generation systems.

(18) Our study could lead to a better understanding of the biological effects of CNTs by deeply elucidating the regulatory effect of CNTs on the body.

(19) This systematic study not only provides direct chemical evidence for the previously reported Lewis base-acid interaction in LiS batteries, but also identifies the governing factor for PS chemisorption capability, which opens a new pathway for realizing practical LiS batteries.

(20) Overall, it is rare to see the application of pyran units into NFA designs and all results indicated the ladder-type *Ph*-DTDP is a promising building block for highly efficient NFA-based PSCs.

(21) Through systematic experiments supported by theoretical calculations, it was verified that the performance of the laser based on a WTe_2 microflake SA is comparable to that of 2D-TMDC-based lasers reported so far.

(22) The study also shows how simplified pure micromechanical modelling schemes can be very helpful in understanding the key underlying physics for a particularly measured behaviour.

(23) In contrast to other polymer-inorganic hybrid materials, not only electrostatic interactions have been taken into account, but rather the precise control of the interface by genetic engineered phages.

(24) G-Fs were directly used as supercapacitor electrodes, demonstrating extraordinary capacitance and outstanding stability and indicating a promising and practical application of flexible electrodes as supercapacitors.

(25) We have demonstrated novel flexible and highly active self-powered force sensor arrays based on highly active composites of lead zirconate titanate (PZT) and polydimethylsiloxane (PDMS).

(26) The study also shows how simplified pure micromechanical modelling schemes can be very helpful in understanding the key underlying physics for a particularly measured behaviour.

(27) Magnetic brightening of dark excitons also provides a route to produce optically observable states with long and widely tunable radiative lifetimes, as well as greater valley stability compared with bright excitons.

(28) VFETs with the optimal geometry exhibited far better performance than that of conventional lateral-channel OFETs.

(29) In summary, this study reports a novel and simple method for increasing the thermal stability of $CH_3NH_3PbI_{3-x}Cl_x$ perovskite solar cells and also increasing their PCEs.

(30) The enhanced radiative and valley lifetimes, combined with the exceptionally strong many-body interactions present in these systems, offer a promising platform to realize correlated exciton states.

(31) Our findings also provide a path forward for obtaining a more complete understanding of IGSCC of important engineering alloys such as stainless steels and Ni-based alloys employed in critical applications.

(32) The physicochemical properties of derived bio-oil confirm it as a valuable low-grade fuel which can be used directly or in a mixture of other conventional fuels in industrial applications but not directly in an engine as a fuel.

(33) In this paper, we demonstrate highly conducting and mechanically stable fiber interconnections for e-textiles using IL/PVDF-coated CNT fiber and adopting a proper pattern design for fiber interconnection.

(34) The precise structural determination of reactive TMO interfaces poised far from equilibrium constitutes a significant advance toward accurate models of catalytic and geochemical processes.

(35) This highly extendable double-cross-linking method and these outstanding results will offer significant progress in porous materials and make the practical applications of transparent flexible aerogel-based superinsulators feasible.

(36) A key merit of the demonstrated prolonged exciton/trion coherences in our work with h-BN-encapsulated material is that they were achieved in a linear optics experiment.

(37) We unambiguously believe that LAD has tremendous potential to replace Li-TFSI/*t*-BP and this innovative strategy of designing a fluorine-containing hydrophobic Lewis acidic dopant will provide an important future direction for developing high-efficiency, hysteresis-less and

stable PSCs.

(38) Conductive PANI-ER/AW composites showed a very low percolation threshold and enhanced mechanical properties.

(39) The central strategy to fabricate gel constructs with precise control of gradient structures by 3D printing of hydrogels with distinct responsiveness shall also apply to other tough hydrogels toward programmed deformations.

(40) The fabrication of the doped MnO_2 ultrathin nanosheets in this study gives hope to the development of an extremely cheap and highly active and stable metaloxide electrode for OER.

(41) The potential applications presented in this work demonstrated the advantageous textural properties of the HPC-G material, while the synthesis strategy presented herein leads to an excellent direction in the rational design of highly porous graphitic carbon materials.

(42) The hydrogel scaffolds presented a well-distributed macroporous structure, high transparency, high swelling ratio, controllable degradability, and excellent biocompatibility, all of which were very eligible for the advanced visualization of wound care.

(43) Another possible field could be the design of chem-FETs based on highly stable SnO that could take advantage of the well-established surface chemistry of oxides and the extreme surface-to-volume ratio of the 2D material.

(44) Small and uniform size, versatile yet sturdy structural features permitting diverse genetic and chemical modifications, and controlled reversible disassembly and re-assembly all represent very attractive characteristics.

(45) Such previously unexplored intriguing property combination not only imparts sufficient exposure and accessibility of NDs but also ensures continuous electron transport and adequate mass transfer through the matrix.

(46) Inevitably, conjoining the advantages inherent in Ç, iron oxide NPs as well as GO, ^{m}Ç-GO-based composite materials offer tremendous promise being mechanically strong, efficient and environmentally friendly enabling Ç to be utilized to its greatest advantage for dye decontamination.

(47) We have shown that oxygen plasma can be used as a very convenient method for the fabrication of tailored nanopores of desired dimension (and probably altered chemical properties) in suspended single-layer graphene, with high precision.

(48) MRR increased with increase in Ce film thickness, and an extremely smooth surface of sub-nanometer roughness was achieved using Ce film of thickness 2.5μm.

(49) The results not only present the fundamental working characteristics of the pyro-phototronic effect based PD, but also provide some guidances for designing novel pyroelectric effect based devices as pyroelectric energy harvesting, IR imaging, fast detecting and smart optoelectronic systems.

(50) Our results indicate that halide perovskites are not only ideal for low-cost and high-efficiency

solar cells, but also possess rich phase change behaviours for switchable optoelectronics.

(51) Dynamic adhesion switching by solvent vapor exposure facilitates rapid and efficient transfer printing of the single-crystalline NWs onto diverse substrates including flexible and biological surfaces without the need for any surface treatment, leading to an extremely high transfer yield of 96%.

(52) We believe the present approach demonstrated a powerful platform for understanding, monitoring, and controlling reaction kinetics.

(53) Furthermore, a collection of favourable quantitative measures, such as small overall device size, small modal volume, small active volume and extremely low threshold power density, could lead to exciting applications in on-chip interconnects with high energy efficiency.

(54) A drastic example is the recrystallization of 95% rolled copper, where the strong cube recrystallization texture was found to be predominantly a consequence of transition band nucleation, which is in good agreement with experimental observations.

(55) Surprisingly the S···S interlayer interactions were also observed to have significant electron sharing, and substantial differences were observed between experiment and theory.

(56) Besides great capacitance, conformability and reliable scalability, the D-MSCs also exhibit long-term and mechanical stability with wide-range working potential.

(57) At last, the practical and real-time application studies of the proposed PW-TENG devices are clearly affirming that the device can be well adapted into wearable clothes, used to sense human movements and also utilized as a self-power source.

(58) Novel fabrication methods such as He-ion beam fabrication produce pores that are almost perfectly cylindrical and should therefore minimize possible artifacts from this source of error.

(59) Furthermore, the vertical architecture is suitable for highly flexible electronics as the inplane cracks in the semiconductor layer have a negligible effect on the vertical current transport.

(60) Also the test frequency has a marginal influence on glass transition temperature.

(61) In the fine fraction, portlandite was also identified as a major phase (10.37 ± 0.03%).

(62) A particularly interesting property is their scalability down to the nanometre scale and the fast and well-understood dynamics of the amorphous-to-crystalline transition.

(63) The robust architecture constitutes a highly porous and integrated conductive network, resulting in fast electrochemical reaction kinetics.

(64) The CTF-1-100W sample, readily fabricated under very moderate conditions, shows a well-defined layered morphology.

(65) Overall, our results uncover the fundamental structure-thermal conductivity link in CVD-grown MoS_2 monolayer highlighting its potential to integrate into modern solid-state devices.

(66) Second, the resulting MMMs show exceptional mechanical performance, far exceeding most of the previous MMMs.

(67) In summary, we reported an effective and facile method to synthesize a highly active and stable HER electrocatalyst by in-situ growth of ultrathin GDY nanosheets on the surface of

MoS_2 nanosheets on conductive 3D CF substrate.

(68) This behavior of PUPCL based material clearly satisfies basic requirements for thermally induced shape memory performance.

(69) The tested bottom ash presents a very low and acceptable risk of collapsibility for non-compacted (0.1%) and compacted (0.001%) configurations.

(70) The 3DG material can be used directly in an easy and scalable manner as a clean water generator with high efficiency under normal (1 sun) and even weaker sunlight (down to 0.25 sun).

(71) Clearly, our studies indicate that photoexcitation can induce bulk polarization through orientation/rational polarization toward decreasing charge recombination in perovskite solar cells.

(72) More interestingly, the energy-storage performance has been greatly improved with the partial substitution of PMW for PHf.

(73) Obviously, this strategy is not restricted to PP, PC, or glass workpieces and will be applied to other insulating substrates with further development.

(74) Most remarkably, the specific capacitance is commendably maintained with 23% decay after 14000 cycles, indicating a high-stable cycling performance.

(75) Most strikingly, however, is that the FG-nups can undergo local and reversible condensation, including transitions from pore-occluding to more open configurations.

Task 18: Analyze the underlined conjuncts (words and expressions showing the writer's varying types and degrees of logical judgment towards the propositions) in the following sentences and group them into the given functional categories.

- Transitional (e.g., besides, in the meantime):

  ____

- Listing (e.g., first, furthermore):

  ____

- Summative (e.g., in conclusion, conclusively):

  ____

- Contrastive (e.g., in contrast, however):

  ____

- Inferential (e.g., if so, predictably):

  ____

- Resultative (e.g., therefore, as a result):

  ____

- Appositive (e.g., for instance, namely):

  ____

(1) Here, we have argued that the orientation of the hydrogen molecules and the structure of the surface are particularly important around pore sizes with widths on the boundary of a mono- to bilayer transition; thus these effects must be included.

(2) Herein, a series of HA nanowire and gelatin composites were synthesized to prepare biomimetic scaffolds for tissue engineering.

(3) There, fiber tension can be adjusted in the manufacturing line to find optimum processing parameters to ensure full impregnation at maximum throughput.

(4) The simple modeling of field enhancement at SWNT tips implied the applicability to the arrays with ~100 SWNTs/μm density, so that the sorting method could eventually lead to the upscaled integration of ultrascaled SWNT transistors for high-performance logic applications.

(5) In the meantime, the entire process is totally solvent-free and is consistent with the requirements of green chemistry.

(6) As a proof of concept, deep-seated draining lymph node networks were visualized by PET imaging with Zr–Gd_2O_3–PEG nanorods, which in the meanwhile also offered strong contrast in MR imaging.

(7) Simultaneously, the phosphorescence lifetime of *m*-BrTCz in crystal is 120 ms, due to the stabilization of the excited triplet states by H-aggregation.

(8) To date, the parameters studied have been promising to achieve films with desirable properties for future application as packaging material.

(9) After that, the interdot/intradot hydrogen bonds and steric protection of semi-ionic C−F bonds can reduce the quenching of RTP by oxygen and stabilize the triplet excitons at room temperature.

(10) PVC is used in various combinations, often as a matrix with a variety of fillers such as clay, nano-$CaCO_3$, or montmorillonite, hence the problems of PVC aging appear there as well.

(11) However, the mechanical properties of WPCs first increased and later decreased with the initial moisture content of the raw material.

(12) After the impact of the supercooled water drops, a rebound occurs, and afterwards smaller secondary drops are formed, which can be easily removed.

(13) Usually, these are accurate to within a few nanometers, but sub-nanometric precisions can be reached by many devices.

(14) Traditionally, the properties of a laser-sintered material would be expected to be significantly worse than those of high performance injection-moulded polymers.

(15) Thereafter, micro-channels on stainless steel using LIPMM with various pulse energies were generated.

(16) Once again, ±45 2D PMC produced higher fatigue run-out stress (69 MPa, or 55 %UTS) than the unitized composites.

(17) At this time, the selection efficiency rate was 92%, and the combustibles content was 99%.

(18) Furthermore, the interplay between small molecules and polymers should be considered carefully to avoid aggregation of the DAE.

(19) In addition, influenced by the electrolytic product, the surface at the ending point was not as smooth as that at the starting point.

(20) Finally, it was hoped that the results of this paper would be useful for the design of microsensors and microactuators.

(21) With the simulated water depth increasing from 0.1 m to 75 m, the austenite content in the weld metals first decreased, then increased, and then decreased.

(22) Thus, AIBN can not only be used for the synthesis of the covalent nanocomposite but also for the preparation of functionalized SiNSs without a polymer matrix.

(23) Additionally, it is possible to make recommendations for the hot pressing process in order to minimize changes in fiber orientation in the future.

(24) On the other hand, intramolecular sheets and, in particular, helices within keratin and silk fibroin chains are responsible for the high extensibility of the composite films.

(25) More importantly, hyperthermia, induced by the photothermal effect of gold nanoparticles, was validated to enhance the efficiency of thrombolysis in vitro and in vivo.

(26) Besides, the driving voltages as well as the efficiency rolloff have been greatly improved.

(27) Second, g-C_3N_4 reduces the intrinsic defect density by passivating the charge recombination centers around the grain boundaries effectively.

(28) In random-nanograined the stress firstly increases steadily and then goes into the slow increase.

(29) The measured magnitude of these fluctuations is all the more unexpected, in particular at ambient temperature and in the absence of complex experimental conditions.

(30) The major advantage of our densely arranged microcavities is the ultrahigh spot density, which allows on the one hand screening of millions of interactions in one experiment and on the other hand allows for the unprecedentedly low consumption of analyte required to screen the synthesized compounds.

(31) Likewise, a strong structure-activity relationship between the integrated current producing C_n products and P content emerges.

(32) Correspondingly, non-radiative lifetime is greatly increased while radiative lifetime remains almost unchanged.

(33) Clearly, we provide a general framework for phase transformation mechanisms in anisotropic materials where the phase boundary movement and ion flux directions are orthogonal.

(34) At last, some important results can be summarized as follows.

(35) In summary, our unique technique has a promising potential to produce multiple channels within a macroporous bioceramic scaffold.

(36) Above all, the ability to define many distinct current levels paves the way for creating high-density memories with 8-bit memory units.

(37) In conclusion, we have investigated ferroelectric superlattices having different periods and lattice strains.

(38) Overall, Bi and Sb doped $Mg_2Si_{1-x-y}Sn_xGe_y$ materials showed high thermoelectric performance.

(39) Taken together, we conclude that the exosomes might have a predomination over ECM in the lineage determination of hMSCs.

(40) To summarize, this study reports the formation of orientation-controlled anisotropic SnS layers on both Si and quartz substrates.

(41) In general, the difference in device performance can not be explained by using traditional arguments derived from photophysical and electrochemical assays.

(42) To conclude, we demonstrated the wafer-scale growth of wrinkle-free graphene single crystals on this thin film.

(43) In short, the o-$NiSe_2$ NCs act as excellent bifunctional HER and OER photocatalysts for the water-splitting PEC cells.

(44) In brief, studies in this paper have proved that the proposed BPs can be used as efficient sensors to real-time and in situ monitor the delamination-dominated damages in laminates without hurting their tensile properties.

(45) All in all, the feasibility of the novel process chain to produce copper bimetal parts was proved.

(46) To sum up, our results developed a method to obtain directed cell migration based on cell communications.

(47) In essence, we have proposed a carbon based single electron electric spaser in the extreme quantum limit

(48) In total the developed measurement setup allows an evaluation of the optical temperature-dependent properties,

(49) In a word, a pure spinel structure LiZnBi ferrite ceramics with various TiO_2 substitution were obtained at low temperatures.

(50) However, the plasma forms at an earlier stage and the plasma size is larger.

(51) In contrast to the scenario for the oscillatory mode under no flow, the critical wavenumber remains nonzero, despite being small.

(52) Nevertheless, such a change gives rise to poor deformation recovery as the breakage of crystallites is irreversible.

(53) Hydrophilic substrates promote repair, while hydrophobic surfaces inhibit it and can, instead, cause the membrane to undergo a structural transformation.

(54) In view of its excellent performance, the PCH material can be an ideal candidate for simple, yet efficient monitoring and encapsulation of uranium in nuclear waste water.

(55) Conversely, when the weight fraction of CNF exceeded the optimal value, the surface layer could be brittle, and the mechanical properties of the specimen decreased.

(56) Nonetheless, there is currently no precise method for ascertaining the presence of GNPs within the exosomes in vivo over long periods.

(57) On the contrary, a sharp interface without nanocrystals or amorphous layers is observed in in-plane textured NiO thin films on Al_2O_3 (0001) substrates.

(58) By contrast, porous analogues of copolymers P3 and P7 show reduced photocatalytic activities.

(59) Notwithstanding their good crystallinity, the analysis verified that the Nb (Ta) distribution in the nanotubes was not uniform along the length of the tubes and also the distance from the surface.

(60) Unfortunately, there are no direct experimental measurements of the full set of elastic coefficients of individual fibers.

(61) Even so, any effective strategy irrespective of the class of materials selected will have to harbor capabilities for lowering the interfacial impedance.

(62) Under these conditions, the inverse relationship imposed by $Ca_3(VO_4)_2$ solubility limits allows higher concentrations of V to accumulate in leachate than those seen under air-excluded conditions.

(63) In this sense, our proposed spaser could be viewed as a multifunctional device.

(64) In this respect, further research must be focused on enhancing biological processes to surge methane production giving the minimum operating costs,

(65) Therefore, the complementary test is no longer appropriate for such cases.

(66) Thus, hardening effects were not captured by the CPFE model considered here but arise naturally in DDP.

(67) Hence, there is an extraordinary bending performance when the actuator is driven by light or electricity.

(68) That would lead to a broader temperature range of criticality and consequently to a broader operational temperature range.

(69) The waviness was eliminated and long scratches were reduced so that a smooth surface was obtained.

(70) As such, other techniques of reducing the bandgap in b-PAs or new 2D materials still need to be investigated to achieve longwave-IR photodetection.

(71) Ultimately, there is a need to develop materials that are elastic while maintaining good electrical performance.

(72) As a consequence, both exhibit broad photoluminescence at room temperature.

(73) To this end, the simulation of the cavitating jet has been conducted to localize the cavitation zone.

(74) As the powder compacts evolved over time, a discrete probability density function of the mean curvature was obtained along with its statistical properties (i.e., mean, variance, and skewness).

(75) The resulting iKEA was superior to conventional methods (e.g., ELISA., Western blotting) in speed and sensitivity.

(76) Hence, the growing alumina scale takes up, for example, yttrium by incorporating yttria particles.

(77) The use of genetically encoded SII as a handhold is superior to those that require post-translational modification (for example, biotin or digoxigenin).

(78) For instance, the single-event (intra-event) analysis probably suffers from deviations in the pore geometry from a perfect cylinder.

(79) Two distinct zones, namely the FZ and the HAZ, are formed.

(80) The approach developed here can be applied to other n-type TMDCs, that is, MoS_2 and $MoSe_2$ while different approaches using n-type dopants would be required to p-type TMDCs, that is, WSe_2 and $MoTe_2$.

(81) In other words, competition between global and local effects imposes that smaller agents dissipate faster to maintain their individuality.

(82) As an example, the fraction of 1W MXene remains important even at high RH with 30% of 1W layers at 95% RH.

(83) To illustrate, thermoplastic polyurethane and polybutylene terephthalate have no individual effect on thermal conductivity.

Task 19: Arrange the sentences in each set logically and coherently into a research article conclusion section.

I

(1) Depending on the laser fluence, we find two types of event.

(2) The mechanism for P-to-AP switching is clearly different from the framework of current AOS and therefore fills a missing piece to achieve optical bipolar switching, as in current-induced STT switching and spin–orbit torque switching.

(3) We believe that this work provides a new route to ultrafast magnetization control by further bridging concepts from spintronics and ultrafast magnetism.

(4) In conclusion, we show that a single femtosecond-laser pulse can induce the reversal of a [Co/Pt] ferromagnetic layer within a common spin-valve structure that contains no rare-earth element.

II

(1) Moreover, the BGNCs can effectively promote rat full-thickness skin wound healing.

(2) Besides, the BGNCs could closely adhere to various tissues.

(3) We envision that the successful construction of the BGNCs not only provides an effective tool for uncontrollable hemorrhage and tissue repair but also offers the possibility for developing bulk inorganic bioactive materials in a nanofibrous, interconnected porous,

resilient, and shape adaptive form.

(4) The *in vivo* studies in rabbit liver and femoral artery hemorrhage models and a rat heart injury model verified that the BGNCs could achieve rapid and effective hemostasis, superior to commercial gelatin hemostatic sponges.

(5) Benefiting from the flexible bioactive glass (BG) nanofibers and stable cellular structure, the resultant bioactive glass nanofibrous cryogels (BGNCs) exhibited high water absorption capacity, fast self-expanding ability, near zero Poisson's ratio, injectability, and superelasticity over 800 compression cycles without plastic deformation.

(6) In parallel, the BGNCs showed sustained ion release, low BCI value, and high adhesion of blood cells.

(7) In summary, we developed superelastic, tissue-adhesive, and bioactive self-expanding cryogels enabling fast hemostasis and effective wound healing.

Ⅲ

(1) Although initially developed for use with 4D-STEM data sets, the method may also be extended to other STEM data sets with associated spectroscopic signals, such as electron energy loss spectroscopy or energy dispersive spectroscopy.

(2) This algorithm is computationally cheap and can be performed on typical consumer hardware.

(3) 4D-STEM is a useful characterization technique to study quantum materials, but its applicability is currently limited by the distortions induced by the intrinsic stability of microscope electronics and cryogenic stages.

(4) It preserves reciprocal space information in the data sets and recovers real space correspondence, making information in four dimensions interpretable.

(5) It can be executed rapidly and thus is potentially applicable to on-the-fly data analysis and visualization for *in situ* cryogenic 4D-STEM experiments to reveal dynamic evolutions of lattice and charge distribution in materials under operational conditions.

(6) This method captures and removes localized distortions by recognizing features common to a reference STEM image and the 4D-STEM data set.

(7) Here, we have developed a workflow combining several algorithms effectively to correct complex distortions in cryogenic 4D-STEM data.

Task 20: Fill in the blanks with appropriate verbs in the right form in the following extracts of research article conclusion sections.

I

Electrochemical Decalcification–Exfoliation of Two-Dimensional Siligene, SixGey: Material Characterization and Perspectives for Lithium-Ion Storage

In summary, we have demonstrated a controlled electrochemical exfoliation of a few-layer siligene (1) ________ Zintl phase $Ca_{1.0}Si_{1.0}Ge_{1.0}$ as a precursor. The successful synthesis was based on the decalcification–intercalation mechanism in a nonaqueous environment, (2) ________ high-quality and high-uniformity 2D SiGe nanosheets with low-level hydrogenation. The obtained siligene is (3) ________ of ultrathin micrometer lateral size flakes with mono- or few-layer thickness and excellent crystallinity. In addition, proposed anodes based on the exfoliated siligene for lithium-ion batteries have been explored. To be specific, the SiGe-integrated LIB (4) ________ a 10% increment of the specific capacity that aligns and/or prevails over the theoretical capacities of double-layer silicene/germanene. Both SiGe_rGOS- and SiGe_MWCNTs-based LIBs (5) ________ similar electrochemical behavior such as low polarization, good cycling stability, and decay of the SEI level after the first galvanostatic discharge/charge cycle. As (6) ________, nanostructuring of microscale electrode materials is considered to be an effective approach to (7) ________ the electrochemical performance of the corresponding LIBs. Therefore, this study will (8) ________ valuable insights into fundamental research of materials chemistry and help to further (9) ________ the remaining challenges for the practical application of 2D Si/Ge-based (nano)structures.

II

Multiresponsive Microactuator for Ultrafast Submillimeter Robots

In summary, we proposed an electric/light-actuated soft microactuator based on nano bimorph films and (1) ________ it into several untethered ultrafast microrobots with exceptional controllability, adaptability, and robustness. Unlike conventional thermal bimorph actuators, this microactuator is composed of multilayer patterned nanofilms and (2) ________ precise and rapid response under low voltages. The same characteristics reappeared with laser irradiation due to the excellent photothermal effect of the aluminum/nitinol films. These responsive performances (3) ________ the microactuator to transform into controllable and fast inchworm-type microrobots actuated by a laser immediately after full freedom. The proposed universal design and microfabrication approach (4) ________ us to access various improved and distinctive robotics configurations simultaneously for faster movement and better controllability. Highly controllable by the laser frequency, the motion speed of the microrobots can (5) ________ 2.96 mm/s (3.66 BL/s) on the polished wafer surface. The excellent movement adaptability of the robot is also verified on other rough substrates. Moreover, directional locomotion can be (6) ________ simply by the bias of the irradiation of the laser spot, and the maximum angular speed reached 167°/s. Benefiting from the simple bimorph film structure, the microrobot could maintain functionalized after being crashed by a payload 67 000-times heavier than its weight. The microrobot with the symmetrical configuration allowed it to (7) ________ its locomotion even in the reversed state, which realized robustness to resist unexpected disturbances to its posture. In comparison with other reported works and creatures, microrobots in this work are capable of moving over 1 BL/s without ratchet surfaces while being untethered, which makes them comparable to the speed of some arthropods in the *Formicidae* family. These untethered microrobots will (8) ________ a promising application prospect in biomedicine and other fields.

In future work, the shape memory effect of the nitinol will be further (9) ________ to enhance the jumping ability and load capacity of the microrobot by taking the advantage of its high power density and to (10) ________ the application scenarios of the robot through programmable actuation temperature. Automatic programming of the microrobot based on the visual aid system and negative feedback control will also be introduced in the future to improve the control accuracy.

Task 21: Conclude an empirical study you have conducted in one to two paragraphs following the outline below, and present it orally in groups.

(1) Briefing research purpose and methods.

(2) Summarizing and evaluating main results.

(3) Indicating implications and/or discussing applications of main results.

(4) Making recommendations or stating a practical need for further study.

References

Unit 1

[1] Popov, G., Mattinen, M., Hatanpää, T., Vehkamäki, M., Kemell, M., Mizohata, K., Räisänen, J., Ritala, M., & Leskelä, M. (2019). Atomic layer deposition of PbI_2 thin films. *Chemistry of Materials*, *31*(3), 1101–1109.

[2] Yu, S. H., Chen, W., Wang, H., Pan, H., & Chua, D. H. C. (2019). Highly stable tungsten disulfide supported on a self-standing nickel phosphide foam as a hybrid electrocatalyst for efficient electrolytic hydrogen evolution. *Nano Energy, 55,* 193–202.

[3] Panáček, A., Kvítek, L., Smékalová, M., Večeřová, R., Kolář, M., Röderová, M., Dyčka, F., Šebela, M., Prucek, R., Tomanec, O., & Zbořil, R. (2018). Bacterial resistance to silver nanoparticles and how to overcome it. *Nature Nanotechnology*, *13*(1), 65-71.

[4] Zhou, G., Wang, W., & Peng, M. (2018). Functionalized aramid nanofibers prepared by polymerization induced self-assembly for simultaneously reinforcing and toughening of epoxy and carbon fiber/epoxy multiscale composite. *Composites Science and Technology*, *168*, 312–319.

[5] Chang, G., Wang, C., Song, L., & Yang, L. (2018). An encouraging recyclable synergistic hydrogen bond crosslinked high-performance polymer with visual detection of tensile strength. *Polymer Testing*, *71*, 167–172.

[6] Rashidi, R., Malekan, M., & Hamishebahar, Y. (2019). Serration dynamics in the presence of chemical heterogeneities for a Cu-Zr based bulk metallic glass. *Journal of Alloys and Compounds*, *775*, 298–303.

[7] Cao, Y., Majeed, M. K., Li, Y., Ma, G., Feng, Z., Ma, X., & Ma, W. (2019). P_4Se_3 as a new anode material for sodium-ion batteries. *Journal of Alloys and Compounds*, *775*, 1286–1292.

[8] Qian, J., Adams, B. D., Zheng, J., Xu, W., Henderson, W. A., Wang, J., Bowden, M. E., Xu, S., Hu, J., & Zhang, J.-G. (2016). Anode-free rechargeable lithium metal batteries. *Advanced Functional Materials*, *26*(39), 7094–7102.

[9] Huong, V. T. H., Duong, N. P., Nguyet, D. T. T., Soontaranon, S., & Loan, T. T. (2019). Local structural change and magnetic dilution effect in (Ca^{2+}, V^{5+}) co-substituted yttrium iron garnet prepared by sol-gel route. *Journal of Alloys and Compounds*, *775*, 1259–1269.

[10] He, Y., Gehrig, D., Zhang, F., Lu, C., Zhang, C., Cai, M., Wang, Y., Laquai, F., Zhuang, X., & Feng, X. (2016). Highly efficient electrocatalysts for oxygen reduction reaction based on 1D ternary doped porous carbons derived from carbon nanotube directed conjugated microporous polymers. *Advanced Functional Materials*, *26*(45), 8255–8265.

[11] Sudrajat, H., Babel, S., Thushari, I., & Laohhasurayotin, K. (2019). Stability of La dopants in $NaTaO_3$ photocatalysts. *Journal of Alloys and Compounds*, *775*, 1277–1285.

[12] Abdeljawad, F., Bolintineanu, D. S., Cook, A., Brown-Shaklee, H., DiAntonio, C., Kammler, D., & Roach, A. (2019). Sintering processes in direct ink write additive manufacturing: A mesoscopic modeling approach. *Acta Materialia*, *169*, 60–75.

[13] Weber, M., Kim, J.-H., Lee, J.-H., Kim, J.-Y., Iatsunskyi, I., Coy, E., Drobek, M., Julbe, A., Bechelany, M., & Kim, S. S. (2018). High-performance nanowire hydrogen sensors by exploiting the synergistic effect of Pd nanoparticles and metal-organic framework membranes. *ACS Applied Materials & Interfaces*, *10*(40), 34765–34773.

[14] Gu, Z., De Luna, P., Yang, Z., & Zhou, R. (2017). Structural influence of proteins upon adsorption to MoS_2 nanomaterials: Comparison of MoS_2 force field parameters. *Physical Chemistry Chemical Physics*, *19*(4), 3039–3045.

[15] Chen, X. L., Dong, B. Y., Zhang, C. Y., Luo, H. P., Liu, J. W., Zhang, Y. J., & Guo, Z. N. (2019). Electrochemical direct-writing machining of micro-channel array. *Journal of Materials Processing Technology*, *265*, 138–149.

[16] Heiber, M. C., Okubo, T., Ko, S.-J., Luginbuhl, B. R., Ran, N. A., Wang, M., Wang, H., Uddin, M. A., Woo, H. Y., Bazan, G. C., & Nguyen, T.-Q. (2018). Measuring the competition between bimolecular charge recombination and charge transport in organic solar cells under operating conditions. *Energy & Environmental Science*, *11*(10), 3019–3032.

[17] Yadav, S. K., Suri, A., Ansari, M. O., & Thomas, S. (2018). Fabrication of graphene oxide and hyperbranched polyurethane composite via in situ polymerization with improved mechanical and dielectric properties. *Polymer Composites*, *39*(8), 2765–2770. https://doi.org/10.1002/pc.24267

[18] He, W., Ahmad, N., Sun, S., Zhang, X., Ran, L., Shao, R., Wang, X., & Yang, W. (2023). Microscopic segregation dominated nano-interlayer boosts 4.5 V cyclability and rate performance for sulfide-based all-solid-state lithium batteries. *Advanced Energy Materials*, *13*(3), 2203703.

[19] Selcuk, S., Zhao, X., & Selloni, A. (2018). Structural evolution of titanium dioxide during reduction in high-pressure hydrogen. *Nature Materials*, *17*(10), 923-928.

[20] Pan, F., Chen, S.-M., Li, Y., Tao, Z., Ye, J., Ni, K., Yu, H., Xiang, B., Ren, Y., Qin, F., Yu, S.-H., & Zhu, Y. (2018). 3D graphene films enable simultaneously high sensitivity and large stretchability for strain sensors. *Advanced Functional Materials*, *28*(40), 180-221.

[21] Ma, T. F., Zhou, X., Du, Y., Li, L., Zhang, L. C., Zhang, Y. S., & Zhang, P. X. (2019). High temperature deformation and microstructural evolution of core-shell structured titanium alloy. *Journal of Alloys and Compounds*, *775*, 316–321.

[22] Dodda, A., Jayachandran, D., Pannone, A., Trainor, N., Stepanoff, S. P., Steves, M. A., Radhakrishnan, S. S., Bachu, S., Ordonez, C. W., Shallenberger, J. R., Redwing, J. M., Knappenberger, K. L., Wolfe, D. E., & Das, S. (2022). Active pixel sensor matrix based on monolayer MoS_2 phototransistor array. *Nature Materials*, *21*(12), 1379–1387.

Unit 2

[1] Shi, J., & Akbarzadeh, A. H. (2019). Architected cellular piezoelectric metamaterials: Thermo-electro-mechanical properties. *Acta Materialia*, *163*, 91–121.

[2] Li, H., Yan, S., Zhan, M., & Zhang, X. (2019). Eddy current induced dynamic deformation behaviors of aluminum alloy during EMF: Modeling and quantitative characterization. *Journal of Materials Processing Technology*, *263*, 423–439.

[3] Breda, A., Coppieters, S., Van de Velde, A., & Debruyne, D. (2018). Experimental validation of an equivalent modelling strategy for clinch configurations. *Materials & Design*, *157*, 377–393.

[4] He, J. F., Guo, Z. N., Lian, H. S., Liu, J. W., Yao, Z., & Deng, Y. (2019). Experiments and simulations of micro-hole manufacturing by electrophoresis-assisted micro-ultrasonic machining. *Journal of Materials Processing Technology*, *264*, 10–20.

[5] Xu, L., Lu, Y., Zhao, C.-Z., Yuan, H., Zhu, G.-L., Hou, L.-P., Zhang, Q., & Huang, J.-Q. (2021). Toward the scale-up of solid-state lithium metal batteries: The gaps between lab-level cells and practical large-format batteries. *Advanced Energy Materials*, *11*(4), 2002360.

[6] Bruno, L. (2018). Full-field measurement with nanometric accuracy of 3D superficial displacements by digital profile correlation: A powerful tool for mechanics of materials. *Materials & Design*, *159*, 170–185.

[7] Shen, X., Wu, Z., Li, J., Kang, J., & Fang, Z. (2019). Phosphor-free white emission from InGaN quantum wells grown on in situ formed submicron-scale multifaceted GaN stripes. *Journal of Alloys and Compounds*, *775*, 752–757.

[8] Wu, Y., Shen, H., Walter, D., Jacobs, D., Duong, T., Peng, J., Jiang, L., Cheng, Y.-B., & Weber, K. (2016). On the origin of hysteresis in perovskite solar cells. *Advanced Functional Materials*, *26*(37), 6807–6813.

[9] Croxford, A. M., Davidson, P., & Waas, A. M. (2018). Influence of hole eccentricity on failure progression in a double shear bolted joint (DSBJ). *Composites Science and Technology*, *168*, 179–187.

[10] Afaneh, T., Sahoo, P. K., Nobrega, I. A. P., Xin, Y., & Gutiérrez, H. R. (2018). Laser-assisted chemical modification of monolayer transition metal dichalcogenides. *Advanced Functional Materials*, *28*(37), 1802949.

[11] Tan, Y. C., & Zeng, H. C. (2017). Defect creation in HKUST-1 via molecular imprinting: Attaining anionic framework property and mesoporosity for cation exchange applications. *Advanced Functional Materials*, *27*(42), 1703765.

[12] Fan, E., Li, L., Wang, Z., Lin, J., Huang, Y., Yao, Y., Chen, R., & Wu, F. (2020). Sustainable recycling technology for Li-ion batteries and beyond: Challenges and future prospects. *Chemical Reviews*, *120*(14), 7020–7063.

Unit 3

[1] Famiani, S., LaGrow, A. P., Besenhard, M. O., Maenosono, S., & Thanh, N. T. K. (2018). Synthesis of fine-tuning highly magnetic Fe@FexOy nanoparticles through continuous injection and a study of magnetic hyperthermia. *Chemistry of Materials*, *30*(24), 8897–8904.

[2] Wang, S., Ding, X., Zhang, X., Pang, H., Hai, X., Zhan, G., Zhou, W., Song, H., Zhang, L., Chen, H., & Ye, J. (2017). In situ carbon homogeneous doping on ultrathin bismuth molybdate: A dual-purpose strategy for efficient molecular oxygen activation. *Advanced Functional Materials*, *27*(47), 1703923.

[3] Lin, C.-L., Arafune, R., Liu, R.-Y., Yoshimura, M., Feng, B., Kawahara, K., Ni, Z., Minamitani, E., Watanabe, S., Shi, Y., Kawai, M., Chiang, T.-C., Matsuda, I., & Takagi, N. (2017). Visualizing type-II Weyl points in tungsten ditelluride by quasiparticle interference. *ACS Nano*, *11*(11), 11459–11465.

[4] Ratzker, B., Wagner, A., Sokol, M., Kalabukhov, S., & Frage, N. (2019). Stress-enhanced dynamic grain growth during high-pressure spark plasma sintering of alumina. *Acta Materialia*, *164*, 390–399.

[5] Eichenberger, L., Haraux, P., Venturini, G., Malaman, B., & Mazet, T. (2019). Structural and magnetic properties of the new $Yb_{1-x}Sc_xMn_6Sn_6$ solid solution. *Journal of Alloys and Compounds*, *775*, 883–888.

[6] Vranic, S., Rodrigues, A. F., Buggio, M., Newman, L., White, M. R. H., Spiller, D. G., Bussy, C., & Kostarelos, K. (2018). Live imaging of label-free graphene oxide reveals critical factors causing oxidative-stress-mediated cellular responses. *ACS Nano*, *12*(2), 1373–1389.

[7] Gambhir, M. L., & Jamwal, N. (2014). *Lab Manual-Building and Construction Materials (Testing and Quality Control)*. McGraw-Hill Education.

[8] Islam, M. A., Sinelnikov, R., Howlader, M. A., Faramus, A., & Veinot, J. G. C. (2018). Mixed surface chemistry: An approach to highly luminescent biocompatible amphiphilic silicon nanocrystals. *Chemistry of Materials*, *30*(24), 8925–8931.

[9] Bronstein, E., Faran, E., & Shilo, D. (2019). Analysis of austenite-martensite phase boundary and twinned microstructure in shape memory alloys: The role of twinning disconnections. *Acta Materialia*, *164*, 520–529.

Unit 4

[1] Klimek-McDonald, D. R., King, J. A., Miskioglu, I., Pineda, E. J., & Odegard, G. M. (2018). Determination and modeling of mechanical properties for graphene nanoplatelet/epoxy composites. *Polymer Composites*, *39*(6), 1845–1851.

[2] Tao, K., Gong, Y., & Lin, J. (2019). Epitaxial grown self-supporting $NiSe/Ni_3S_2/Ni_{12}P_5$ vertical nanofiber arrays on Ni foam for high performance supercapacitor: Matched exposed facets and re-distribution of electron density. *Nano Energy*, *55*, 65–81.

[3] Li, L., Sun, L., Dai, Z., Xiong, Z., Huang, B., & Zhang, Y. (2019). Experimental investigation on mechanical properties and failure mechanisms of polymer composite-metal hybrid materials processed by direct injection-molding adhesion method. *Journal of Materials Processing Technology*, *263*, 385–395.

[4] Bastola, A. K., & Li, L. (2018). A new type of vibration isolator based on magnetorheological elastomer. *Materials & Design*, *157*, 431–436.

[5] Tomków, J., Rogalski, G., Fydrych, D., & Łabanowski, J. (2018). Improvement of S355G10+N steel weldability in water environment by Temper Bead Welding. *Journal of Materials Processing Technology*, *262*, 372–381.

[6] Wang, X., Wang, J., Gao, Z., Xia, D.-H., & Hu, W. (2018). Fabrication of graded surfacing layer for the repair of failed H13 mandrel using submerged arc welding technology. *Journal of Materials Processing Technology*, *262*, 182–188.

[7] Boborodea, A., Brookes, A., & O'Donohue, S. (2018). Polyethylene characterization by analytical temperature rising elution fractionation with evaporative light scattering detector. *Polymer Testing*, *72*, 172–177.

[8] Chang, P.-Y., & Doong, R. (2019). Microwave-assisted synthesis of SnO_2/mesoporous carbon core-satellite microspheres as anode material for high-rate lithium ion batteries. *Journal of Alloys and Compounds*, *775*, 214–224.

Unit 5

[1] Wang, Q., Wu, W., Zhang, J., Zhu, G., & Cong, R. (2019). Formation, photoluminescence and ferromagnetic characterization of Ce doped AlN hierarchical nanostructures. *Journal of Alloys and Compounds*, *775*, 498–502.

[2] Tian, J., Xie, S.-H., Borucu, U., Lei, S., Zhang, Y., & Manners, I. (2023). High-resolution cryo-electron microscopy structure of block copolymer nanofibres with a crystalline core. *Nature Mate*rials, 22(6), 786–792.

[3] Tébar-Soler, C., Martin-Diaconescu, V., Simonelli, L., Missyul, A., Perez-Dieste, V., Villar-García, I. J., Brubach, J.-B., Roy, P., Haro, M. L., Calvino, J. J., Concepción, P., & Corma, A. (2023). Low-oxidation-state Ru sites stabilized in carbon-doped RuO_2 with low-temperature CO_2 activation to yield methane. *Nature Materials*, *22*(6), 762–768.

[4] Liu, W., Zhang, C., Alessandri, R., Diroll, B. T., Li, Y., Liang, H., Fan, X., Wang, K., Cho, H., Liu, Y., Dai, Y., Su, Q., Li, N., Li, S., Wai, S., Li, Q., Shao, S., Wang, L., Xu, J., … Wang, S. (2023). High-efficiency stretchable light-emitting polymers from thermally activated delayed fluorescence. *Nature Materials*, *22*(6), 737–745.

[5] Yang, S., Feng, X., Xu, B., Lin, R., Xu, Y., Chen, S., Wang, Z., Wang, X., Meng, X., & Gao, Z. (2023). Directional self-assembly of facet-aligned organic hierarchical super-heterostructures for spatially resolved photonic barcodes. *ACS Nano*, *17*(7), 6341–6349.

[6] Chen, H., Zheng, J., Ballestas-Barrientos, A., Bing, J., Liao, C., Yuen, A. K. L., Fois, C. A. M., Valtchev, P., Proschogo, N., Bremner, S. P., Atwater, H. A., Boyer, C., Maschmeyer, T., & Ho-Baillie, A. W. Y. (2022). Solar-driven co-production of hydrogen and value-add conductive polyaniline polymer. *Advanced Functional Materials*, *32*(52), 2204807.

[7] Alon, H., Stern, C., Kirshner, M., Sinai, O., Wasserman, M., Selhorst, R., Gasper, R., Ramasubramaniam, A., Emrick, T., & Naveh, D. (2018). Lithographically patterned functional polymer–graphene hybrids for nanoscale electronics. *ACS Nano*, *12*(2), 1928–1933.

[8] Li, J., Li, T., Jia, Y., Yang, S., Jiang, S., & Turng, L.-S. (2018). Modeling and characterization of crystallization during rapid heat cycle molding. *Polymer Testing*, *71*, 182–191.

[9] Wang, X., Ma, C., Li, C., Kang, M., & Ehmann, K. (2018). Influence of pulse energy on machining characteristics in laser induced plasma micro-machining. *Journal of Materials Processing Technology*, *262*, 85–94.

[10] Li, H., Gordeev, G., Wasserroth, S., Chakravadhanula, V. S. K., Neelakandhan, S. K. C., Hennrich, F., Jorio, A., Reich, S., Krupke, R., & Flavel, B. S. (2017). Inner- and outer-wall sorting of double-walled carbon nanotubes. *Nature Nanotechnology*, *12*(12), 1176–1182.

[11] De, S., & Aluru, N. R. (2018). Energy dissipation in fluid coupled nanoresonators: The effect of phonon-fluid coupling. *ACS Nano*, *12*(1), 368–377.

[12] Smeets, P. J. M., Cho, K. R., Sommerdijk, N. A. J. M., & De Yoreo, J. J. (2017). A mesocrystal-like morphology formed by classical polymer-mediated crystal growth. *Advanced Functional Materials*, *27*(40), 1701658.